Introduction to Mixed-Signal, Embedded Design

Introduction to Mixed-Signal, Embedded Design

Alex Doboli
State University of New York
Stony Brook, NY, USA

Edward H. Currie
Hofstra University
Hempstead, NY, USA

 Springer

Alex Doboli
Department of Electrical Engineering
State University of New York
Stony Brook, NY 11794, USA
adoboli@ece.sunysb.edu

Edward H. Currie
Department of Computer Science
Hofstra University
Hempstead, NY 11549, USA
edward.currie@hofstra.edu

A previous version of this book was published by The Cypress University Alliance in 2007

ISBN 978-1-4939-3944-2   ISBN 978-1-4419-7446-4 (eBook)
DOI 10.1007/978-1-4419-7446-4
Springer New York Dordrecht Heidelberg London

Mathematics Subject Classification (2011): 94CXX, 94-04

©Springer Science+Business Media, LLC 2011
Softcover reprint of the hardcover 1st edition 2011

Printed on acid-free paper

Springer is part of Springer Science+Business Media(www.springer.com)

Contents

List of Figures

List of Tables

Preface

This textbook was developed for upper-level undergraduate, and first-year graduate level, curricula. It addresses three important aspects of embedded mixed-signal systems: (a) the defining characteristics of embedded applications, (b) embedded mixed-signal architectures, and (c) top-down design and design activities for developing performance-satisfying, optimized implementations. Although the authors have attempted to make the material presented here as self-contained as possible, the student will find it helpful to take advantage of the bibliographies at the end of each chapter, and the companion laboratory manual available online.

Embedded applications involve: (i) interfacing to analog signals and digital data, (ii) functionality sensing, control, actuation, data computation, and data communication, functionality; and (iii) design and performance constraints that must be satisfied by the system implementation. In addition to considering the general challenges of designing mixed-signal, embedded systems, various analog and digital interfaces and interfacing modules, for example, interfaces to temperature sensors, tachometers, and LCDs, filters, analog-to-digital converters (ADCs), quantizers, interrupt subsystems, and digital communication components based on standard protocols, e.g., SPI, UART, and I2C, are discussed in detail.

The topics discussed include the hardware and software used to implement analog and digital interfaces, for example, $\Delta\Sigma$ ADC topologies and circuits, various filter structures, amplifiers and other signal-conditioning circuits, PWMs, timers, I/O ports, interrupt service routines (ISRs), high-level communication primitives (APIs), firmware routines, and data structures for handling multiple, but similar, peripheral devices.

The authors have chosen Cypress Semiconductor's Programmable System on (a) Chip (PSoC) to illustrate many of the key points developed in this text. This choice was based largely on the fact that PSoC provides a broad range of the various components of a typical mixed-signal embedded system, for example, A/D and D/A converters, UARTs, $\Delta\Sigma$ modulators, filters, programmable gain amplifiers, instrumentation amplifiers, comparators, DTMF dialer, counters, timers, digital buffers, digital inverters, LCD and LED support, sleep timers, watchdogs, MUXs, PWMs, random sequence generators, flash temperature sensors, eand so on, and all in a single chip.

Detailed design examples are presented throughout the text as illustrative examples of embedded systems, such as, interfacing to temperature sensor, tachometer and fan, tachometer ISR, SPI and UART implementations, SPI- and UART-based task communications, and ISR for decimation in $\Delta\Sigma$ ADCs. Recommended exercises are also provided at the end of each chapter.

The embedded functionality treated here can be divided into four broad categories: (1) control dominated systems, which are specified as finite state machines (FSM), (2) data-dominated applications expressed as acyclic dataflow graphs (ADFG), (3) multitasking systems defined as task graphs, and (4) multimode systems specified as control and dataflow graphs. In addition, depending on the semantics of execution with respect to time, embedded mixed-signal systems are continuous-time, discrete-time, and event-driven (reactive) systems.

There are different types of embedded functionalities presented, for example, temperature controller systems, stack-based operations, subroutines for unsigned multiplication, bit manipulations, sequence detector applications, scalar products of two vectors, and several examples of communicating tasks. Regarding analog functionality, the material includes comparators, instrumentation amplifiers, filters, $\Delta\Sigma$ ADCs, differential and common mode amplification, and several signal conditioning front-ends.

This text emphasizes the importance of performance constraints and requirements in determining the optimal implementation of an embedded application. The performance attributes considered are cost, time-to-market, size and weight of the implementation, real-time constraints, and data accuracy. Constraints can be classified into global constraints, if they refer to the overall system, and local constraints, if they are related to the individual modules and subsystems. Constraint transformation is the relating of the global constraints to the local constraints, including the correlations between analog and digital constraints. Performance profiling is often used for computing the performance-criticality of functional blocks.

Several specific constraint transformation procedures, such as, relating the required processing accuracy/bandwidth of the input signals to the signal-to-ratio requirement and sampling frequency of the ADC, and to the latency constraint of the digital processing are presented. Other examples include correlating the maximum bit rate of a serial input to the processing latency of the processing algorithm, studying the impact of the memory access mechanism and time on the total execution time of an application, estimating and reducing the timing overhead of data communication, handling predefined timing limits for the tachometer ISR, and improving system performance by architecture customization to the application's characteristics.

This textbook also specifies the performance attributes that describe analog and mixed-signal modules, for example, the nonidealities of continuous-time analog blocks (e.g., OpAmp finite gain, poles, zeros, input and output impedances, distortion, offset, power supply rejection ratio, saturation, slew rate, and circuit noise), the nonidealities of switched capacitor blocks (e.g., the nonzero switch resistance, channel charge injection, clock feedthrough, and finite OpAmp gain), and the concepts of quantization noise power, signal-to-noise ratio (SNR), dynamic range (DR), and power spectrum density in ADCs. Finally, the text models the dependency of the ADC performance on circuit nonidealities (e.g., jitter noise, switch thermal noise, integrator leakage, and OpAmp bandwidth, slew rate, saturation and noise).

The final chapter provides descriptions of two case studies, which reiterate the design flow for embedded systems, and ends by introducing some of the current challenges related to design automation for analog and mixed-signal systems.

LEGAL NOTICE

In this textbook the authors have attempted to teach the techniques of mixed-signal, embedded design based on examples and data believed to be accurate. However, these examples, data, and other information contained herein are intended solely as teaching aids and should not be used in any particular application without independent testing and verification by the person making the application. Independent testing and verification are especially important in any application in which incorrect functioning could result in personal injury, or damage to property.

For these reasons the authors and Cypress Semiconductor Corporation make no warranties, express or implied, that the examples, data, or other information in this volume are free of error, that they are consistent with industry standards, or that they will meet the requirements for any particular application.

THE AUTHORS AND CYPRESS SEMICONDUCTOR EXPRESSLY DISCLAIM THE IMPLIED WARRANTIES OF MERCHANTABILITY AND OF FITNESS FOR ANY PARTICULAR PURPOSE, EVEN IF THE AUTHORS AND CYPRESS SEMICONDUCTOR CORPORATION HAVE BEEN ADVISED OF A PARTICULAR PURPOSE, AND EVEN IF A PARTICULAR PURPOSE IS INDICATED IN THE TEXTBOOK. THE AUTHORS AND CYPRESS SEMICONDUCTOR ALSO DISCLAIM ALL LIABILITY FOR DIRECT, INDIRECT, INCIDENTAL, OR CONSEQUENTIAL DAMAGES THAT RESULT FROM ANY USE OF THE EXAMPLES, EXERCISES, DATA, OR OTHER INFORMATION.

Acknowledgment

Completion of this work would not have been possible without the participation of a number of people each of whom offered continual support and encouragement throughout the project.

Cypress Semiconductor participants included: George K. Saul (Vice President of the PSoC Business Unit at Cypress), Steve Gerber (Product Marketing Director), Heather Montag (Senior Technical Writer), Harold Kutz (Design Engineering Director), Patrick Kane (Director of Cypress' University Alliance Program), Dennis Sequine (Senior Applications Manager) and David Van Ess (Principal Applications Engineer).

Particular appreciation is expressed to David Van Ess, who can truly be said to be the "Engineer's engineer". Dave's prolific writings on "all things-PSoC" served as a rich resource upon which we drew deeply, often and with reverence.

Pengbo Sun, Varun Subramanian, Christian Ferent and Ying Wei, graduate students at State University of New York at Stony Brook, helped the authors develop several of the examples in the book. Yang Zhao helped by improving the figures.

And finally, our heart felt thanks for the many helpful suggestions provided by the students who participated in the first graduate level, mixed-signal and embedded design course at Stony Brook to use these materials.

The book is based in part on the following material courtesy of Cypress Semiconductor Corporation: PSoC Mixed Signal Array (Technical Reference Manual, Document No. PSoC TRM 1.21, 2005), PSoC Express (Version 2.0, 2006), PSoC Designer: Assembly Language User Guide (Spec. #38-12004, December 8 2003), PSoC Designer: C Language Compiler User Guide (Document #38-12001 Rev.*E, 2005), PSoC EXPRESS, Driver Author Guide (Version 2.0, 2006, available online), Motor Control with PSoC (presentation, 2006), Pseudo Random Sequence Generator (CY8C29/27/24/22/21xxx Data Sheet, September 21 2005), 8-bit and 16-bit Pulse Width Modulators (CY8C29/27/24/22/21xxx, CY7C6xxxx, and CYWUSB Data Sheet, October 3 2005), SPI Master (CYC29/27/24/22/21xxx, CY7C6xxxx, and CYWUSB Data Sheet, October 3 2005), SPI Slave (CYC29/27/24/22/21xx, CY7C6xxxx, and CYWUSB Data Sheet, October 3 2005), UART (CYC29/27/24/22/21xxx, CY7C64215, and CYWUSB Data Sheet, October 3 2005), DelSig8 v3.2, 8 Bit Delta Sigma ADC (Application Note, Oct. 3 2005), and the following application notes: AN224, AN2032, AN2036, AN2038, AN2041, AN2091, AN2112, AN2120, AN2129, AN2132, AN2148, AN2166, AN2168, AN2209, AN2216, AN2218, AN2233a, AN2269, AN2281, AN2283, AN2315, AN2325, AN2351, AN2326, AN2329, AN2332, AN2333, AN2367.

Chapter 1

An Overview of Mixed-Signal, Embedded System Design

This chapter is intended to serve as an introduction to embedded mixed-signal systems and discusses:

- Top-down design flow for developing performance-efficient implementations

- Characteristics of embedded systems with respect to interfacing, functionality, and design/performance constraints and presents mixed-signal architectures for embedded applications

- Top-down successive refinement design including specification, simulation, implementation by tradeoff analysis, testing and debugging, and validation

As a simple and introductory case study, a fan controller is discussed that illustrates the most important issues regarding the heterogeneous nature of embedded systems (involving various kinds of sensing and actuation devices, analog circuits, digital hardware, and software), and the top-down process for determining the design and performance requirements of the implementation

- Section 1 presents types of embedded applications, and their defining characteristics with respect to interfacing, functionality, and design/performance requirements.

- Section 2 introduces popular architectures used in embedded implementation, for example, microprocessors and microcontrollers, digital signal processors (DSP), very large instruction word (VLIW) processors, and reconfigurable architectures.

- Section 3 discusses the design activities of top-down design methodologies.

- Section 4 provides some predictions about the future of the embedded system market.

- Section 5 details the case study for an embedded fan controller design.

- Section 6 summarizes the chapter by enumerating the discussed topics.

- Section 7 presents further readings on embedded systems.

A. Doboli, E.H. Currie, *Introduction to Mixed-Signal, Embedded Design*,
DOI 10.1007/978-1-4419-7446-4_1, © Springer Science+Business Media, LLC 2011

1.1 Embedded Applications

Embedded systems are quietly permeating day-to-day life. They serve as entertainers, cooks, message carriers, and it is hoped soon as drivers, maids, butlers, nurses, and much more. Embedded systems are ubiquitous and unavoidable, and their importance will continue to increase as new ways of incorporating electronic intelligence emerge in every aspect of our daily lives.

A typical embedded processing flow includes data (signal) acquisition by different kinds of sensors, data processing and storing, and finally data communication to external devices and process control actuation. Figure 1.1 illustrates some of the popular signal sensing and actuation devices, including sensing units, for example, temperature sensors, sonars, tachometers, switches, position sensors, microphones, RFID readers, and actuation devices, for example, motors and speakers. Moreover, embedded systems can operate either as stand-alone systems, or, in conjunction with other embedded systems or computing devices. In the latter case, embedded systems may incorporate wired and wireless links for data communication with the other computing devices. Finally, various display units often display data during an embedded system's operation.

The generic embedded processing flow is instantiated in many popular applications. For example, the embedded controller in a "smart shoe" adjusts the shoe's cushioning level to a runner's weight and running style. Or, the embedded microcontroller inside a tennis racquet improves the hitting of a tennis ball by amplifying the electrical current generated by the piezoelectric fibers at the ball's impact. Other popular applications include many other consumer goods, such as, TV sets, microwave ovens, "smart" pens, cellular phones, and personal digital assistants.

More complex applications involve large networks of interconnected embedded systems (e.g., in automobiles, aircraft and ships). Although smaller, they are still very complex applications, (e.g., robots). In addition to sensing and control at the level of the individual systems, these applications require networking of the systems to provide more comprehensive sensing of the environment and coordinated control. Embedded systems in automobiles are traditionally used to control basic functions such as air intake and fuel, ignition and combustion, exhaust, and transmission systems [33]. Other embedded systems supervise the air bags and air conditioning, or provide navigational information.

Depending on their purpose, embedded systems can be differentiated as follows:

1. *Embedded systems as controllers*: In this case, embedded controllers are stand-alone systems for supervising the behavior and adjusting the characteristics of the monitored parameters. The systems possess interfacing capabilities to specific sensors and actuators. Moreover, embedded controllers compute the control signals for the supervised entities by using a control algorithm operating on the sensed input signals and the system state. The computational complexity of the executed algorithms is, in general, low to medium. However, many applications require satisfaction of timing (real-time) constraints, and safety-critical functioning of the system, which among others includes accurate signal conversion. Many designs ought to be low cost. Some of the typical applications are fan controllers, vending machine controllers, pacemakers, robots and automotive controllers.

2. *Embedded systems as specialized interfaces*: Such embedded systems function as complex interfacing devices to computing systems (e.g., high performance desktop computers) that lack the necessary interfaces. The required "embedded" functionality may include interfacing to analog signals as well as to networks and networking devices, for example Ethernet cards. These embedded systems not only supply the missing hardware (e.g., the analog circuits) but also relieve the computing system from any data processing related to

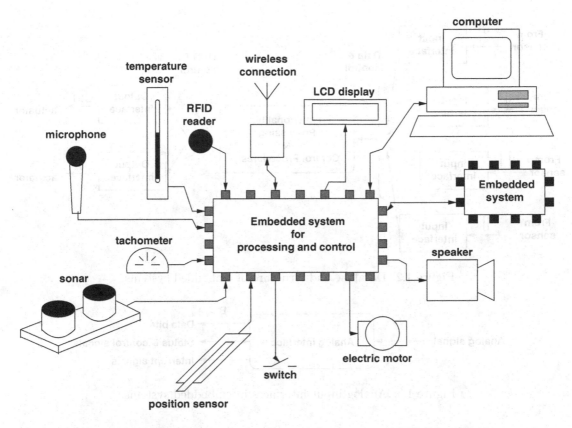

Figure 1.1: Popular sensing and actuation devices in embedded applications.

interfacing, such as data conversions, formatting, encoding, decoding, etc. The computational complexity of these applications is, in general, low to medium, but it can become high for high-speed signals and data links. In such implementations, cost has to be low, size/weight small, throughput fast, and energy consumption minimal. Popular applications include data acquisition, networking cards, and interfacing to peripherals, such as the mouse and capacitive sensors.

3. *Embedded systems as customized coprocessors*: These systems operate as application-specific coprocessors for a computing system (e.g., desktop computer). Their task is to relieve the main system of computationally demanding tasks for which the embedded coprocessor provides an efficient implementation. In data-intensive applications, such as digital signal processing, graphics, multimedia, image processing, and so on, coprocessors offer data processing functions that are much faster than those of a general-purpose computer. For example, data processing using the video coprocessor of the Philips PNX 1300 media processor [9] is about ten times faster than using its CPU core. Similarly, coprocessors can be customized for other computing requirements too, for example low-energy processing for mobile applications. The cost and size of embedded coprocessors must be kept low, considering that they are part of larger systems.

4. *Networks of embedded systems*: Networks of embedded systems are traditionally used in automotive applications in which a fairly large number of sensors, actuators, and embedded microcontrollers are networked together. The systems execute control algorithms in

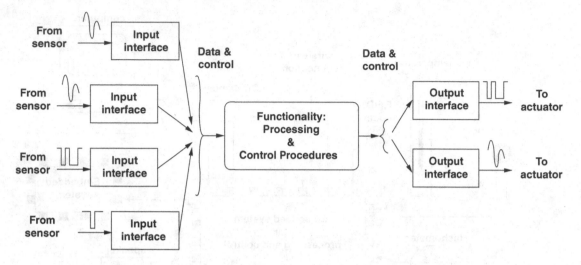

Figure 1.2: Dataflow and structure of embedded systems.

Figure 1.3: Analog input interfaces in embedded systems.

addition to the networking functionality. Therefore, their operation is subject to timing and safety-critical constraints. Wireless sensor networks have become a promising platform for many applications related to infrastructure management and security, healthcare and environment protection. One of the challenges is to develop low-cost, physically small integrated circuits that incorporate processors, memory, interfaces to various sensors, and wireless transceivers. The energy consumption of these chips needs to be as low as possible. In addition, energy-efficient communication protocols are needed, including algorithms for data routing and data aggregation. Finally, scalable, distributed, control techniques are required to ensure efficient operation of the network.

Figure 1.2 introduces the three defining aspects of embedded applications: (a) the required interfacing to sensors, actuators, and data links, (b) the embedded functionality for processing and control, and (c) the design and performance requirements that must be satisfied by the embedded system's implementation. Table 1.1 summarizes the characteristics of the four main types of embedded systems.

A. Interfaces

A common feature of many embedded systems is that they can sense and process a wide variety of electrical signals, and generate control/output data for a diversity of devices. This characteristic stresses the importance of providing an embedded system with *interfaces* to many different types of analog and digital data. The interfaces can be divided into two main categories, depending on the nature of the involved signals:

Table 1.1: Summary of embedded systems' characteristics.

Embedded Systems Types	Interfaces			Data processing	Requirements and constraints
	Sensing	Actuation	Data communication		
Controllers	Wide range	Wide range	-	Control	Cost, accuracy, timing, safety-critical
Specialized interfaces	Wide range	-	Various speed	Conversion, formatting, encoding, etc.	Cost, size, accuracy, timing, power consum.
Co-processors	-	-	High-speed	Data-intensive customized processing	Cost, size, speed, power consum.
Networks of embedded systems	Wide range	Wide range	Various speed	Control, conversion, encoding, etc.	Cost, size, safety-critical, power consum.

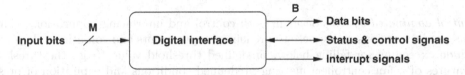

Figure 1.4: Digital input interfaces in embedded systems.

1. *Analog interfaces* convert analog signals into digital signals for processing by the embedded processors, and vice versa, digital signals from the processors are converted into analog signals. Analog inputs of different kinds (e.g., voltage, current, charge, phase, and frequency) are produced for an embedded system by thermistors, thermocouples, pressure sensors, strain gauges, passive infrared detectors, ultrasonic receivers, GPSs, proximity sensors, velocity sensors, an so on [24]. Also, diverse analog signals must be generated by an embedded system for speech synthesizers, step motors, displays and actuators. Figure 1.3 graphically illustrates an analog input interface that converts an analog signal into B data bits. In addition, this interface generates status, control, and interrupt signals that are used by the embedded processing and control algorithms.

Interfacing to these signals requires analog and digital circuits, including signal conditioning circuits, analog filters, analog-to-digital converters, digital-to-analog converters, analog multiplexers, comparator circuits, sampling circuits, amplitude detectors, frequency detectors, modulation/demodulation circuits, analog multipliers and mixers. For example, in a "smart shoe," the currents sensed by capacitive sensing are amplified and converted into digital data and then control signals for the embedded processor.

Because analog signals are of different kinds (e.g., sine, square, triangular, differential inputs, etc.) they determine the type of interfacing circuits required. In addition, analog signal characteristics, e.g., range, frequency, precision, noise, speed of variation, impose specific interface performance requirements, some of which are illustrated in Figure 1.7, and discussed in the rest of this chapter.

2. *Digital interfaces* link embedded systems to external peripherals, other embedded systems, and digital networks. Figure 1.4 illustrates the concept of a digital interface that creates the "transition" from M digital input bits to B output data bits. In addition, the interface generates status, control, and interrupt signal used by the embedded functionality. Digital interfaces are of two basic types: serial, and parallel. Serial interfaces transmit one data bit at a time. Parallel interfaces transmit several bits (e.g., one byte) simultaneously. Digital interfaces usually follow a standard communication protocol, such as I2C, FireWire, SPI, UART, or USB for serial communication, and PCI Bus and ARM Bus for parallel communication. Wireless protocols such as IEEE 802.11 and Bluetooth protocols are becoming increasingly important.

B. Functionality

With respect to the nature of their data processing, embedded systems range from those that produce single control actions in response to a certain event (e.g., a fan controller system) to more complex systems that perform multimode data processing, as in the case of network routers. Embedded systems can be classified into the following four types, depending on the type of processing involved:

1. *Control dominated systems* occur in both control and interfacing applications. These systems produce actions in response to predefined input events and conditions. Typical events include exceeding, or falling below, predefined threshold values (e.g., the threshold temperatures of a fan controller) meeting predefined conditions and expiration of preset time intervals.

   Control-dominated systems are described graphically as finite state machines (FSM) [28] with arcs that define the transition between states. Unlabeled arcs indicate that the transition occurs unconditionally at the next iteration. For labeled arcs, the transition occurs at the next iteration only if the associated condition, as expressed by the arc label, is true. Otherwise, the FSM remains in the same state. These conditions are based on the inputs (including data and events). Output data are generated for a state transition using the inputs and the current FSM state. FSMs have an initial state that is entered upon resetting the FSM.

   Figure 1.5(a) illustrates a FSM with three states. The FSM remains in $state_1$ provided that no event occurs because both output transitions from $state_1$ are controlled by events. For example, $event_1$ causes the transition to $state_2$. The system remains in this state until the next iteration, when it returns to $state_1$. The output signal, *out*, is asserted upon switching to $state_2$. $state_1$ is the initial state of the FSM.

2. *Data-dominated systems* are usually signal (data) processing applications in which the embedded system operates as a specialized coprocessor. The functionality involves a large amount of computations (e.g., additions, multiplications, divisions, shift operations, etc.) executed iteratively in loops, and on large amounts of data stored in arrays. Digital signal processing algorithms, such as digital filtering, are typical examples.

   Data-dominated systems are described as acyclic dataflow graphs (ADFGs) in which the nodes define the data processing operations, and arcs indicate the sequencing of the nodes.

Each node is executed as soon as its predecessor is finished. The execution of a node takes a predefined number of time units, measured in terms of clock cycles.

Figure 1.5(b) depicts an ADFG with four nodes and the operations that correspond to each node as indicated. The execution time of a node is also shown in this figure. The execution of $node_1$ is followed by the execution of $node_2$ and $node_3$. Finally, $node_4$ is executed only after both $node_2$ and $node_3$ have finished their respective executions.

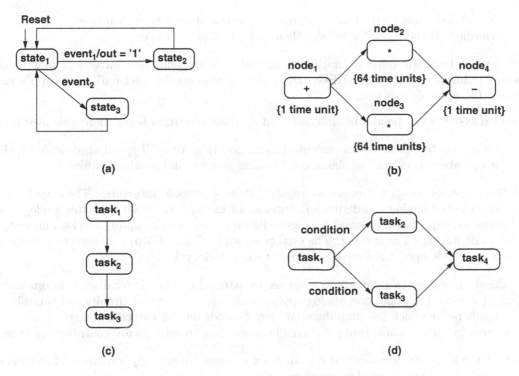

Figure 1.5: Functionality types for embedded systems.

3. *Multitasking systems* are embedded systems with complex functionality, for example JPEG and MPEG algorithms for multimedia, and face detection/recognition in image processing. These systems include concurrently executing tasks that exchange data. Each task is an independent entity with its own inputs, states, and outputs.

   Multitasking systems are expressed as task graphs in which nodes correspond to tasks, and arcs represent the data communicated between the tasks. A task is executed only after all its predecessors are finished, and the data have been communicated from them to the task. The execution model for task graphs assumes that tasks are executed only once for each traversal of the graph.

   Figure 1.5(c) presents the task graph with three tasks that are executed in sequence.

4. *Multimode systems* provide the complex data processing functionalities that are influenced by the specific operation conditions (operation mode). Depending on certain conditions

(e.g., input events and/or the computed values) multimode systems execute only certain parts (modes) of their overall functionality.

Multimode systems are described as data and control graphs. The graph nodes describe tasks, unlabeled arcs present data communications between tasks, and conditional arcs (i.e., arcs labeled with logic expressions) express the functionality that is executed only if the corresponding condition is true. Examples of multimode systems are adaptive filters and complex networking modules.

Figure 1.5(d) shows a dual mode system. If the condition is true then tasks *1*, *2*, and *4* are performed. If the condition is false then tasks *1*, *3*, and *4* are executed.

Note that this textbook refers primarily to embedded applications that have their functionality expressed either as FSMs or as ADFGs. Discussion of multitasking and multimode applications is beyond the scope of this text.

Embedded systems can be also be differentiated by their activation (execution) rule over time:

1. *Continuous-time systems* are executed continuously in time. Typical applications include analog subsystems, such as signal conditioning systems and continuous filters.

2. *Discrete-time systems* execute at regular, that is periodic intervals. These systems include clocked analog and digital subsystems, for example switched capacitor analog blocks (switched capacitor filters and analog-to-digital converters), sampling blocks, and sampled data digital signal processing. The system shown in Figure 1.5(b) is a discrete-time system because node operations are performed at each clock cycle.

3. *Event-driven systems* operate in response to external events. Typical applications include digital control systems, and analog applications, e.g., comparator circuits and variable gain amplifiers for which the amplification gain depends on the external control signals. The system in Figure 1.5(a) is event-driven because state transitions are controlled by events.

Note that the majority of embedded mixed-signal systems are actually a mixture of continuous-time, discrete-time, and event-driven subsystems.

C. Design and performance requirements

With respect to their processing speed and memory size, embedded systems can be inferior to desktop architectures, which deploy more aggressive design solutions. However, the architecture of embedded systems must take into account a much larger variety of design and performance requirements and constraints, as briefly summarized in Table 1.1.

1. *Low cost*: The majority of embedded system designs have a low-cost constraint.

2. *Short time-to-market*: Most of the embedded applications have a short time-to-market requirement, which demands high productivity design flows that are largely based on computer-aided design (CAD) tools. This requirement can trade off some of the achievable design quality for a faster design process.

3. *Small size and weight*: Many embedded designs have "tight" size and weight requirements, including the batteries needed for powering the system, displays, keyboards, sensors, and actuators. This influences the packaging options for the chips, for example package size and number of pins.

4. *Real-time constraints*: Many embedded applications impose timing, or timing related, requirements for their implementations. Real-time constraints demand that certain functionalities of the design are time constrained: that their execution ends before a preset time limit, starts after a certain time, or ends within a given length of time. Real-time constraints are *hard*, if their violation causes the failure of the system functioning, and *soft*, otherwise.

The *latency* constraint requires that a system's execution time (latency) be shorter than an imposed limit. Figure 1.6(a) defines the latency constraint for an ADFG by requiring that the difference between the ending time of $node_4$ and the starting time of $node_1$ is less than the constraint T:

$$T_{node_4}^{ending} - T_{node_1}^{starting} \leq \Delta T \tag{1.1}$$

The *throughput* constraint requires that the processing of one data token takes less than a predefined value.

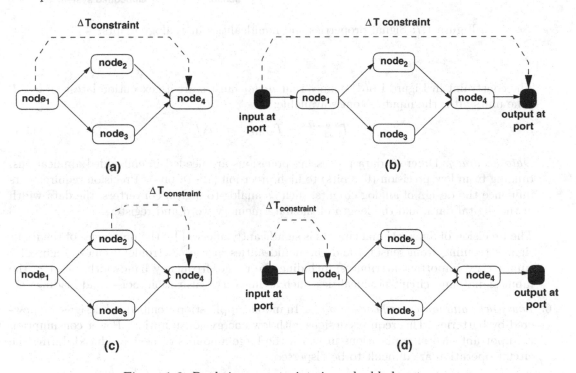

Figure 1.6: Real-time constraints in embedded systems.

The *timing* constraint in Figure 1.6(b) states that the interval between the time T_{avail}, when the input data are available to the system, and T_{prod}, when the corresponding output data are produced, must be less than the constraint ΔT:

$$T_{prod} - T_{avail} \leq \Delta T \tag{1.2}$$

In Figure 1.6(c), this constraint requires that the difference between the time when $node_4$ is finished and $node_2$ starts must be less than ΔT:

$$T_{node_4}^{ending} - T_{node_2}^{starting} \leq \Delta T \tag{1.3}$$

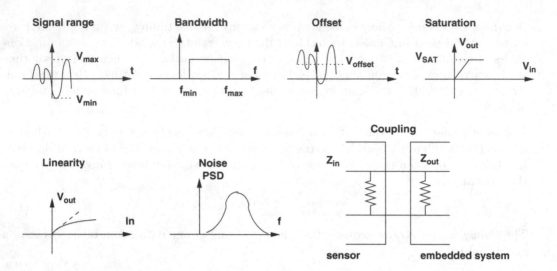

Figure 1.7: Signal properties and nonidealities in analog circuits.

The constraint in Figure 1.6(d) defines that $node_1$ must start its execution latest after ΔT time units after the input becomes available:

$$T^{starting}_{node_1} - T_{avail} \;=\; \Delta T \tag{1.4}$$

5. *Data accuracy:* Different data processing precisions are needed in embedded applications, ranging from low precision (6–8 bits) to high precision (12–16 bits). Precision requirements influence the design of analog circuits, such as analog-to-digital converters, the data width of the digital data, and the length of the data memory word and registers.

   The precision of analog signal circuits is significantly affected by the properties of the input signals (coming from sensors) and the nonidealities of the electronic circuits. Figure 1.7 summarizes important attributes, including the range, frequency bandwidth, offset of the input signal, and circuit nonidealities, such as linearity, saturation, noise, and loading.

6. *Low power and low energy consumption:* In mobile applications, embedded designs are powered by batteries. This requires designs with low energy consumption. Power consumption is important also in applications in which the large amounts of heat produced during the circuit operation are difficult to be dispersed.

7. *Safety:* A large number of embedded designs must guarantee correct operation for any kind of input data and in any environmental conditions. For example, this is important for medical devices (e.g., pacemakers), but also in other safety-critical applications, such as airbag controllers, and car suspension controllers. This sets additional requirements for testing, debugging, and validating the embedded designs.

8. *Robustness:* Embedded systems may operate in harsh environments, including environments with high noise levels, large temperature variations and humidity. Robustness defines the capacity of an embedded system to correctly operate in such environments. For example, robustness might relate to a circuit's operation in the presence of noise, including input noise, circuit noise, and coupling noise. The impact of noise on the quality of a design is reflected in various performance metrics (e.g., signal to noise ratio and noise margins).

9. *Flexibility in developing new applications*: Single embedded designs are rarely developed, instead designs usually pertain to families of applications. These applications share significant portions of their hardware and software designs. It is desirable that a given embedded system design - while satisfying the given specification requirements, is still flexible enough in incorporating new functionalities and in being customized for new performance constraints.

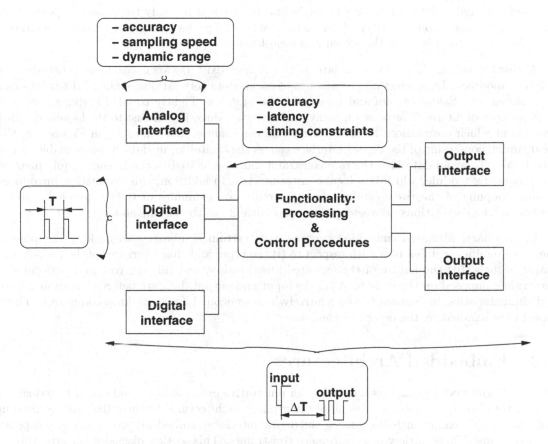

Figure 1.8: Embedded systems design and performance requirements.

Figure 1.8 illustrates different design and performance requirements in an embedded mixed-signal system, and some of the interactions between them. The constraints can be classified into *global constraints*, if they are at the level of the entire system (e.g., accuracy, speed and power consumption), and *local constraints*, if they characterize a particular module (e.g., the constraints of the analog interface). For example, the figure shows a global timing constraint between data bits being available at the digital interface and the corresponding outputs being produced. This defines the maximum overall execution time of the input interface, processing, and output interface modules. These are local constraints that characterize the modules. The process of relating the global constraints to the local constraints is called *constraint transformation*, and is important for a top-down design methodology, as detailed in Section 1.3 and in the following chapters of this textbook.

Another example of constraint transformation relates to mapping the global accuracy constraint of the system into local requirements of the analog and digital modules. For example, the analog interface might be constrained by accuracy constraints, which limits the ratio between the power of the useful signal and noise, a metric called the signal-to-noise ratio. The bandwidth of the signal and the accuracy constraints impose limits on the sampling speed of the interface, which, in turn, constrains the speed of the data processing module. In addition, the interface must have the capability to process the entire signal range, which defines the dynamic range and also the sampling speed of the interface circuit. Finally, the required data accuracy defines the bit width of the digital data, which affects the hardware cost and possibly the processing execution time also. In conclusion, a variety of correlated design and performance constraints characterize an embedded system both at the global and local levels.

A third example shows the correlations that might exist between the local constraints of different modules. Depending on the amount of constraints correlations, embedded systems can be grouped into *tightly coupled* and *loosely coupled systems*. Tightly coupled systems involve a large number of interactions between many module constraints, in contrast to the loosely coupled systems in which correlations involve few module constraints. For example, in Figure 1.8, the local timing constraint of the digital interface causes the digital input data to be available at the rate $1/\Delta T$. To avoid data loss, the rate constraint must be related to the latency requirement of the processing module, which is a local constraint also. In addition, this correlation introduces a loose coupling of the two modules. Analog circuits are examples of tightly coupled systems because of the correlations between voltages and currents at the circuit nodes.

In summary, although embedded systems have certain similarities to desktop computers, there are significant differences with respect to (i) the types and characteristics of their input and output signals, (ii) the nature of the executed functionality, and (iii) the cost and performance constraints imposed on the system. Also, the input and output data are heterogeneous in nature and characteristics, in contrast to the relatively homogeneous I/Os in desktop computers. These aspects are handled by the design methodology.

1.2   Embedded Architectures

The PSoC architecture [12] was selected as an illustrative example of an embedded mixed-signal architecture because it incorporates all of the primary architectural features that are required in embedded applications, including analog and digital interfaces, embedded processing, and support for addressing a large variety of performance constraints. This section discusses the structure of embedded mixed-signal architectures, their main subsystems, and some of the popular embedded architectures.

Mixed-signal, embedded architectures consist of both digital and analog modules. In addition, compared to desktop processor architectures, such architectures are resource-constrained. This means that their data widths are shorter, the clock frequencies are slower, the data and program memory spaces are smaller, and the embedded CPUs have less hardware resources, e.g., fewer registers, no pipeline stages, no on-chip cache memory and no floating point operations. However, embedded architectures interface resources to connect to a rich set of peripherals, have customized circuitry allowing high performance execution of certain operations, and support for implementing interrupt service routines and multitasking execution.

The following is a summary of the main subsystems of embedded mixed-signal architectures:

- *CPU*: The CPU supervises the functioning of the entire system, performs data processing, and participates in interfacing with peripherals. Many embedded CPUs have simple architectures, which reduces their cost and power consumption, and simplifies the complexity of

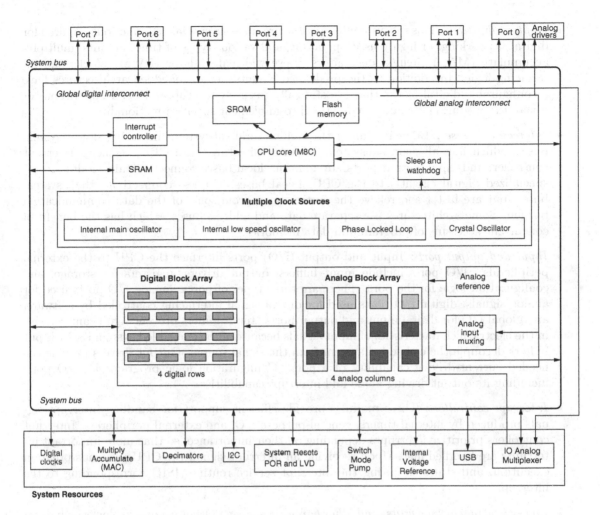

Figure 1.9: PSoC mixed-signal SoC architecture [12].

their software tools, including compilers, simulators, and emulators. Table 1.2 summarizes the main types of CPUs in embedded architectures.

- *Memory system*: Different kinds of on-chip and off-chip memories store programs and data of embedded applications. Both nonvolatile (flash, EPROM) and volatile memories (RAM) make up the relatively small memory space of embedded architectures, which places constraints on the footprint of the operating system and size of the application software that can be supported.

- *Analog subsystem*: This subsystem includes analog and mixed-signal circuits used primarily to interface the CPU core to analog peripherals. The available analog circuits include various types of filters, analog-to-digital converters, digital-to-analog converters, programmable gain amplifier circuits, analog output drivers, analog multiplexers, and comparator circuits. Newer embedded architectures also contain on-chip receiver-transmitter circuits for wireless communication. In addition to these blocks, some embedded architectures have reconfigurable analog circuits that can be customized for application-specific operation.

- *Digital subsystem*: This subsystem has digital circuits with specialized functions required for digital networking, or highly used operations and various types of timer circuits, multiply-accumulate (MAC) circuits, decimators, pulse width modulator (PWM) blocks and interfacing logic. In addition to the predefined circuits, some embedded architectures have reconfigurable digital blocks that are statically (at system setup and before operation) or dynamically (during operation) configured to implement specific functionality.

- *Interconnect buses*: Buses link all of the architectural subsystems including slower system buses, which link all main subsystems of an architecture, such CPU, memory, interrupt controller, and input/output ports. In addition, local buses connect certain modules (e.g., specialized digital circuits) to the CPU. Local buses offer less connectivity than system buses, but are faster and relieve the system buses from some of the data communication burden. Some architectures use separate data and address buses, which has the benefit of concurrent accessing of program and data memories.

- *Input and output ports*: Input and output (I/O) ports interface the CPU to the external peripherals. I/O ports include input buffers, output drivers, registers for storage, and configuration logic for the port. There are three types of ports: analog I/O ports used for analog signals, digital I/O ports used for digital signal interfacing controlled by software, and global I/O for digital input and output ports. Input buffers prevent significant changes in the electrical characteristics of input signals because of the embedded system itself. Input buffers decouple the embedded system from the signal source. Output drivers provide the needed current-driving capability of a port. Configuration logic programs an I/O port, including its output driving mode and interrupt capabilities.

- *Interrupt controllers*: These modules provide efficient support for handling interrupt signals produced by internal timers, general-purpose I/O, and external peripherals. Interrupt controllers prioritize interrupts depending on their importance, so that more important interrupts are handled first. Upon receiving an interrupt signal, the CPU suspends program execution, and starts executing the interrupt service routine (ISR) corresponding to the interrupt.

- *Power supply, clocking units, and other hardware support*: Other hardware blocks include an internal main oscillator, internal low-speed oscillator, external crystal oscillator, sleep and watchdog timer modules, phase-locked loop circuit, interfacing circuits (e.g., I2C controller, and USB controller), internal voltage reference, and switch mode pump.

Table 1.2 lists the three main processing unit types used in embedded applications: microprocessor and microcontroller, digital signal processors (DSPs), and reconfigurable architectures.

Table 1.2: Popular CPU types used in embedded applications.

Architecture types	Speed (MHz)	Memory	Interfacing	Data width (bits)
Microprocessors & microcontrollers	12–100	4 k-64 k program., 128bytes-2 k data	serial, UART, I2C,ADC,SPI	8-32
DSP	160–200	16 k-8 M	serial, DMA, I2C, PWM, etc.	16-40
Reconfigurable	80–400	0-8 M	programmable	programmable

This table also summarizes their main characteristics: processing speed, available memory space, interfacing capabilities, and power consumption. These units are detailed next. More examples of embedded architectures can be found in the related literature [32, 39, 40]. Table 1.3 introduces several popular general-purpose microprocessors and microcontrollers.

Table 1.3: Popular microprocessors and microcontrollers.

Model	Speed	On-chip memory	Interfacing	Data width
ARM7 (ARM)	90 MIPS @ 100MHz	4G addressable	-	32 bits
DSP56800 (Freescale)	35 MIPS @ 70MHz	28k-152k	serial, CAN	16 bits
MCS51(8051) (Intel)	1 MIPS @ 12MHz	4k bytes ROM, 128 bytes RAM	full duplex serial	8 bits
PIC 24FJXX (Microchip)	16 MIPS @ 32MHz	64k-128k flash 8192 bytes RAM	SPI, UART I2C, parallel	16 bits
80C552 (Philips)	1.36 MIPS @ 24MHz, 0.91 MIPS @ 16MHz	8k bytes ROM, 256 bytes RAM	parallel I/Os, UART, I2C	8 bits
MAXQ3120 (Maxim)	8MIPS @ 8MHz	16k flash, 256 words RAM	I/Os, LCD driver UART, ADCs	16 bits
PPC (IBM)	pipelined RISC	32k data cache, 32k prog. cache	timers, memory & coprocessor interf.	32 bits
PSoC (Cypress)	24MHz	64k flash up to 2k RAM	I/Os, I2C, UART, PWM, ADC	8 bits
XScale (Intel)	superpipelined RISC	32k instr.cache, 32k data cache	coprocessor interf., media processing	32 bits

ARM7 [2] is a microprocessor core used in portable devices, advanced user interfaces and automotive applications. This processor has lower power consumption than other processors, and provides reasonably high computing speed. The CPU is a 32-bit RISC processor with three-stage pipelined execution. Sixteen of the thirty-seven 32-bit registers are accessible to developers. ARM cores have seven operation modes: (1) user mode (normal program execution mode), (2) data transfer mode, (3) IRQ mode (used for interrupt services), (4) supervisor mode (for operating system support), (5) abort mode (when aborting data and instruction fetch), (6) system mode (provides privileged mode for user), and (7) undefined mode (when fetching an undefined instruction). Besides regular instructions for data processing, data transfer, and branch, the instruction set also includes instructions specialized for interaction with the coprocessor (e.g., coprocessor data transfer, coprocessor data operation, and coprocessor register transfer) and software interrupts. ARM Thumb extension reduces the instruction length to 16 bits, hence reducing the memory size and cost of an embedded system. ARM10 improves instruction execution by branch prediction, caches and parallel load/store unit. ARM Jazelle has an extension for Java-bytecode execution.

DSP56800 [3] is a microcontroller for digital signal processing in wireless and wireline communication, voice and audio processing, high-speed control, and many other applications. The microcontroller is based on a 16-bit Harvard architecture, which can be interfaced to flash and

RAM, and a large set of external peripheral circuits (e.g., serial communication interfaces) pulse width modulator circuits, analog-to-digital converters, watchdog timers, programmable I/Os, and controller area networks (CAN). This microcontroller architecture is optimized for efficient execution of multiply-accumulate instructions, such that the execution unit, memory, and peripherals can operate in parallel, that is in one instruction cycle the processor can execute one instruction prefetch, one multiplication involving two 16-bit operands, one 36-bit addition, two data moves, two address pointer updates, sending and receiving full-duplex data, and timer counting. It can separately address one program memory and two data memories, by using 16-bit addresses. The DSP56800 microcontroller has two operation modes for reducing power consumption. In the wait mode, the core is shut down while the peripherals and the interrupt controller continue to operate. The stop mode saves more power than the wait mode by deactivating the clock oscillator. This microcontroller has an *on chip emulation unit* that allows a user to examine the core, peripheral registers, and memory. Breakpoints can also be set on the program and data memories, and single-stepping through program execution is supported.

PIC24FJXXGA006 [8] integrates a 16-bit microcontroller with on-chip flash, timers, and a large variety of interfacing circuits for example ADC, comparators, PWM, parallel ports for master and slave operations, and serial communication modules for UART, I2C, and serial peripheral interface (SPI). The CPU (PIC24 core) is a modified Harvard architecture with 16-bit data path and 24-bit address path. The chip has 64 k–128 k bytes of flash. However, the total addressable memory space includes 8 M bytes of program memory and 64 kbytes of data memory. On-chip analog circuits include a 10-bit ADC and comparators with programmable I/O configurations. Three modes of operation are provided to reduce power consumption. On-the-fly clock switching allows the device clock to be changed under software control. Doze mode reduces CPU speed during slow operations such as serial communications. The instruction-based, power-saving mode suspends all operations, or shutsdown the core leaving only the peripherals active.

Philips 83C552 [4] is a single-chip 8-bit microcontroller with an analog-to-digital converter, high-speed outputs, and pulse width modulation circuit. The microcontroller is an 80C51 CPU core capable of addressing the 8 kbytes of on-chip ROM and 256 bytes of on-chip RAM. Memory can be extended externally up to 64 kbytes of ROM and 64 kbytes of RAM. The chip also includes a 10 bit analog-to-digital converter, PWM circuit, watchdog timers, 16-bit timers/counters, I2C serial I/O port, full duplex UART circuit, five 8-bit parallel I/O ports, and one 8-bit parallel input port, all connected by an 8-bit system bus. For reducing power consumption, the chip has two low power modes. In the idle mode, the CPU is stopped but RAM, timers, serial ports, and the interrupt controller are operating. In the power-down mode, only RAM remains active.

MAXQ3120 [6] is a mixed-signal controller that integrates on a single chip a RISC processor, program/data memory, analog-to-digital converters, timers/counters, watchdog timers, LCD driver, USART, and interfacing circuits for PWM. The core is a 16-bit RISC processor operating at 8 MHz with 16 registers, 16-bit instruction words and 16-bit address lines. The controller incorporates 16-bit $\Delta\Sigma$ ADCs, including the related $sinc^3$ filters for decimation. This chip also has 32 general-purpose I/O pins, an LCD driver, two serial USARTs for serial communication, a multiply-accumulate circuit, 16-bit programmable timers/counters, PWM support, and one programmable watchdog timer. The processor clock can be slowed down by a factor of 2, 4, 8, and 256, or stopped to reduce power consumption.

PPC 405FX [10] is a 32-bit RISC processor with a 5 stage pipeline. The processor has thirty-two 32-bit registers, 16 Kbyte instruction cache, 16 Kbyte data cache, hardware circuits for fast multiplication and division, and can address 4 Gbytes of external memory. In addition, programmable interval timers, watchdog timers, and fixed interval timers are also available.

Higher computing performance is achieved by coupling the processor to a dedicated coprocessor by the auxiliary processor unit interface, or by connecting memory by the on-chip memory interface.

Table 1.4: Popular digital signal processors (DSPs).

Model	Speed	On-chip memory	Interfacing	Data Width (bits)
ADSP-219x (Analog Devices)	160 MHz	16 k-128 k	serial, JTAG programmable I/Os	16
PNX 1330 (Philips)	VLIW, 5 instr.simult.	16 k data cache, 32 k instr. cache	CCIR, I2C, SSI, S/PDIF	32
SHARC (Analog Devices)	333 MHz	6 M ROM, 2 M SRAM	DMA, S/PDIF, PWM, DAI, DPI	32/40
TMS320C6000 (Texas Instruments)	VLIW, 8 FU	512 k program, 512 k data	DMA, serial, timers, expansion bus	32
LSI40x (ZSP400) (VeriSilicon)	200 MHz	96 k-252 k	serial, JTAG, programmable I/Os	16/32

Table 1.4 summarizes the main attributes of some popular digital signal processors. *PNX 1330* [9] is a high-performance media processor. It is meant for real-time, high-quality audio and video processing in video phones, digital TV sets, desktop computers, and so on. The processor includes a very large instruction word CPU core, and several multimedia-oriented coprocessors that can operate independently or in parallel with the CPU. The CPU and the coprocessors are linked by an internal "data highway" bus that also connects the external SDRAM and external peripheral devices. Coprocessors are customized for multimedia input, output, and processing: the (1) *video in unit* interfaces the media processor to input data from a CCIR 601/656 compliant device, such as a digital camera. (2) The *image coprocessor* performs image scaling and filtering, and (3) the *variable length decoder* decodes the Huffman-encoded video frames, for example in high-rate MPEG-1 and MPEG-2 video streams. (4) the *enhanced video out unit* produces 8-bit CCIR 656 video data streams and (5) the *Sony/Philips Digital Interface Out* (S/PDIF) module outputs one-bit, high-speed data streams. The (6) *I2C interface* provides multimaster, multislave interfacing at rates of up to 400 kbits/sec. (7) The *synchronous serial interface* provides interfacing to analog modem and ISDN front-end devices. Finally, (8) *audio in* and (9) *audio out* coprocessors are interfaces to input and multi-channel output stereo digital audio data. The CPU core is a VLIW processor that can simultaneously execute five operations in each clock cycle. The CPU core has a data width of 32 bits, and can address a linear address space of up to 512 M words. It has 128 full-purpose, 32-bit registers, 16 k of data cache, and 32 k of instruction cache. Power consumption is reduced by two features: in the global power-down mode, all internal clocks are shut down, SDRAM is set to a low-power, self-refreshing mode, and the majority of peripherals are shut down. In addition, the processor can selectively power down peripheral devices.

SHARC - ADSP 2136x [13] is a 32-bit processor for high-quality signal processing, including audio processing, medical imaging, speech recognition, communication, and so on. The processor architecture is a super Harvard architecture with parallel data transfers and arithmetic operations. The clock frequency is 333 MHz. The chip has 6 Mbytes of on-chip ROM and 2 Mbytes of SRAM. The DSP includes DMA, PWM, and interfaces such as the digital audio interface (DAI), serial peripheral interface, serial port for point-to-point multiprocessor communication, S/PDIF, and

JTAG for test access. The digital peripheral interface (DPI) consists of interrupt controller, two wire interface part, signal routing unit, timers, and UART. Two low power modes are available: the PLL circuit frequency can be programmed in software, and the clock to the peripherals can be disabled.

TMS320C6000 [16] is a more general-purpose processor consisting of the CPU core, peripherals, and memory. The CPU core is a VLIW processor with eight functional units that operate in parallel. Functional units are organized into two identical sets, each set including one logic unit for arithmetic and logic operations, one shifting unit, one multiplier, and one data address unit for data transfers between the registers and memory. The CPU core has 32 registers organized as two register files. Registers are 32 bits. On-chip peripherals include DMA controller, external memory interface, power-down logic, serial ports, expansion bus, and timers. The chip has 32-bit addresses. The on-chip memory includes a 512 k bits instruction memory and a separate 512 k bit data memory. Instruction memory words are 256-bits, and data memory words can be 8, 16, or 32-bits. For low-power processing, there are three power-down modes for the TMS320C6000 processor. The first power-down mode blocks the internal clock at the boundary of the CPU, thus preventing most of the CPU logic from switching. The second power-down mode stops the on-chip clock structure at the output of the phased lock loop (PLL) circuit. In addition, the third power-down mode also stops the entire internal clock tree, and disconnects the external clock structure from the PLL circuit.

Table 1.5: Popular reconfigurable architectures.

Model	Speed (MHz)	Memory	Interfacing	# gates	Max. # pins
Altera MAX 7000 (Altera)	125	0	PCI, JTAG programmable I/Os	600-5 k	208
Altera FLEX 10k (Altera)	up to 200	41 k RAM	PCI, JTAG	250 k	470
Xilinx Spartan-II (Xilinx)	80	0	PCI, programmable I/Os	40 k	224
Xilinx Vertex (Xilinx)	400	8 M RAM	serial, JTAG, Ethernet, audio CODEC	4,074 k	808

Table 1.5 lists the performance attributes for four popular reconfigurable architectures. *MAX 7000* [1] is based on EEPROM-based programmable logic devices (PLDs). The basic programmable unit is a macrocell. Each macrocell consists of a programmable logic *AND* plane, a fixed logic *OR* array, and configurable registers. Hence, the macrocell structure is well suited for implementing logic functions described as sums of products. Sixteen macrocells form a logic array block (LABs), and several LABs exist in a MAX 7000 chip. In addition to the macrocells, each LAB contains 16 shareable expanders that can be used to extend the functionality of any of the LAB's macrocells. Depending on the circuit types, the number of LABs is between 2 and 16, which is equivalent to having between 32 to 256 macrocells, or between 600 to 5000 digital gates. The maximum clock frequency is $\approx 125\ MHz$ depending on the circuit type. The architecture includes a reconfigurable interconnect connecting the macrocells, and programmable I/O ports. Interfaces are for PCI devices and JTAG, but the programmable PLDs can implement a much larger variety of interfacing procedures. To save power consumption, the macrocells can operate

at a power consumption of 50% or lower (eight modes in total), and their slew rate (SR) can also be lowered so that less signal coupling results during digital switching.

Spartan [15] is a reconfigurable architecture based on configurable logic blocks (CLBs). The functionality of a CLB is set by programming three lookup tables (LUTs), two multiplexer circuits, and two flip-flops. A LUT is a logic function generator: two of a CLB's LUTs can implement four-input functions, and the other LUT can realize a three input function. CLBs are organized in two-dimensional arrays, the size of the array varying from 10×10 to 28×28. Hence, Spartan architectures can include between 5k to 40k gates, between 360 to 2016 flip-flops, and between 77 to 224 programmable I/Os. The architecture includes three kinds of interconnect: CLB routing channels along the rows and columns, IOB routing channels connecting the I/Os and the CLB routing channels, and the global routing channels for the clock and high fanout signals. The CLB's flip flop can be also used as on-chip RAM. The architecture includes a low power-down pin which reduces the power supply current to about 100 μA but at the expense of losing the flip-flop content.

1.3 Top-Down Design Flow

In essence, the goal of embedded system design is to customize the hardware and software components of a design in order to meet the specific functional, cost, and performance needs of the application, and optimize its characteristics for its operating environment. The idea of customizing an embedded system design for an application is very important, because it then achieves the specified performance requirements at relatively low costs compared to desktop computers. On the negative side, customization may result in less flexibility in terms of efficient reuse of an embedded system design for a different application. For example, setting a tight cost constraint for the embedded fan controller (e.g., total cost below $5) might require the use of low-end microcontrollers or smaller-size reconfigurable chips. In contrast, for implementing complex control algorithms such as those used in server systems, if the total cost margin is set higher (e.g., $20) a DSP-based architecture might be used to meet the real-time constraints of the algorithms. In the customization process, a large number of cost–size–performance tradeoffs are identified, expressed and analyzed to determine the specific attributes of the hardware circuits and software modules of the implementation.

Figure 1.10 presents a top-down design flow for embedded system design. The flow starts from an abstract system specification developed using the functional, interfacing, cost, and performance requirements for the system. Then, the design is incrementally refined during the top-down design process by continuously adding new implementation details to the design. This process transforms the global requirements of the system into local requirements for the subsystems and modules. For example, earlier refinement steps determine the number of processing cores, the partitioning of functionality to the cores, the structure of the memory subsystem and the number of buses. Later refinement steps determine the implementation details of the building blocks, such as the topology of the analog circuits, circuit transistor dimensions, logic level design of the digital circuits, among others. Each refinement step optimizes the design by conducting a *tradeoff analysis* that considers different design solutions and their impact on the overall system performance. Thus, the design flow should incorporate a *performance evaluation mechanism* to evaluate the quality of a design decision in terms of its impact on the system performance, and a *modeling procedure* to express the defining attributes (e.g., speed, power consumption, etc.) of the building blocks in the design.

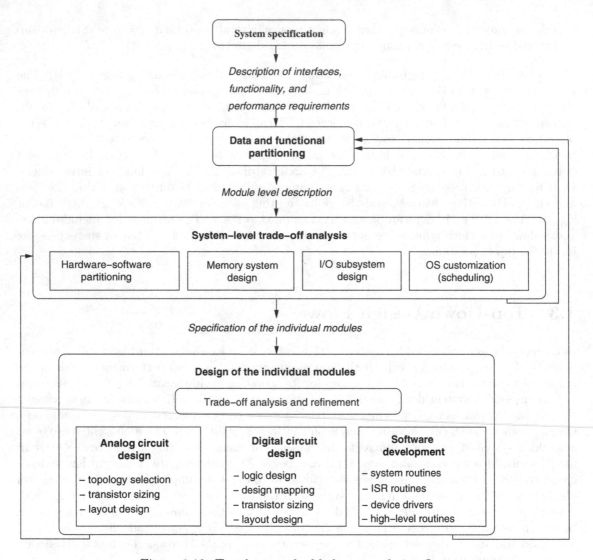

Figure 1.10: Top-down embedded system design flow.

1. *System specification* is the process of describing the interfacing, functionality, and performance constraints of an embedded system. System specification uses, in most cases, a simulatable notation, for example a specification language. This has the advantage of being able to simulate the specification to verify its correctness and completeness. Specifications are at different levels of abstractions, such as concept level, algorithm level, block (module) level, and circuit (component) level. High level specifications offer more design opportunities at the expense of a more sophisticated design flow.

2. *Functional partitioning* step reorganizes a system specification into subsystems and modules that perform a specialized functionality, such as a specific type of interfacing function or data processing. The identified modules may undergo separate, top-down, design procedures. For example, data processing modules, such as digital filters, are obviously designed in a different way than analog processing modules, such as analog-to-digital converters. The output of this step is a set of descriptions presenting the functionality and interfaces

of all modules. Functional partitioning can be performed multiple times in a design flow, at different levels of abstraction and for different subsystems.

3. *System-level tradeoff analysis* maps the system-level performance and design requirements into requirements for the building blocks. Hence, this step realizes constraint transformation. Numerous design tradeoffs are formulated and analyzed during this step (e.g., cost vs. speed, speed vs. power consumption and bandwidth vs. accuracy vs. speed). Possible tradeoff analysis steps include deciding whether a module should be implemented in hardware or software, an activity known as hardware/software partitioning, or deciding on the main attributes of the memory subsystem, such as memory size, number of memory modules, access time required, and so on needed to meet the timing requirements of the system. At the I/O level, this step determines the number of I/O ports, their communication protocols and mapping of input and output signals to ports. Tradeoff analysis also includes refining the operating system (OS) or firmware level routines (e.g., scheduling of tasks executing by the same processor, and arbitration of module access to system and local buses). It is important to stress that at the end of system-level tradeoff analysis, modules can be individually designed, because the interfaces, functionality, and performance constraints of each module are now known.

4. *Design of the individual modules* is the activity of implementing the modules of the system, such as different interfacing modules, data processing modules (e.g., video and audio processors), the memory module, and the bus communication subsystem. Modules can include hardware circuits, and software routines of different natures, e.g., drivers, interrupt service routines (ISR), as well as, control and data access routines. Complex modules undergo a top-down design process in which their design is refined by partitioning each module's functionality into sub-modules. Design tradeoff analysis is utilized to identify the design requirements of the submodules and define their design.

5. *Analog circuit design* is the process of implementing the analog and mixed-signal circuits of a design. Popular analog circuits include operational amplifiers, transconductor amplifiers, comparator circuits, analog multiplexers, integrator circuits, and sample-and-hold circuits and many more. Analog circuits are implemented as either continuous-time circuits or as switched capacitor circuits. Analog circuit design includes the selection of the circuit topology, transistor sizing, and circuit layout design.

6. *Digital circuit design* of customized digital circuits for the embedded design includes interfacing circuits and customized data processing hardware, for example various digital filters, decimator circuits, encoders, for example Huffman encoders for multimedia applications to name just a few types. The design methodologies used depend on the complexity of the circuits. More complex digital subsystems (e.g., a universal asynchronous receiver transmitter module or a video processor) are subject to separate top-down refinement, including data and functional partitioning step, followed by tradeoff analysis, and module design. For simple circuits, the first step is to develop the logic design for combinational and sequential circuits, followed by mapping these designs to basic digital gates such as NAND and NOR gates, inverters, multiplexers, decoders, flip-flops, and registers. This is followed by transistor sizing for basic gates, and circuit layout design.

7. *Software development* creates application and system software. Application software includes methods for data processing, control, and graphical user interfaces (GUIs). System software development addresses real-time operating systems (RTOS), middleware, and networking.

8. *Circuit modeling* is the process of characterizing the behavior of hardware circuits, including electrical behavior, silicon area, and number of pins. This step is very important for correctly predicting the system's performance, because electronic circuit behavior can vary significantly from their ideal counterpart. Circuit models predict the main nonidealities of analog and digital circuits, such as propagation delay, power consumption, noise margins for digital circuits, and finite gain, poles/zeros, bandwidth, harmonic distortion, and circuit noise for analog circuits. Different kinds of nonidealities, and therefore different kinds of models, have to be contemplated at different levels of abstraction in a top-down design flow. Early design steps, and hence steps at higher levels of abstraction, use coarser circuit models, because second- and third-order circuit nonidealities have a minor impact on system performance. In contrast, later design steps employ more detailed circuit models, in as much as a more precise insight into circuit behavior is needed. Other kinds of nonidealities include bus delay, power supply voltage variations, substrate coupling, sensitivity to temperature, additional harmonics of input signals, and so on.

9. *Software characterization* develops models predicting performance attributes of software routines (e.g., speed, required memory, memory access patterns and power consumption). Software characterization is difficult because of data-dependent programming constructs, such as if statements, case statements, and loops. The performance of these constructs is determined by the specific data values, and therefore can be determined precisely only at execution time.

Table 1.6: Embedded market predictions (Source: BCC Research Group)[17].

	2004 ($ billions)	2009 ($ billions)	AAGR % (2004–2009)
Embedded software	1,641	3,448	16.0
Embedded hardware	40,539	78,746	14.2
Embedded boards	3,693 billions	5,95	10.0
Total	45,873 billions	88,144	14.0

10. *Performance evaluation* is the step of finding the performance attributes of the system and its modules. This step is critical in predicting the performance of different design options without actually building, testing and measuring the designs. Performance evaluation can be based on analytical expressions such as system models, processor models, circuit macromodels, etc., that predict system performance. Models are simple to use in design, but their development requires significant effort, and their prediction accuracy is not very high. The alternative is to base performance predictions on simulating a simplified description of the system. The drawback to this approach is that the simulation time can become lengthy. This is prohibitive when it is necessary to analyze a large number of design alternatives.

Designing embedded systems requires significant cross-disciplinary knowledge, including system design, computer architectures, microprocessors and microcontrollers, specification languages, real-time operating systems, analog and digital circuit design, programming languages, and compilers. This textbook emphasizes the correlations between these disciplines, and stresses the four main aspects in top-down, refinement-based design, (i) specification, (ii) tradeoff analysis, (iii) modeling, and (iv) performance evaluation.

1.4 Embedded Systems Market

BCC Research Group [17] has estimated the worldwide embedded system market to be approximately $45.9 billion in 2004, and suggested that it will grow to $88,144 billion by 2009. They have also predicted that this dramatic growth of the embedded systems market is due to four primary reasons: (1) the popularity of embedded systems and systems on chip for mobile consumer products and automotive applications, (2) the growth of an ultra-wide band Internet, (3) the availability of integrated functionality which increases the value of embedded products significantly above the value of the components, and (4) the emergence of large markets for specialized computing systems that are part of more complex systems (not necessarily electronic systems). Currently, integrated chips for embedded systems are the majority of semiconductor sales: according to the *World Semiconductor Trade Statistics Blue Book*, embedded processors represent 94% of the total semiconductor sales (5 billion processors) in contrast to only 6% for desktop PC processors, including Pentium and PowerPC processors.

Embedded systems are envisioned by many to be at the core of the next IT revolution. Projections for future applications include embedded systems in healthcare, Internet-friendly consumer appliances, sensor networks for environment monitoring for security and protection, etextiles and support for elderly people and people with disabilities.

1.5 Embedded Design Example: Fan Control System

The section provides a brief illustration of the main characteristics of a simple(r) embedded system, and the steps applied to develop its implementation. This application is an embedded fan controller that monitors the temperature inside a computer, and controls the speed of a fan to ensure that the electronic circuits do not overheat. This example introduces the many facets of embedded system design, including system specification, specification simulation, design implementation/testing and debugging. The following chapters elaborate these issues in greater detail, based on design solutions in [18] and [11].

Table 1.7: Fan speed control function depending on the measured temperature.

Temperature (°C)	Fan Speed
$-55 - +25$	Off
$+25 - +35$	Low
$+35 - +50$	Ramp
$+50 - +130$	High

The fan controller adjusts the fan's speed based on the sensed temperature. The controller selects one of four predefined fan speed values, (i.e., off, low, ramp, and high) as shown in Table 1.7. If the temperature stays below 25°C then the fan remains off. If the temperature is in the range 25°C to 35°C, then the fan rotates at a low speed. For temperatures ranging from 35°C to 50°C, the fan's speed must increase linearly with the temperature, following the ramp dependency:

$$Speed_{fan} = K \frac{room\ temperature}{10} \tag{1.5}$$

where K is a constant.

Finally, if the temperature exceeds 50°C then the speed of the fan is set to high. An LCD displays the measured temperature, and the fan speed that is set by the embedded fan controller. The fan's speed is measured in rotations per minute (RPMs). The accuracy of the control procedure with respect to the sensed temperature should be less than *1%*. In addition, the cost and the size of the implementation should be as small as possible, based on the fact that the controller ought to be usable in low-cost products.

1.5.1   Description of the Fan Controller System

The starting point in designing the signal processing flow of the system is the description of the fan controller: (i) a temperature-sensing module continuously delivers temperature readings to the processing unit, (ii) then the processing subsystem computes the required RPM value for the sensed temperature, and (iii) generates the control (actuation) signals that set the RPM value of the fan motor to the computed RPM value, and (iv) finally, the LCD unit displays the temperature and RPM values. The linear processing flow is illustrated in Figure 1.11.

According to this algorithm, the controller apparently performs only a linear succession of processing steps, from the input to the output. The caveat is that there is not a one-to-one correspondence between the desired RPM values and the control signals that have to be produced for the fan motor. Instead, the controller must continuously adjust the values of the control signals in a feedback loop until the resulting RPM values of the fan correspond to the values in Table 1.7. Therefore, the linear processing flow of the fan controller system must be extended to the feedback structure shown in Figure 1.12. The complete processing flow of the fan control system is now as follows. (i) A temperature sensor provides continuous temperature readings; (ii) a tachometer measures the rotational speed of the fan motor (i.e., shaft speed) and provides the value to the processing unit, (iii) the processing unit reads the inputs coming from the temperature sensor and tachometer, and accordingly adjusts the control signals to the motor to increase or decrease its RPM value as required by the description in Table 1.7; and finally (iv) the LCD displays the temperature and RPM values. The continuous adjustment of the control signals sets the rotational speed of the fan to the RPM values specified in Table 1.7.

Note that the controller design has to solve an important "mismatch" between the nature of the input signal sensed by the temperature sensor and the type of processing conducted by the processing unit. The sensed signal is *continuous time*, that is the signal has a well-defined value at each instant of time. In contrast, the digital processing unit operates at *discrete time units* with a time period defined by the system clock. It is a defining characteristic of mixed-signal embedded systems to consist of heterogeneous parts, some of which are defined and operate in continuous-time, others in discrete-time. The mixed-signal system must incorporate interfacing

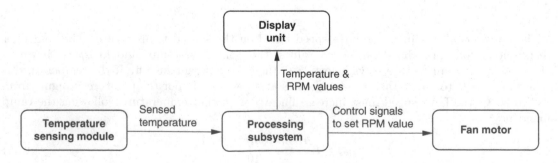

Figure 1.11: Linear signal-processing flow.

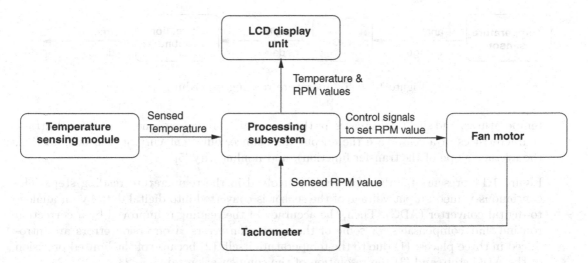

Figure 1.12: Feedback signal-processing flow (concept-level description).

mechanisms between continuous-time data/processing and discrete-time data/processing. In the present example, the processing unit samples the continuous-time temperature sensor output at uniform instances of time, which makes the fan controller a *sampled system*.

The block structure in Figure 1.13 is based on the processing flow in Figure 1.12.

- *Temperature sensor*: The temperature sensor used in this example provides an output voltage proportional to the temperature. Temperature sensors are small, cheap, and accurate enough for normal temperature ranges (e.g., -55°C to 130°C [5]). Other possibilities are to sense the temperature using thermistors, thermocouples, or RTDs, but these would complicate the design.

  The behavior of a sensor is characterized by its *output voltage versus temperature* transfer function, which provides the output voltage of the sensor for a specific temperature value. The temperature range that can be measured by a sensor depends on the supply voltage of the sensor, for example lowering the supply voltages reduces the temperature range that can be measured. The accuracy of the measured temperature is higher for ambient

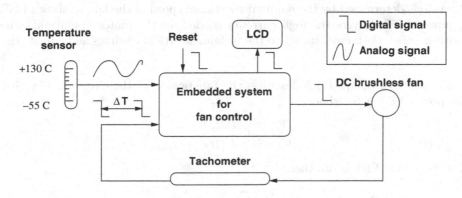

Figure 1.13: Block structure for the embedded fan control system (concept-level description).

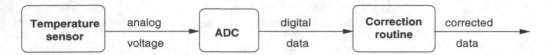

Figure 1.14: Temperature reading procedure.

temperatures, and slightly decreases for the temperature range extremes [5]. Other electrical characteristics of a sensor are the output impedance, quiescent current, sensor gain (i.e., the average slope of the transfer function), and nonlinearity [5].

Figure 1.14 presents the data processing involved in the temperature reading step. The continuous-valued output voltage of the sensor is converted into digital data by an analog-to-digital converter (ADC). Then, the accuracy of the reading is improved by a correction routine that compensates for some of the conversion errors. Conversion errors are introduced in three places: (1) due to the temperature itself, (2) because of the limited precision of the ADC step, and (3) the resolution of the compensation table [5, 23].

- *Fan*: The fan used in this example is a DC "brushless" fan with a duty cycle α, and therefore speed, controlled by the output of the embedded controller. The system must control the speed of the fan by modifying its duty cycle.

The behavior of a DC motor can be expressed, with acceptable accuracy, by the circuit shown in Figure 1.15(a) [18]. Voltage V is provided by the power source, current I is the resulting current, resistance R the armature coil resistance, and E_{bemf} the voltage generated across the motor due to the back electromotive force. T represents the torque, and ω the shaft speed of the fan motor. The following equations, referred to as the transducer equations, describe the motor's behavior [18]:

$$V = I R + E_{bemf} \tag{1.6}$$

$$E_{bemf} = K_V \omega \tag{1.7}$$

$$T = K_M I \tag{1.8}$$

The constants R, K_V, and K_M depend on the motor model and manufacturer. More details about the functioning and characteristics of electric motors can be found in [27, 34].

Figure 1.15(b) illustrates the electric circuit for operating and controlling the fan motor. This circuit includes a pulse width modulator circuit, and a power FET [14]. PWM circuits produce waveforms with a programmable duty cycle, α. For the fan controller design, the duty cycle is determined by the required rotational speed of the fan, as shown in Table 1.7. The power FET delivers the high currents needed for the motor operation. The control algorithm senses the temperature and the tachometer RPM readings and adjusts the PWM's duty cycle.

The following set of equations describes the functioning of the circuit in Figure 1.15 [18]. If the power FET is on, then:

$$V_i = I R_s \tag{1.9}$$

$$V_m = I (R_S + r_{dson}) \tag{1.10}$$

and if the power FET is off, then:

$$V_i = 0 \tag{1.11}$$

$$V_m = V - E_{bemf} \tag{1.12}$$

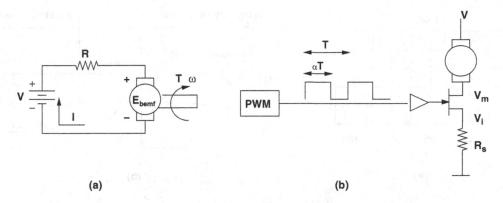

Figure 1.15: DC motor model and operation [18].

Based on the these equations, the following relationships are obtained for the average torque T and the average shaft speed ω, as functions of the PWM duty cycle α,

$$V_i^{average} = I\,R_s\,\alpha \tag{1.13}$$

$$V_m^{average} = I\,(R_s + r_{dson})\,\alpha + (V - E_{bemf})\,(1 - \alpha) \tag{1.14}$$

and

$$T_{average} = K_m\,I^{average} = K_m\,\frac{V_i^{average}}{R_s} \tag{1.15}$$

$$\omega_{average} = \frac{E_{bemf}}{K_v} = \frac{\frac{V_i^{average}\,(R_s + r_{dson})}{R_s\,(1-\alpha)} + V - \frac{V_m^{average}}{1-\alpha}}{K_v} \tag{1.16}$$

Expression (1.16) relates the PWM duty cycle α to the average shaft speed $\omega_{average}$ of the DC motor.

- *Tachometer*: The tachometer measures the rotational speed ω of the fan, that is the shaft speed, and generates input signals that are used for computing the fan speed. Then, using Equation (1.16), the PWM duty cycle α is modified so that the fan speed satisfies the requirements expressed in Table 1.7 for different temperature ranges. If the instantaneous speed ω is larger or smaller than the required value then the needed adjustment for the duty cycle α is computed using expression (1.16).

- *LCD*: The LCD displays the values of the temperature measured by the temperature sensor and the fan rotational speed ω measured by the tachometer.

Before starting the design of the actual embedded fan controller, the designer has to decide on the sensing and actuation devices to be used in the application, including the types of the temperature sensor, fan, tachometer, and LCD. This step is important because it defines the nature of the input and output signals to the embedded fan controller, and the performance requirements for the system design.

The selected temperature sensor was the LM20 sensor [5] because of its small cost and simple interfacing requirements. Figure 1.16(a) presents the output voltage versus the temperature transfer function of the sensor [5] because the power supply voltage influences the temperature range that can be measured. Therefore a supply voltage of 2.7 V – 5.5 V is required for this

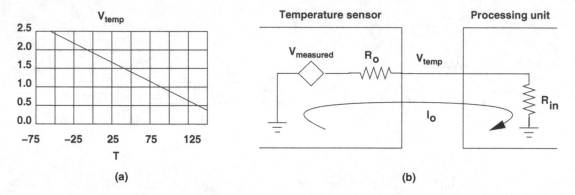

Figure 1.16: Temperature sensor.

application. Note that the curve is predominantly linear, but has a small parabolic curvature at the extremes of the temperature range. The temperature error is 1.5°C at normal temperature, and 2.5°C at the extremes. This error is corrected by the calibration routine shown in Figure 1.14. The following equation relates the sensor output voltage and the measured temperature [5],

$$T = -1481.96 + \sqrt{2.1962 \times 10^6 + \frac{1.8639 - V_o}{3.88 \times 10^{-6}}} \tag{1.17}$$

The sensor has a small output impedance ($\approx 160\Omega$) simplifying the interfacing of the sensor to the embedded processor which has very high input impedance. The equivalent circuit is shown in Figure 1.16(b). Because the current I_o is extremely small, the voltage V_{temp} at the processor pins is the same as the voltage $V_{measured}$ for the sensed temperature.

The type of fan determines the control signals that must be produced by the fan controller, for example the digital output signals for defining the drive frequency, drive polarity, spin-up time, and so on. Due to its small size, a three wire, brushless DC fan was chosen for this design. The fan's supply voltage is 5 V, and the power FET IRF7463 [14] was selected to deliver the high current to the motor.

The tachometer generates successive digital pulses that are used to compute the rotational speed of the fan. The type of the tachometer used defines the embedded processor interfacing procedure. For example, the tachometer used in the design has a five-step interfacing procedure, as follows: (1) Upon receiving the first signal from the tachometer, the processor must set the duty cycle α of the fan to high so that it rotates at constant speed. (2) Upon receiving the next tachometer signal, the processor waits for a predefined time for the tachometer reading to settle. (3)–(4) The third and fourth tachometer pulses are used to compute the time difference between the pulses. The RPM of the fan is based on this time difference. The controller reconnects to the fan motor so that the fan control algorithm can resume. (5) After receiving the fifth pulse, the control can proceed with a new tachometer reading and repeat this five-step procedure.

Five pins are needed for interfacing the devices to the controller: one pin for the temperature sensor, one for the tachometer, one pin for the fan, and two for the LCD. The LCD module displays outputs of the embedded system using a standard I2C bus interface. A supply voltage of 5 V can be used for all circuits.

All of the external device details are incorporated into Figure 1.17. This figure presents a refined block structure of the fan control system, including the external resistors and capacitors for

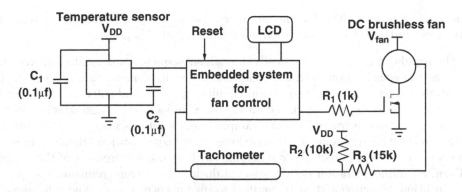

Figure 1.17: Interfaced fan controller system (concept-level description).

connecting the devices to the embedded controller. This step completes the process of defining the design requirements for the embedded fan controller because all of its interfaces and functionality are now completely determined.

Note the importance of the application requirements on the implementation decisions made during the development process. In this case, the cost and size of the fan controller had to be minimized. Therefore, low-cost devices were selected for temperature sensing, fan motor control, and display. These devices also require a very simple interfacing structure, which keeps the cost low for the circuits that interface the sensing and actuation devices to the processing unit. All electronic circuits can be powered by a single 5 V power supply, which also keeps the system cost and size low. The control algorithm was kept as simple as possible.

This application does not involve any special speed or reliability requirements. In contrast, the implementation of the cooling system for a high-end computer server, or a server farm, although starting from a signal processing flow similar to that of the fan controller, will require totally different implementation decisions and therefore, it is important to maximize the system's reaction speed to events and its reliability to failures. For example, if the temperature rises above a dangerous limit, then the controller must react quickly and shut down the server, or server farm. Also, if several of its parts fail to operate normally the system must remain functional. These requirements will change not only the type of the sensing and actuation devices used, but also the block structure of the system, for example by having redundant parts, and the functionality of the control algorithm to incorporate handling of special events.

1.5.2   Design of the Fan Controller System

The implementation process for the embedded fan controller follows a top-down design and validation flow. The purpose of the flow is to help correct implementation of the system, while minimizing its cost and time-to-market time.

The top-down methodology starts from the system description, and first develops a high-level specification of the system functionality and constraints. Having a high-level specification for the fan controller helps to achieve the goal of building correct implementations because the specification can be verified either by simulation or rapid prototyping before the system is actually built. Any errors in the system description are eliminated at this point.

Then, the top-down design flow proceeds through subsequent implementation steps that successively refine the specification until the complete system implementation is created. The final implementation consists of analog and digital circuits, memory, and software routines. Each refinement step focuses on a specific implementation issue, and analyzes possible solutions that each gives different performance tradeoffs. For example, some solutions may be faster, other require less hardware or software and some might have lower power consumption. Refinement steps select the solutions with the best potential for satisfying the low-cost requirement of the fan controller design. Then, all implementation details selected during the current refinement step are added to the specification, which can then be verified again for correctness. The refinement process continues until all of the implementation details have been determined, and the implementation specification has been produced.

The last step of the top-down flow is building the physical system and testing/debugging it. Figure 1.18 illustrates the entire design flow. The top-down methodology used the following four steps for the fan controller design.

1. *Embedded controller specification*: The specification encodes the complete description of the embedded system, including its interface, functionality, and design/performance requirements. This step usually uses specification notations, i.e., specification/programming languages (C++, SystemC, VHDL, UML, MATLAB, etc.) or graphical notations such as Simulink and PSoC Express. The majority of the specification notations support the development of descriptions that can be simulated, or emulated, for rapid prototyping. In addition validation of the system description correctness, specifications also enable the analysis of the effectiveness of different sensing and actuation devices.

2. *Specification testing and debugging*: The correctness of the system description is validated by simulating the specification for a set of input signals. The set must be sufficiently large to cover all situations that can occur during operation. Testing reveals the presence of errors in a specification and debugging locates the errors and corrects them.

3. *System implementation*: This step concentrates on the implementation of the system hardware and software. Hardware implementation includes designing the analog and digital interfacing circuits including the analog-to-digital converter for interfacing the temperature sensor, and the PWM circuit for actuating the fan motor and selecting the processing units, on-chip and off-chip memories, power control circuits, and timers. Software implementation consists of tasks such as developing the drivers for interfacing the peripherals, the middleware routines (including sensor error compensation routines), the functions of real-time operating systems such as task synchronization/communication/scheduling, and the control algorithm of the system.

4. *Physical implementation, testing, and debugging*. The hardware and software designs are implemented, tested, and debugged. The implementation must be tested for a set of input signals that represent all of the possible errors that can actually occur. The experimental setup includes the controller circuit, temperature sensor, fan, tachometer, LCD, power source, and an oscilloscope for signal measurement. In addition, a monitoring environment is needed for observing program execution that displays the register and memory contents, execution traces, stepwise program execution, and defines the breakpoints in a program.

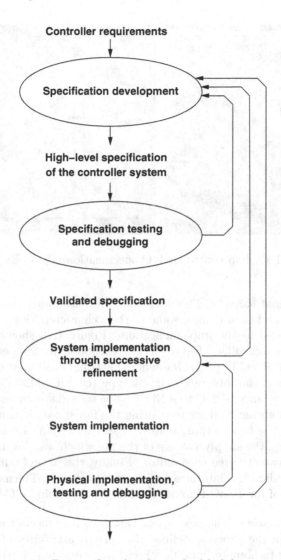

Figure 1.18: Design flow for the embedded fan controller.

The four design steps for the fan controller are as follows.

1. Embedded fan controller specification. This specifies the system functionality using PSoC Express' graphical notation[11].

PSoC Express' description philosophy is to express a system as a netlist of predefined modules that are the basic building blocks (e.g., temperature sensors, fans, tachometers, etc). PSoC Express' library modules are parameterized, so that a designer can customize the parameters depending on the application requirements. For example, the range of the temperature sensor can be defined based on the expected temperature range of the application. PSoC Express' interface also provides an easy way to describe the control algorithm, and for mapping the input and output signals to the pins of a single-chip implementation. This makes PSoC Express easy to use by less experienced designers.

Figures 1.19 and 1.20 illustrate the specification steps in PSoC Express:

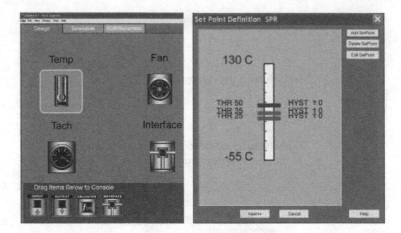

Figure 1.19: Fan controller I/O specification (PSoC Express).

- *Controller inputs and outputs*: This step identifies the input and output modules of an embedded system, and then defines some of their characteristics depending on the nature of the devices utilized in the implementation. Figure 1.19 shows the input and output modules of the fan controller. Input "Temp" denotes the temperature sensor, and is a predefined device in PSoC Express. Its defining attributes are part of the specification, and include the following: the integrated circuit type (circuit LM20 [5] in this case), and the covered temperature range, -55°C to +130°C. The second system input is called "Tach". It represents the tachometer used for measuring the fan speed. It is also a predefined device in PSoC Express. The first output, named "Fan" is the fan. Its attributes are related to the fan selected, e.g., the supply voltage of the fan which was 5 V in this case, and the fan type for example, two or three wire motor. Finally, the second output describes the LCD connected by a standard I2C interface to the processing unit. The module called "Interface" is an instantiation of the PSoC Express module for describing I2C-based interfaces.

- *System netlist description*: The next specification step connects the modules into the system netlist, and, in the process, defines the related attributes of the input and output modules. First, the tachometer module "Tach" is "connected" to the "Fan" block, because the tachometer measures the fan speed. This is achieved in PSoC Express by setting a dedicated attribute of the module "Tach", called *AssociatedFan,* to point to the module "Fan". Next, the three threshold temperatures of the sensor "Temp" are defined as specified in Table 1.7. This is achieved in PSoC Express by adding the *Interface Valuator* module to the specification. The valuator module is named "SPR". The input to the "SPR" module is set to the block "Temp". Three threshold values are defined for the valuator block corresponding to the four temperature ranges in Table 1.7. The resulting valuator description is shown in Figure 1.19. Hysteresis could also be specified for the module to denote regions where changes in the input values do not modify the output. This avoids fast switching of the control signals due to noise. This completes the description of the functional link between the temperature sensor "Temp" and the valuator block "SPR".

  At this point, note that the system structure is not completely defined. The connection between the sensed temperature and the fan control has not been characterized.

- *Control algorithm*: The last specification step describes the control algorithm of the fan controller. This completes the system structure description by relating the four temperature

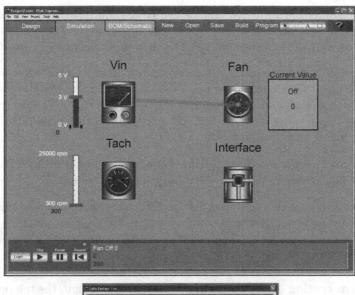

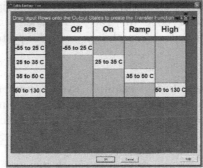

Figure 1.20: System structure and table lookup transfer function (PSoC Express).

ranges in Table 1.7 to the required speed of the fan. In PSoC Express, this is achieved by defining the transfer function of the module "Fan". The transfer function was specified as a table lookup, but other specification constructs are also available (e.g., priority encoder).

Table lookup transfer functions explicitly define the output states of a module for every combination of its input values. Any input combination can have only one associated output state, and only one input combination can be active at any time. Input combinations that do not have any associated output states do not change the system state. For the fan controller example, the valuator block "SPR" is first designated as the input to the "Fan" module. Figure 1.20 illustrates the complete table lookup transfer function for the system.

Thus, the embedded fan controller specification developed with the graphical interface of PSoC Express has the following important characteristics:

1. The specification is at a *high level*, that is few circuit details are present. The specification is based on abstract parametric modules, which are defined mainly by their functionality. There is no need to indicate the physical structure of the modules. The library modules can be interconnected without having to consider interfacing requirements (e.g., the I2C interfaces), or electrical loading constraints (e.g., impedance matching).

Graphical notations are intuitive and descriptive, require little learning effort, and are easy to use. The specification is easy to change, for example for analyzing other types of temperature sensing devices such as an RTD, or a thermocouple.

2. The specification is *heterogeneous*. It incorporates modules that operate continuously in time, and routines that are executed upon the occurrence of certain events. The temperature sensor, tachometer, and fan operate continuously in time. The control algorithm and the LCD execute only after an event, e.g., the exceeding of the temperature threshold value, or displaying a new value, respectively.

3. The specification is *comprehensive*. It defines the input and output interfaces, functionality, and design requirements of the embedded fan controller system.

The examples throughout this textbook further illustrate that specifications of mixed-signal embedded systems are heterogeneous and comprehensive. In addition, using high-level specifications, instead of low-level descriptions, enhances the design options that can be analyzed, and simplifies the specification process.

2. Specification testing and debugging. To verify the design, the temperature values are changed to cover the entire range of -55°C to +130°C. For each temperature setting, the value of the output "Fan" and "SPR" modules must change accordingly to conform to the specification in Table 1.7.

3. System implementation. In this example, system implementation went through three successive refinement steps, each step being guided by the goal of minimizing the cost of the final implementation. The following refinement steps have been used in the design process.

- *Functional partitioning*: Functional partitioning identified the functional modules (i.e., blocks) of the implementation, and defined the *block structure* of the embedded fan controller system. Hence, functional partitioning mapped the system specification to the blocks.

  Each functional block has well-defined interfaces and functionality, but there are still few implementation details available for the block. Also, there is little resource sharing between blocks, such as shared hardware circuits or software routines. Therefore, functional partitioning establishes the starting point in devising the controller architecture by defining the signal and dataflow of the architecture, and the functionality of the architectural blocks.

- *Implementation domain selection*: This step refines the functional blocks in the system architecture by deciding which of their sub-functions are to be implemented as continuous-time, and which as discrete-time, processing units. The units can employ analog circuits, digital circuits, and software routines. Each block is designed separately, and meet the requirements set for the block interfaces and functionality.

  For example, this step determines whether the fan control block that fixes the duty cycle, α, should be a software routine, or a customized digital circuit. Using a software routine is cheaper solution, but it might provide slower control of the fan than a dedicated hardware circuit.

- *Circuit design and software development*: This refinement step completes the implementation process by designing the related analog and digital circuits, and developing the software routines for the fan control algorithm, drivers, and interrupt service routines. This step is then followed by the building of the physical implementation of the system.

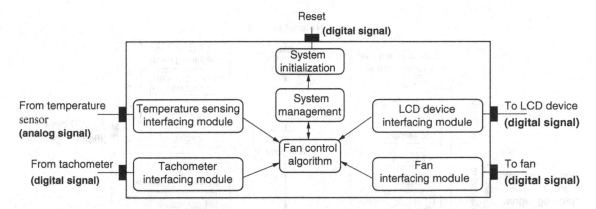

Figure 1.21: Block-level design after functional partitioning.

The three refinement steps are detailed next for the fan controller.

A. *Functional partitioning.* Figure 1.21 presents the block-level structure of the embedded fan controller. Starting from the system specification in Figures 1.19 and 1.20, the description is organized into blocks for interfacing the system to the sensors and actuators, and implementing the fan control algorithm (in Table 1.7). In addition, one block supervises the functioning of the entire system. Now, the controller architecture and the inputs, outputs, and functionality of each block in the architecture have been determined.

The architecture consists of the following blocks.

- *Temperature sensor interfacing block:* - converts the analog voltage produced by the temperature sensor into a digital, unsigned, 8-bit data representing the temperature. This provides the required accuracy for temperature sensing. This block also compensates for the systematic errors that can occur during signal conversion, for example voltage offsets, and ADC precision.

- *Tachometer interfacing block:* calculates the rotational speed of the fan (shaft speed) using the digital signal output of the tachometer. The digital signal is utilized to generate an interrupt signal serviced by a corresponding routine. The interval between two consecutive interrupt signals is measured, and used to calculate the number of rotations per minute (RPM) of the fan.

- *Fan interfacing block:* controls the fan operation by starting/stopping the fan and setting its duty cycle, α, which is determined by the length of the "high time", and "low time" of the block's digital output signal.

- *LCD interfacing block:* connects the LCD to the embedded system through a standard I2C interface. The LCD displays the measured temperature and fan speed.

- *Fan control algorithm:* implements the control algorithm described in Table 1.7. This module takes unsigned, digital inputs from the temperature sensing and tachometer interfacing blocks. It produces digital control signals for the fan interfacing block, and data and control signals for the LCD interfacing block.

- *System management block:* supervises the operation of the entire embedded system. Upon system reset, including at boot time, it (1) initializes the system configuration, i.e., sys-

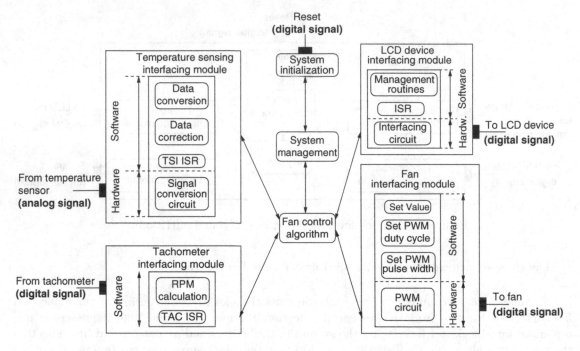

Figure 1.22: Refined block-level design after domain selection.

tem registers, interrupt service routine table, etc., (2) powers up analog circuits (e.g., the analog-to-digital converters of the temperature sensor interfacing block), buffers, and the tachometer interface, (3) creates the structures for interfacing the blocks, and (iv) starts the execution of the fan control algorithm.

B. Implementation domain selection. Once the specification of each module was completed, the design process further refined the composing modules. The resulting design is shown in Figure 1.22. This figure details the four interfacing modules presenting their hardware and software components:

- *Temperature sensor interfacing module:* includes hardware circuitry for converting the analog input voltage into one-byte, unsigned, digital data representing the measured temperature. The software module includes three layered routines that bridge the gap between the unsigned data in the buffers of the conversion circuit and the data input to the control algorithm. The signal conversion hardware includes analog-to-digital converters (ADC) designed to offer a low-cost, required speed and precision for data conversion, and robust operation. The software component includes an interrupt service routine executed each time the embedded processor receives an interrupt from the signal conversion hardware. The routine called *TSI ISR* performs digital processing and then moves the converted data from the data buffer of the converter to a reserved memory address. The *data correction* routine processes the read data to compensate for offset/resolution errors, and translates the corrected value into the corresponding temperature value. This routine is called by the fan control algorithm. Assembly code programming is introduced in Chapter 2, and ISR programming in Chapter 3. Chapter 9 presents the signal conversion module, in detail.

- *Tachometer interfacing module:* consists of two software routines that compute the RPM values based on the digital interrupt signal from the tachometer. ISR routine *TAC ISR* cyclically executes a five state routine for interfacing with the tachometer: Upon receiving the first interrupt signal, it disconnects the pin "connected" to the fan and sets it "high", so that the fan rotates at constant speed which is required for correct reading by the tachometer. Upon receiving the next interrupt signal, the second state of the ISR waits for the reading to settle, after switching the input signal multiplexing circuit. The third ISR state measures and stores the first edge of the tachometer timer. The fourth state measures the second edge of the tachometer timer, and calculates the time difference between the two edges. Then, it reconnects the pin to the fan, so that fan control can resume. The fifth state waits until the next interrupts are received, after which, it loops back to the first ISR state. Routine *RPM calculation* computes RPM based on the time difference calculated in the fourth ISR state. This routine is called by the control algorithm. Chapter 3 details the ISR implementation.

- *Fan interfacing module:* includes hardware for generating the digital control signal for the fan, and software routines that allow easy control of the module hardware from the control algorithm. The module hardware is a pulse width modulator, a digital circuit for setting the duration of high and low levels of the digital signal for the fan. These durations are defined as multiples of a timer interval. The low level routine, *Set PWM pulse width*, is used to configure the PWM control registers. Routine *Set PWM duty cycle* computes the length of time length between the high and low levels for a required duty cycle of the fan. Finally, the routine *Set Value*, which is called from the fan controller algorithm, finds the necessary duty cycle using the measured value of the temperature. Chapter 4 presents the design of customized digital circuits (e.g., PWM) by programming reconfigurable digital blocks.

- *LCD interfacing module:* consists of digital hardware circuit and software routines for interfacing the fan control algorithm to the display unit. The LCD and digital hardware circuit are connected at the bus level. The LCD acts as a slave module controlled by the digital hardware circuit of the interface. This circuit generates all control data for the display and transmits the data serially, bit by bit. The LCD generates an interrupt each time it receives a data byte, signaling the interface module that it can continue its execution, for example by executing the ISR routine for transmitting the next byte. The management routines initialize and set up the data buffers that hold the data transmitted by the ISR routine and displayed on the LCD.

C. Circuit implementation and software development. The design was implemented using the CY8C27443 PSoC chip [12] to provide a low-cost implementation of the embedded controller. PSoC is a system on chip (SoC) that includes a wide variety of specific functional blocks, such as configurable (i.e., programmable) analog and digital circuits, an 8-bit microcontroller, on-chip program and data memory, and timers. Chapters 3 to 7 describe PSoC in detail.

Figure 1.23 shows the circuit structure of the fan controller:

- *Temperature sensor interfacing module:* includes both analog and digital circuits. The analog circuits are a gain stage to amplify the input voltage from the sensor, and an ADC to convert the voltage into 8-bit data. Different ADC types could be used [22], but a first–order $\Delta\Sigma$ converter was selected because of its simplicity, robustness, and low cost. The digital circuits include a buffer that stores the ADC output, and a decimator that performs the digital processing required for the analog-to-digital conversion process. Chapters 6 and 7 discuss analog circuit design, including gain stages, and Chapter 9 offer details on the functioning and design of $\Delta\Sigma$ ADCs. Once a data byte is available for a new temperature

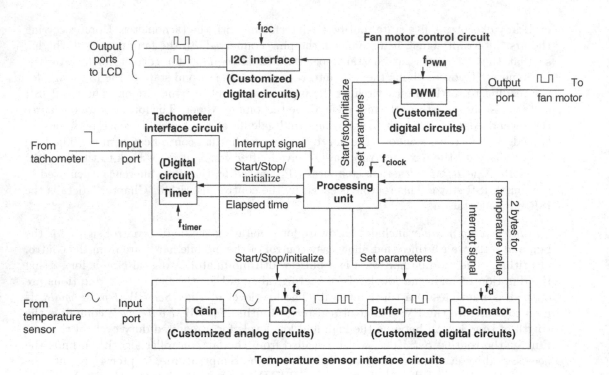

Figure 1.23: Circuit level implementation after domain selection.

reading, the module produces an interrupt for the processing unit, and the temperature value is transferred.

The hardware required is the following: one analog PSoC block for the gain stage, one analog PSoC block for the ADC, and one digital PSoC block for the decimator.

The software routines were developed in assembly and C code: the TSI ISR was written in assembly code to ensure fast access to the decimator's registers, and the data correction and data conversion routines were written in C.

- *Tachometer interfacing module:* uses a digital timer to measure the interval between consecutive interrupt signals from the tachometer.

  The hardware implementation required is one digital PSoC block for the timer.

  The software routines include assembly code for the TAC ISR to provide access to the interrupt mechanism of the chip, and C for the RPM routine.

- *Fan interfacing module:* one digital circuit was required for the pulse width modulator.

  This required one digital PSoC block for the PWM.

  The fan-related ISR was developed in assembly code, and the management routine in C.

- *LCD interfacing module:* required one digital PSoC block for the I2C interface with the LCD. The driver routines were written in assembly code.

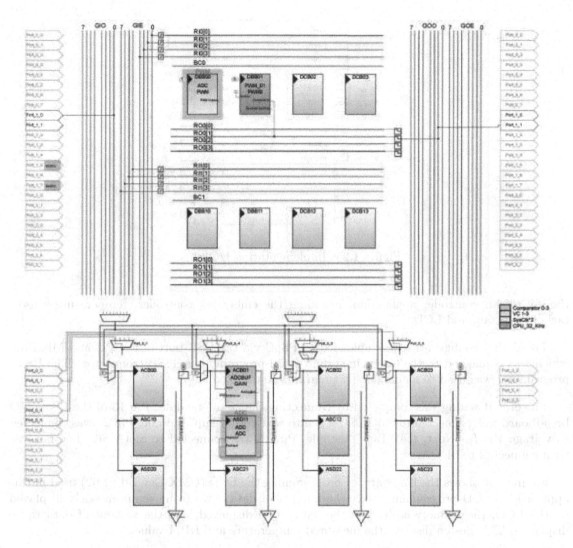

Figure 1.24: PSoC implementation (after circuit implementation and software development).

- *Processing unit* - executes the control algorithm and also the system management routine. In addition, it starts/stops, and sets the parameters of the two sensing modules, the fan interfacing module, and the LCD. The PSoC's on-chip microcontroller executes these functions. In addition, the implementation requires 21 bytes of data RAM, and 804 bytes of program memory.

Figure 1.24 illustrates the PSoC implementation of all analog and digital circuits. The chip has 28 pins. Because the number of available pins is much larger than the number of the controller inputs and outputs, the pin assignment was set to default, and is illustrated in Figure 1.25. Chapter 3 provides more details on PSoC's I/O ports.

4. Physical implementation, testing, and debugging. After completing the hardware and software design, the PSoC Express Development Kit (CY3210-ExpressDK) was used for

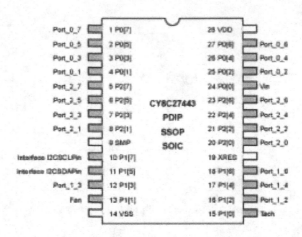

Figure 1.25: Implementation I/O pins.

the entire fan controller application, including the embedded controller, temperature sensor, tachometer, fan, and LCD.

The design configuration was uploaded to PSoC via the USB port of the PC and MiniProg which was connected to the target in-system serial programming (ISSP) connector. The PSoC programmer was used to automatically program the targeted chip.

The circuit wiring was completed by connecting the "Fan" module. Pin 13 of the PSoC I/O breadboard was connected to SlotA_S4. Then to wire the "Temp" module, Pin 24 was connected to voltage $Vref_1$. Next, the "Tach" module, Pin 15 was connected to SlotA_S6. The fan was then connected to Slot A.

Figure 1.26 shows the flowchart for programming the LCD (CMX Dev Bd v1.02) used in this application. LCD programming fixes the number of data bytes to be simultaneously displayed on the LCD, the memory address of the data to be displayed, and the amount of data to be displayed. This design displays the measured temperature and RPM value.

The system was operated by changing the value of voltage $Vref_1$ and monitored using the LCD. The duty cycle of slot A was set to 100%. The higher the duty cycle, the hotter the heater gets.

Microsoft's HyperTerminal can be used for more comprehensive monitoring of the system's operation. The on-board LCD can display only the values of three registers. However, there are four registers of primary interest. Monitoring was accomplished by connecting the serial port COM1 to the development board. Using HyperTerminal, the serial communication speed was set to 38,400 bits per second. Then, the data transfer parameters were selected, for example ASCII Setup, Echo typed characters locally for the ASCII Setup. Figure 1.27 shows HyperTerminal's graphical user interface.

The following steps were necessary to monitor the system. First, the system was reset by pressing Master Reset. As a result, the message "CMX Dev Bd v1.02" was displayed on HyperTerminal. Next, the duty cycle of slot A was set to 100% by typing the command "Slot A 100". HyperTerminal returned the current parameters of slot A. Third, the I2C address and the count value were defined by the command "Cird 04 04". After entering this command, HyperT-

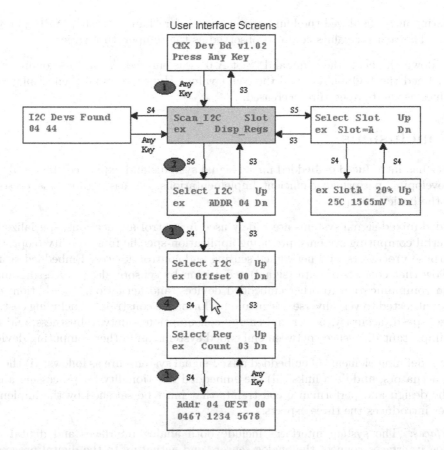

Figure 1.26: LCD device operation for monitoring and debugging.

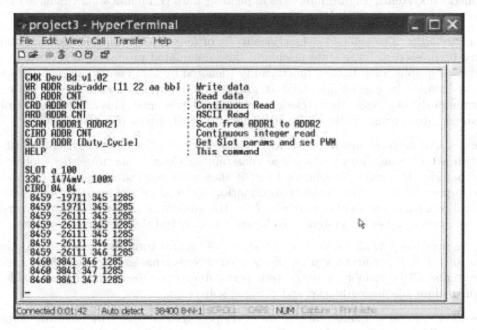

Figure 1.27: Fan controller monitoring using HyperTerminal.

erminal continuously displayed the four register values, for "Fan", "Tach", "SPR", and "Temp", respectively. The register values could be observed as the temperature varied.

To turn down the heat, the command "Slot A 0" was entered. Then, the command "Cird 04 04"which defined the I2C address and the count value. HyperTerminal then displayed the four register values as the temperature decreased.

1.6   Conclusions

This chapter has introduced embedded mixed-signal systems and explained the top-down design flow for developing performance-efficient implementations. A case study was presented that illustrated the basic concepts.

Embedded mixed-signal systems are widely used in control applications, specialized interfacing to powerful computing systems, providing application-specific functionality (coprocessors) to general-purpose processors, and networked sensing and control devices. Embedded systems realize a dataflow that consists of data (signal) acquisition by sensors, data processing and storing, digital data communication to other embedded devices, and actuation. In addition, embedded systems are subjected to very diverse design and performance constraints, including cost, time–to–market, size, speed, accuracy, power and energy consumption, safety, robustness, and flexibility. These are important differences between embedded systems and other computing devices.

The three defining elements of embedded mixed-signal systems are as follows: (i) the interfaces to sensors, actuators, and data links, (ii) the embedded functionality for processing and control; and (iii) the design and performance constraints that must be satisfied by the implementation. The chapter introduces the three aspects.

- *Interfaces*: The system interfaces include both analog interfaces and digital interfaces. Analog interfaces connect the analog sensors and actuators to the digital processor of the embedded systems. Digital interfaces implement data communication channels by various mechanisms, such as SPI, UART and I2C communication standards.

- *Functionality*: The embedded functionality is of four types: control-dominated functionality (expressed as finite state machines) data-dominated processing described by acyclic dataflow graphs, multitasking functionality (denoted by task graphs) and multimode operation (defined by control and dataflow graphs). In addition, the functionality is executed continuously, at discrete times (clocks), or triggered by events. This differentiates embedded systems into continuous-time, discrete-time, and event-driven (reactive) systems.

- *Design and performance constraints*: Implementations of embedded systems must satisfy a rich set of design and performance constraints. Constraints are either global, if they characterize the entire system, or local, if they refer only to subsystems or modules of the implementation. Constraint transformation is the process of mapping (translating) the global constraints to local constraints. Also, the amount of existing constraint correlations differentiates embedded systems into loosely coupled and tightly coupled systems.

This chapter has introduced the main features of architectures for embedded applications, and some of the popular architectures. Embedded mixed-signal architectures have the following subsystems: CPU, memory, input/output ports, interrupt subsystem, programmable digital blocks, programmable analog blocks, and clocking subsystem.

Embedded systems' implementations are developed systematically by following a top-down design flow. The goal is to address the imposed design and performance requirements by customization of the hardware and software to the application's needs. The design flow conducts a successive, and iterative, refinement process starting from a high-level description of the embedded system until completing the implementation. This includes specification, functional partitioning, system tradeoff analysis (including constraint transformation), design of the individual modules, analog circuit design, digital circuit design, software development, circuit modeling, software characterization (profiling), and performance evaluation.

The second part of this chapter discussed an embedded fan control system case study and the issues that arise in the top-down design flow of the fan controller. The controller modifies the fan speed according to the temperature sensed by an analog temperature sensor. In addition, the implementation must provide good accuracy for the sensed temperature, have low cost and small size.

Additional, related topics discussed in this chapter included:

- A description of the fan controller (including the sensing and control principles).

- Development of the system specification based on the graphical notation of PSoC Express' environment.

- Functional partitioning and tradeoff analysis for implementation domain selection. The trade-off analysis presented also includes constraint transformation for mapping the global system constraints to local constraints of the subsystems. Functional partitioning determines the signal processing flow of the implementation, for example the block structure (each type of block has a basic functionality), block interfaces, and block connections. Implementation domain selection determines whether the system blocks should be continuous-time, discrete-time, or event-driven; analog or digital; and implemented in hardware or software.

- The physical implementation, testing, and debugging step is based on the actual analog circuits, digital circuits, and software routines of the final design. For the PSoC architecture, the designs were implemented by programming the reconfigurable analog and digital blocks, connecting the sensors and actuators, programming the LCD, and monitoring the system operation with HyperTerminal.

1.7 Further Readings

There is an abundant amount of literature on embedded system design and various kinds of embedded applications. F. Vahid and T. Givargis [39], and P. Marwedel [30] discuss important theoretical aspects related to embedded system design, such as mathematical (formal) description of embedded systems, specification languages for embedded design, and algorithms for automated design and optimization of embedded systems.

A second category of material focuses on presenting architectures and hardware circuits for embedded systems, including microprocessors, microcontrollers, and DSPs used in embedded applications. T. Noergaard [32] provides a comprehensive discussion of embedded system architectures encompassing hardware building blocks, processors, memory types, I/Os, buses, and device drivers. R. Haskell [26] and J. Valvano [40] present designs for Motorola's 68HC12 family of microcontrollers, including the microcontroller architecture, interfacing, interrupts, and programming in assembly code. T. Morton [31] also discusses embedded system design using the Motorola's 68HC11 and 68HC12 microcontrollers, but with more focus on embedded programming, including assembly code programming, assembly program design and structure, real-time

I/O and multitasking, real-time multitasking kernels, and C language programming. K. Short [38] presents embedded system design using Intel's 80C188EB microprocessor with emphasis on microprocessor architecture and hardware design. M. Predko [35] discusses embedded design using the PICmicro processor and details the PICmicro architecture, interfacing, software development, debugging, and various embedded applications using the microcontroller. J. Hamblen and M. Furman [25] discuss digital design using Altera's MAX 7000 and FLEX 10 k programmable devices. R. Ashby [20] describes the PSoC mixed-signal, system- on-chip architectures, and provides many tips and tricks for using PSoC in real-world design.

A third category of work focuses on embedded software. D. Lewis [29] and M. Barr [21] present embedded programming in C and C++ programming languages, data representation, mixing C and assembly code, memory management, concurrent programming, scheduling, and system initialization. Hardware and software coverification techniques are described by J. Andrews [19].

Other temperature-sensing based embedded systems are presented by M. Raaja in [36] and [37]. These systems use thermocouples and RTDs. Fundamentals of electric motors are discussed by A. Hughes in [27] and C.-M. Ong [34], among others. R. Ohba presents many interesting applications and ideas for sensor-based embedded systems in [33], and J. Fraden explains the principles and functions of many popular sensing devices [24].

1.8 Recommended Exercises

1. List five embedded mixed-signal systems in your home/garage and five embedded mixed-signal systems in your school library, cafeteria, and grocery store.

2. Identify five popular sensing devices, five actuation devices, (other than those discussed in the chapter) and five embedded applications that could use these sensing and actuation devices.

3. What other embedded control applications could use the same signal flow as the embedded fan controller? What is different in these applications?

4. Consider two embedded systems present in your home, and analyze the type of interfaces of each. For each interface, state whether it is an analog interface or a digital interface. Identify which part of each system is analog, and which is digital.

5. Select three embedded systems in your home, and identify what are the most important constraints for each of them. Is it cost, size, weight, reliability, speed, easy to use, and so on?

6. Identify other applications that would involve both continuous and discrete time signals and processing. What kind of interfacing procedures must exist between the two domains? Would it be possible to modify the system structure so that all signals and processing belong to a single domain, for example to the continuous- or discrete-time domains?

7. Discuss the structure of a modified fan controller subject to the constraint that should the ambient temperature exceed the threshold value of 100°C, the controller would immediately react by shutting down the system.

8. Modify the structure of the system in Figure 1.13 so that it reads data from more sensors and controls multiple fans. Discuss the advantages and limitations of the system structures.

9. The fan controller system discussed in this chapter could vary the duty cycle of the PWM module based on the measured torque, T, instead of measuring the shaft speed ω, as was the case in Section 1.5. Develop the signal processing flow that would implement this kind of control. What kind of functional blocks does the design involve? What hardware circuits and software routines are needed for the implementation? How does the performance and cost of your design relate to those of the implementation in Section 1.5?

10. How would the system implementation change if the PWM module is implemented in software instead of hardware as was the case in Section 1.5? What advantages and problems do you anticipate in this case?

11. Specify and simulate the fan controller system using the C programming language. Develop separate modules for the temperature sensor, fan, tachometer, processing unit, and LCD. Find a solution to mimic the continuous time on a clocked (discrete-time) computer.

12. Consider an embedded system capable of autonomously parking an automobile. What kind of sensing and actuation devices would your system use? Are such devices currently available? What would be their defining characteristics? Describe the signal processing flow for the system. What performance requirements must the parking system address? Create a block-level description of the design.

13. Develop the signal processing flow and the block structure for a home monitoring system. The system's functionality must include the following: (1) activating/deactivating the security

alarm, (2) closing the garage doors at night, (3) turning lights on/off, and (4) watering the flower garden. Identify the design and performance requirements of this application. What sensing and actuation devices are needed? What are the building blocks of the system?[1]

14. Design the block structure of an intelligent garage door opener. The system must automatically open/close the garage door, issue specific warning signals if (a) the car position is too close to the garage walls, (b) the garage door cannot be closed, or (c) the humidity inside the garage exceeds a specified threshold value. What are the design requirements for this application? What sensing and actuation devices should be used to minimize the implementation cost and maximize the reliability of the implementation?[2]

---

[1] *Acknowledgment:* This exercise was inspired by the article by J. Benjamin, M. Benjamin, Multifunctional home control system. *Circuit Cellar*, 187:14-20, February 2006.

[2] *Acknowledgment:* This exercise was inspired by the article by V. Lick, "Intelligent automatic garage door opener. *Circuit Cellar*,179:80-85, June 2005.

Bibliography

[1] MAX 7000 Programmable Logic Device Family, Data Sheet, DS-MAX7000-6.7, *Altera Corporation*, September 2005.

[2] ARM 1136JF-S and ARM 1136J-S, Technical Reference Manual, Revision r1p1, ARM DDI 0211H, ARM Limited.

[3] DSP56800, Family Manual, 16-Bit Digital Signal Controllers, DSP56800FM, Rev. 3.1, Freescale Semiconductor, November 2005.

[4] 80C552/83C552 Single-Chip 8-Bit Microcontroller with 10-Bit A/D, Capture/Compare Timer, High-Speed Outputs, PWM, Data Sheet, September 3 2002, Philips Semiconductors.

[5] LM20 2.4V, 10μA, SC70, micro SMD Temperature Sensor, National Semiconductor, DS100908, 1999.

[6] High-Precision ADC Mixed-Signal Controller, Maxim Integrated Products, Maxim Dallas Semiconductor, Rev 1; 8/05.

[7] Intel XScale Microarchitecture for the PXA255 Processor, User's Manual, March 2003.

[8] PIC24FJ1286GA, General Purpose 16-bit Flash Microcontroller, DS39747A, Microchip Technology, 2005.

[9] PNX 1300 Series Media Processors, Data Book, Preliminary Specification, Philips Semiconductors, February 15 2002.

[10] PPC 405 Fx Embedded Processor Core, User's Manual, IBM, SA 14-2764-00, January 28 2005.

[11] PSoC Express, Version 2.0, Cypress Semiconductor, 2006.

[12] PSoC Mixed Signal Array, Technical Reference Manual, Document No. PSoC TRM 1.21, Cypress Semiconductor Corporation, 2005.

[13] ADSP-2136x SHARC Processor, Hardware Reference Manual, Analog Devices, Revision 3, May 2006.

[14] SMPS MOSFET IRF7463, International Rectifier, PD-93843A, March 30 2000.

[15] Spartan-3 FPGA Family: Complete Data Sheet, Product Specification, DS099 April 26 2006, Xilinx Inc.

[16] TMS320C6000, Technical Brief, Texas Instruments, February 1999.

[17] RG-299R Future of Embedded Systems Technology, Business Communications Company, INC., 25 Van Zant Street, Norwalk, CT 06855, publisher@bccresearch.com.

[18] Motor Control with PSoC, presentation, Cypress Semiconductors, 2006.

[19] J. Andrews, Co-Verification of Hardware and Software for ARM SoC Design, New York: Elsevier, 2005.

[20] R. Ashby, *Designer's Guide to the Cypress PSoC*, New York: Elsevier, 2005.

[21] M. Barr, Programming Embedded Systems in C and C++, Sebastopol, CA: O' Reilly, 1999.

[22] M. Basinger, PSoC Device Selection Guide, Application Note AN2209, Cypress Semiconductors, May 12 2005.

[23] E. Denton, Application Brief. Tiny Temperature Sensors for Portable Systems, National Semiconductor Corporation, 2001.

[24] J. Fraden, *AIP Handbook of Modern Sensors. Physics, Design and Applications*, Woodbury, NY: American Institute of Physics, 1993.

[25] J. O. Hamblen, M. D. Furman, *Rapid Prototyping of Digital Systems*, Boston: Kluwer Academic Publishers, 2001.

[26] R. E. Haskell, *Design of Embedded Systems Using 68HC12/11 Microcontrollers*, Upper Saddle River, NJ: Prentice Hall, 2000.

[27] A. Hughes, *Electric Motors and Drives. Fundamentals, Types and Applications*, New York: Elsevier, Third edition, 2006.

[28] E. Lee, P. Varaiya, *Structure and Interpretation of Signals and System*, Reading MA: Addison-Wesley, 2003.

[29] D. W. Lewis, *Fundamentals of Embedded Software. Where C and assembly meet*, Upper Saddle River, NJ: Prentice Hall, 2002.

[30] P. Marwedel, *Embedded System Design*, Boston, MA: Kluwer Academic Publishers, 2003.

[31] T. Morton, *Embedded Microcontrollers*, Upper Saddle River, NJ: Prentice Hall, 2001.

[32] T. Noergaard, *Embedded Systems Architecture. A Comprehensive Guide for Engineers and Programmers*, New York: Elsevier, 2005.

[33] R. Ohba, *Intelligent Sensor Technology*, New York: John Wiley, 1992.

[34] C.-M. Ong, *Dynamic Simulation of Electric Machinery using MATLAB/SIMULINK*, Upper Saddle River, NJ: Prentice Hall, 1998.

[35] M. Predko, *Programming and Customizing PICmicro Microcontrollers*, Upper Saddle River, NJ: McGraw-Hill, 2002.

[36] M. G. Raaja, RTD Temperature Measurement, Application Note AN2120, Cypress Microsystems, September 12 2003.

[37] M. G. Raaja, Measuring Temperature Using a Thermocouple, Application Note AN2148, Cypress Microsystems, March 16 2004.

[38] K. L. Short, *Embedded Microprocessor Systems Design. An Introduction Using the INTEL 80C188EB*, Upper saddle River, NJ: Prentice Hall, 1998.

[39] F. Vahid and T. Givargis, *Embedded System Design. A Unified Hardware/Software Introduction*, New York: John Wiley, 2002.

[40] J. Valvano, *Embedded Microcomputer Systems. Real Time Interfacing*, Thomson, Third edition, London: 2007.

Chapter 2

Microcontroller Architecture

The chapter presents the main characteristics of a microcontroller instruction set, and discusses programming techniques in assembly language for several applications. It also defines the instruction set architecture of PSoC's M8C microcontroller.

The instruction set of the M8C microcontroller consists of instructions for (i) data transfer, (ii) arithmetic operations, (iii) logic operations, (iv) execution flow control, and (v) other miscellaneous instructions.

The microcontroller provides ten addressing modes, which result from combining four basic modes: immediate, direct, indexed, and indirect addressing. Each addressing mode defines a specific tradeoff among the execution time of the code, the required program memory, and the flexibility in modifying the code.

The PSoC memory space includes SRAM for storing data, and nonvolatile memory for storing the program code and the predefined subroutines used in booting up the architecture, accessing the flash memory that holds the application program, and circuit calibration. In addition, the register space stores the status and control information of the embedded mixed-signal architecture.

Several applications illustrate programming techniques in assembly language, and discuss the performance of these solutions. The applications include data block transfer, stack operation, unsigned data multiplication, calling assembly routines from programs in high-level programming languages, bit manipulations, and sequence detectors.

The chapter has the following structure:

- Section 1 presents M8C's instruction set and addressing modes.

- Section 2 explains PSoC's SRAM and ROM subsystems.

- Section 3 presents chapter conclusions.

2.1 Microcontroller Architecture

PSoC's M8C microcontroller is based on an eight-bit Harvard architecture with separate data and address buses. Figure 2.1 shows the architecture of the M8C, that is the microcontroller structure, its internal registers and external memory space [2].

A. Doboli, E.H. Currie, *Introduction to Mixed-Signal, Embedded Design*,
DOI 10.1007/978-1-4419-7446-4_2, © Springer Science+Business Media, LLC 2011

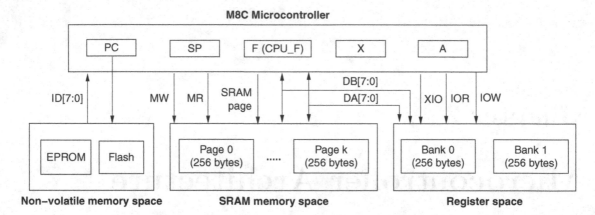

Figure 2.1: The M8C microcontroller structure [2].

The M8C has five internal registers.

- A Register (Accumulator) an eight bit, general-purpose register used by instructions that involve data transfer, arithmetic and logical operations, jump instructions, and so on.

- X Register (Index Register) an eight bit register that can be used either as a general-purpose register, similar to register A, or for implementing certain addressing modes, for example source-indexed and destination indexed addressing.

- F Register (Flag Register), also referred to as the CPU_F register an eight bit, nonaddressable register located at address x,F7H.[1] This register stores the various flag and control bits of the microcontroller utilizing the following bit structure.

  - Bit 0 is the GIE (Global Interrupt Enable) bit and determines which of the external interrupts are enabled, or disabled.
  - Bit 1 is the ZF (Zero Flag) bit, and Bit 2, the CF bit (Carry Flag). Both bits are set by certain types of data transfers, or by data processing instructions.
  - Bit 4 is the XIO bit (IO Bank Select), determines the active register bank.
  - Bits 6 and 7 are the PgMode bits (Page Mode). They control the accessing of data stored in SRAM. More details are offered in Subsection 2.2.

- SP Register (Stack Pointer) an eight bit register that points to the top of the stack. The SRAM page of the stack is pointed to by the eight-bit STD_PP register, in the register space.

- PC Register (Program Counter) stores the 16-bit program memory address, representing 64 K of memory space, and points to the next instruction to be executed.

The PSoC memory space consists of three distinct memory regions:

- *Nonvolatile memory space* consisting of the permanent read-only memory (EPROM) and flash memory, that are used to store the program code to be executed by the M8C processor. The size of the flash memory space can be up to 64 K words, inclusive. The address of a memory word is pointed to by the PC register. The data read is returned via a dedicated eight-bit bus, called ID[7:0], as shown in Figure 2.1.

---

[1] An 'x' before the comma in the address field indicates that this register can be read or written to, no matter what bank is used.

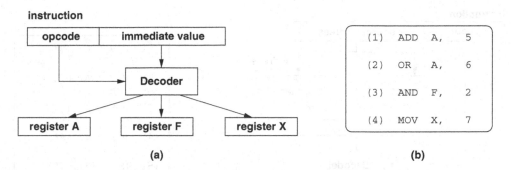

Figure 2.2: Source-immediate addressing mode.

- *SRAM space* stores both global and local variables, and the stack implementation. The maximum size of the SRAM space is 2048 words, inclusive, and the number of pages is limited to eight. Because each page is 256 bytes long, eight-bit address words are required to access an SRAM page. The accessing of pages is controlled by the "control bits" in the CPU_F (CPU Flags) and CUR_PP (Current Page Pointer) registers. The SRAM and register spaces share the same eight-bit address and data buses, DA and DB, respectively. Control signals MR and MW indicate a memory read and a memory write, respectively.

- *Register space* consists of the registers needed to control PSoC's resources, for example the digital and analog reconfigurable blocks, SRAM and interrupt system. Additional details about the individual registers are provided as PSoC's programmable hardware is discussed later in the text.

  The M8C architecture includes two register banks that are selected by bit four (bit XIO) of the CPU_F register. Having two register banks is useful for *dynamic hardware reconfiguration* (i.e., when the hardware is reconfigured during execution, to provide different functionality and/or performance). Each of the banks stores the control information for a configuration mode.

(Subsection 2.2 provides additional details about the PSoC memory system.)

2.1.1 Microcontroller Addressing Modes

The microcontroller addressing modes impose different conventions, that is rules for generating the address used in accessing SRAM. Address modes range from very simple address generation rules, such as utilizing the physical address of the word, specified in a field of the instruction word, to the implementation of more complex rules that use index registers, or indirect addressing based on pointers.

Different addressing modes impose different tradeoffs with respect to the (i) flexibility in addressing memory, (ii) number of instructions required to prepare a memory address, (iii) number of clock cycles required to access memory, and (iv) the number of registers utilized by an addressing mode. Simple addressing modes, for example those specifying the operand value, or indicating the physical address as part of the instruction word, offer faster memory access, execute in fewer clock cycles, and require fewer instructions to generate the address. The major disadvantage of such addressing schemes is their inflexibility, because they impose a rigid mapping of data to memory words. Unless consistent addressing guidelines are followed, and reused in each new application, it is difficult to link modules with each other.

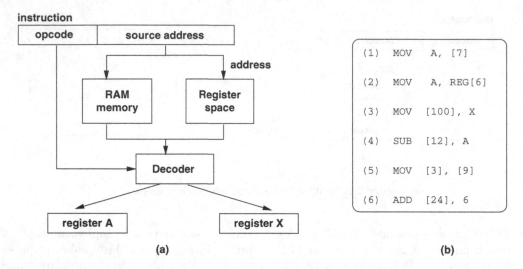

Figure 2.3: Source-direct and destination-direct addressing modes.

The M8C microcontroller supports the following addressing modes.

A. Source-Immediate Addressing

Figure 2.2(a) illustrates source-immediate addressing. Instructions using this mode include a field that contains the value of one of the instruction operands. The other operand, and the result, are kept in registers A, X, and F, respectively.

Figure 2.2(b) shows four instructions that use source-immediate addressing. The first instruction adds the value 5 to the value in register A, and the result is stored in register A. The second instruction performs a bitwise OR operation on the bits of register A and the mask '00000110'. As a result, bits 1 and 2 of register A are set to the value '1' and the rest of the bits in register A are unchanged. The third instruction describes a bitwise AND operation between register F, the register holding the flag and control bits, and the mask '00000010'. This instruction resets all bits of register F except for bit 1, which is left unchanged. The fourth instruction loads the value "7" into register X.

B. Source-Direct and Destination-Direct Addressing

Figure 2.3(a) describes source-direct and destination-direct addressing. The source address field of the instruction, for source-direct addressing, contains the address of a location in either the SRAM space, or the register space. For example, the first instruction in Figure 2.3(b), MOV A, [7], points to the SRAM cell at the physical address 7. The value found at this address is loaded into register A. The second instruction is MOV A, REG[6] and refers to the register at address 6 of the register space. Its contents are loaded into the A register.

Destination-direct addressing uses a field of the instruction word to store the destination address. A destination location can be in SRAM or in the register space. The third instruction in Figure 2.3(b) uses destination-direct addressing. The value in the X register is loaded in SRAM

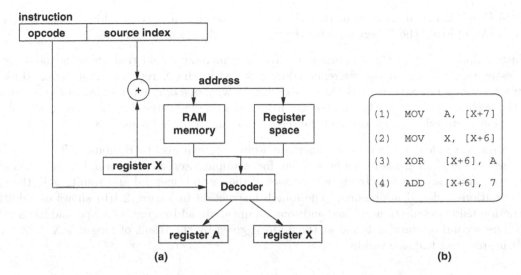

Figure 2.4: Source-indexed and destination-indexed addressing modes.

at the physical address 100. Similarly, the fourth instruction subtracts the value in register A from the value at SRAM address 12, and the result is stored at SRAM address 12.

Source-direct and destination-direct addressing are combined in instructions that include two address fields, one to specify the address of the source and the other, the destination address. This combined mode is called destination-direct source-direct addressing [2]. Instruction five, in Figure 2.3(b), illustrates a "move" instruction using source-direct and destination-direct addressing. In this example, the value in memory address 9 is copied into location with the memory address 3.

Source-immediate and destination-direct addressing can also be employed simultaneously by an instruction. One field of the instruction stores the value of one of the source operands, using source-immediate addressing. A second field contains the address of the destination, which for certain instructions also represents the destination-direct address of the second operand. Instruction six, in Figure 2.3(b), illustrates the addition instruction in which the source operand is accessed by immediate addressing (the operand's value is 6), the second operand is at the SRAM address 24, and the result is stored at address 24. Destination-direct addressing is used to determine the destination.

C. Source-Indexed and Destination-Indexed Addressing

Figure 2.4(a) is a summary of the characteristics of source-indexed and destination-indexed addressing. For source-indexed addressing, the instruction includes a field for the source index that represents the relative displacement of the addressed memory location with respect to a base address stored in the index register X. Hence, the address of the accessed memory location is obtained by adding the content of the X register and the source index field. The address can point to a location in either the SRAM space, or to the register space.

The address of the memory location is loaded into the A or X register, depending on the value of the instruction opcode. Figure 2.4(b) shows two instructions using source-indexed addressing.

MOV A,[X+7] loads the A register with the value of the memory cell at address register X + 7. MOV X,[X+6] loads the X register with the value at address register X + 6.

Instructions using destination-indexed addressing include a field that specifies the offset of the destination related to the reference address stored in the X register. Instruction three in Figure 2.4 shows an exclusive OR (XOR) instruction with one operand located in the A register and the second operand at the address given by the value "register X + 6". The result of the instruction is stored at the address of the second operand, that is "register X + 6".

Destination-indexed and source-immediate addressing can also be combined. These instructions use one of their fields to define a value, for example, used as an operand in an arithmetic operation, and a second field to store the offset of the second operand and result, as in the case of destination- indexed addressing. The fourth instruction in Figure 2.4(b) shows an addition instruction using destination-indexed and source-immediate addressing. One operand has a value of 7. The second operand is found at the address given by the result of "register X + 6". The result is stored at the same address.

D. Source-Indirect and Destination-Indirect Postincrement Addressing

Instructions using source indirect postincrement addressing are often used in transferring blocks of data, and include a field containing the SRAM address of a pointer to the source data. After executing the instruction, the value of the pointer is incremented, so that the next datum in the block can be accessed. The pointer is always located in the current memory page pointed to by the CUR_PP register at the address in the register space. This pointer references data located in the SRAM page pointed to by the MVR_PP register. (Additional details are given in the sections on MVI instructions, the only instructions that use this mode.)

Destination-indirect postincrement addressing uses a similar addressing scheme for the destination. The instruction field contains the address of a pointer used to refer to the destination. After executing this instruction, the value of the pointer is incremented. The pointer is always located in the current SRAM page pointed to by the CUR_PP register. The destination is located in the SRAM page pointed to by the MVW_PP register. MVI instructions are the only ones that use this addressing mode. (Additional details are given in the sections on MVI instructions.)

The mechanism for these two addressing modes is summarized in Figure 2.5.

2.1.2 Instruction Set

The M8C instruction set includes instructions for (i) data transfer, (ii) arithmetic operations, (iii) logical operations, (iv) execution flow control, and (v) other miscellaneous instructions. These instructions use source-immediate, source-direct, source-indexed, source-indirect-postincrement, destination-direct, destination-indexed, and destination-postincrement addressing.

A. Instructions for Data Transfer

Instructions for data transfer include the following instructions: MOV, MVI, SWAP, POP, PUSH, ROMX, and INDEX instructions. The instructions are detailed next.

A.1 MOV and MVI instructions

MOV instructions transfer data among the A, SP, and X registers, SRAM space, and register space. Table 2.1 lists the different kinds of *MOV* instructions. For each instruction, the table

Table 2.1: **MOV** instructions [2].

Instruction	Semantics	Opcode	Bytes	Cycles
MOV X,SP	X ← SP	0x4F	1	4
MOV A,expr	A ← expr	0x50	2	4
MOV A,[expr]	A ← SRAM[expr]	0x51	2	5
MOV A,[X+expr]	A ← SRAM[X+expr]	0x52	2	6
MOV [expr],A	SRAM[expr] ← A	0x53	2	5
MOV [X+expr],A	SRAM[X+expr] ← A	0x54	2	6
MOV [expr$_1$],expr$_2$	SRAM[expr$_1$] ← expr$_2$	0x55	3	8
MOV [X+expr$_1$],expr$_2$	SRAM[X+expr$_1$]← expr$_2$	0x56	3	9
MOV X,expr	X ← expr	0x57	2	4
MOV X,[expr]	X ← SRAM[expr]	0x58	2	6
MOV X,[X+expr]	X ← SRAM[X+expr]	0x59	2	7
MOV [expr],X	SRAM[expr] ← X	0x5A	2	5
MOV A,X	A ← X	0x5B	1	4
MOV X,A	X ← A	0x5C	1	4
MOV A,REG[expr]	A ← REG[expr]	0x5D	2	6
MOV A,REG[X+expr]	A ← REG[X+expr]	0x5E	2	7
MOV [expr$_1$],[expr$_2$]	SRAM[expr$_1$] ← SRAM[expr$_2$]	0x5F	3	10
MOV REG[expr],A	REG[expr] ← A	0x60	2	5
MOV REG[X+expr],A	REG[X+expr] ← A	0x61	2	6
MOV REG[expr$_1$],expr$_2$	REG[expr$_1$] ← expr$_2$	0x62	3	8
MOV REG[X+expr$_1$],expr$_2$	REG[X+expr$_1$] ← expr$_2$	0x63	3	9

shows the opcode of the instruction, the number of bytes occupied by the instruction, and the number of clock cycles required to execute the instruction.

The semantics of the instructions and the addressing mode to be used are as follows.

- *MOV X,SP:* The value of the SP register is loaded into the X register.

- *MOV A,expr:* The value of expr is loaded into the A register. This instruction uses immediate addressing for accessing the source value to be loaded into the register.

- *MOV A,[expr]:* The value found at address expr in SRAM is loaded into the A register (source-direct addressing).

- *MOV A,[X+expr]:* The value in the SRAM, at the address given by the value "X register + expr", is loaded into the A register (source-indexed addressing).

- *MOV [expr],A:* This instruction copies the value in the A register to the SRAM cell at address expr (destination-direct addressing).

- *MOV [X+expr],A:* The value in the A register is stored in the SRAM at the address "X register + expr" (indexed addressing for the destination).

- *MOV [expr$_1$],expr$_2$:* Value expr$_2$ is loaded into SRAM at address expr$_1$ (source-immediate and destination-direct addressing).

- *MOV [X+expr$_1$],expr$_2$:* Value expr$_2$ is stored in SRAM at address "X register + expr$_1$" (source-immediate and destination indexed addressing).

- *MOV X,expr:* The X register is loaded with the value expr (source-immediate addressing).

- *MOV X,[expr]:* The value in SRAM, at address expr, is copied into the X register (source-direct addressing).

- *MOV X,[X+expr]:* The X register is loaded with the value found in SRAM at address "X register + expr" (source-indexed addressing).

- *MOV [expr],X:* The content of the X register is stored in SRAM at the address expr (destination-direct addressing).

- *MOV A,X:* The A register is loaded with the value in the X register.

- *MOV X, A:* The value in the A register is copied into the X register.

- *MOV A,REG[expr]:* This instruction loads the A register with the value in register expr, which is in the register space.

- *MOV A,REG[X+expr]:* Using the source-indexed addressing mode, the A register is loaded with the value in the register at address "X register + expr".

- *MOV REG[expr], A:* Register expr is loaded with the value of the A register.

- *MOV REG[X+expr],A* - The value in the A register is copied into the register pointed to by the value of "X register + expr" (destination indexed addressing).

- *MOV REG[expr$_1$],expr$_2$* - Value expr$_2$ is loaded into the register selected by expr$_1$ (destination-direct addressing).

- *MOV REG[X+expr$_1$],expr$_2$* - Value expr$_2$ is stored into the register pointed by expression "X register + expr$_1$" (destination indexed addressing).

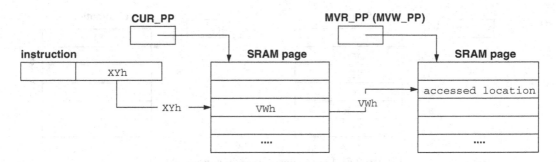

Figure 2.5: MVI instructions.

If the value zero is loaded into the A register, the ZF flag (Zero Flag) is set and the CF flag (Carry Flag) remains unchanged.

MVI instructions implement data transfers using source indirect and destination indirect addressing. Table 2.2 shows the two types of MVI instructions. In contrast to MOV instructions, the expr field of a MVI instruction is a pointer to a SRAM cell, for example the contents of SRAM at address expr is an address of another SRAM cell.

- *MVI A,[expr]* loads the A register with the contents of the memory cell pointed to by [expr], and [expr] is incremented. The SRAM cell is in the SRAM page pointed to by the MVR_PP register. The last three bits of the MVR_PP (MVR_PP[2:0]) register at address 0,D4H select the SRAM page for MVI instructions (source indirect addressing).

- *MVI [expr],A* stores the value found in the A register in the SRAM cell referred to by the pointer at address expr and the pointer is incremented. The SRAM cell is located in the SRAM page pointed to by the MVW_PP register. The last three bits of the MVW_PP register, that is bits 2-0, (MVW_PP[2:0]), at address 0,D5H, select the SRAM page for the MVI instruction (destination indirect addressing).

The CF flag is unchanged by these instructions. If the value zero is loaded into register A, the flag ZF is set. Figure 2.5 illustrates the source indirect post-increment and destination indirect post-increment addressing used by MVI instructions. The address field of the instruction points to a location in the current SRAM page selected by the CUR_PP register. In the figure, this location is pointed to by XYh. Then, the content of the selected location, shown as value VWh in the figure, is used as a pointer to a memory location in the SRAM page selected by the MVR_PP register for MVI A, [expr] instructions, and by the MVW_PP register for MVI [expr],A instructions. This is the memory location used in the data transfer.

Table 2.2: MVI instructions [2].

Instruction	Semantics	Opcode	Bytes	Cycles
MVI A,[expr]	A ← SRAM[SRAM[expr]] SRAM[expr] ← SRAM[expr] + 1	0x3E	2	10
MVI [expr], A	SRAM[SRAM[expr]] ← A SRAM[expr] ← SRAM[expr] + 1	0x3F	2	10

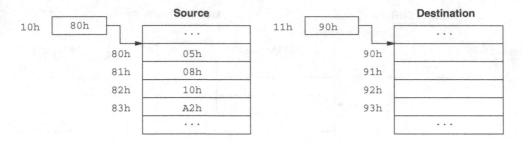

Figure 2.6: Data vector transfer example.

Characteristics of MOV and MVI instructions

As shown in Table 2.1, there is a variety of ways in which data can be transferred in a system. Immediate, direct, indexed, and indirect post-increment addressing have specific execution times that are expressed in terms of clock cycles. Each requires different amounts of memory for storing their code, expressed in bytes, and some modes involve the A and X registers. Finally, the four addressing modes offer specific flexibility and benefits in terms of reusable, assembly language, source code that can be incorporated into new applications.

With respect to execution times, accessing the A register requires the shortest time, followed by X register accesses, register space accesses, and finally SRAM accesses, which are the slowest. Immediate addressing is the fastest, followed by direct addressing, and indexed addressing. MOV [expr$_1$],[expr$_2$] requires the longest execution time, viz., ten clock cycles, because both source and destination are located in SRAM, and involve direct addressing.

Example (Execution time for addressing modes). The following example illustrates the performance characteristics of the four addressing modes. The goal is to transfer a data vector of size 4 from address A in the SRAM to address B in memory. In this example, a data value is one byte long, and the four data values are stored at consecutive memory locations. Figure 2.6 is a graphical representation of the operations to be performed. Table 2.3 shows the performance characteristics of the four description styles, namely, the execution time in clock cycles, nonvolatile (flash) memory requirement in bytes, corresponding number of instructions, and the flexibility in the development of new programs.

If the four values to be transferred are known in advance, immediate addressing can be used for data transfer. Figure 2.7(a) shows the source code for this example. Each of the MOV instructions uses source-immediate addressing to point to the value to be transferred, and destination-direct addressing to point directly to the location that the data are to be copied to in memory. The data transfer is fastest in this case, requiring only 32 clock cycles, which is approximately three times less than the number of clock cycles required for the slowest case, that is using indirect post-increment addressing. Also, the flash memory requirement is the smallest of the four cases, with only 12 bytes of memory being required to store the four instructions. However, this method provides little flexibility, which can be a significant disadvantage. If other values have to be transferred, or if data are stored at different memory locations, then the code has to be modified.

Figure 2.7(b) shows the assembly language source code required for the data transfer if source and destination-direct addressing are used. The flash memory requirements are still small (only

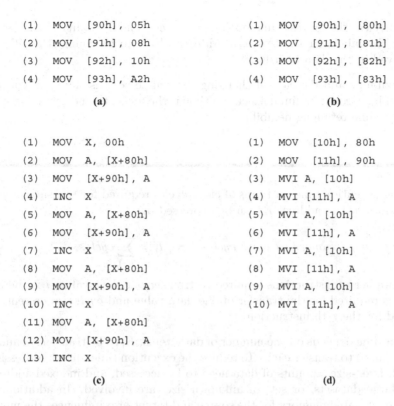

```
(1)  MOV  [90h], 05h              (1)  MOV  [90h], [80h]
(2)  MOV  [91h], 08h              (2)  MOV  [91h], [81h]
(3)  MOV  [92h], 10h              (3)  MOV  [92h], [82h]
(4)  MOV  [93h], A2h              (4)  MOV  [93h], [83h]

              (a)                               (b)


(1)  MOV  X, 00h                  (1)  MOV  [10h], 80h
(2)  MOV  A, [X+80h]              (2)  MOV  [11h], 90h
(3)  MOV  [X+90h], A              (3)  MVI  A, [10h]
(4)  INC  X                       (4)  MVI  [11h], A
(5)  MOV  A, [X+80h]              (5)  MVI  A, [10h]
(6)  MOV  [X+90h], A              (6)  MVI  [11h], A
(7)  INC  X                       (7)  MVI  A, [10h]
(8)  MOV  A, [X+80h]              (8)  MVI  [11h], A
(9)  MOV  [X+90h], A              (9)  MVI  A, [10h]
(10) INC  X                       (10) MVI  [11h], A
(11) MOV  A, [X+80h]
(12) MOV  [X+90h], A
(13) INC  X

              (c)                               (d)
```

Figure 2.7: Assembly code for data vector transfer.

12 bytes), but the execution time is slightly longer than for the first case. Forty clock cycles are required for the data transfer and although the flexibility is somewhat better, overall it is still poor. Although this code is independent of the data values that are transferred, it will have to be changed if different memory addresses and regions for source and destination are involved in the transfer.

Figure 2.7(c) shows the source code using indexed addressing. The X register stores the index of the data involved in the transfer. This register is initialized to zero, and incremented after each transfer. Because of the instructions available, a data value has to be copied from the source location to the A register, and then moved from the A register to the destination location. This requires thirteen instructions, which occupy 22 bytes of flash memory, almost double the amount needed for the first two cases. Although the execution time is also much longer (viz., 68 clock cycles) which is more than twice the time for the case in Figure 2.7(a), this case affords greater flexibility. If different memory regions are involved in the transfer, then only the base addresses of the regions (i.e., 80h for the source, and 90h for the destination) have to be updated. This is easily achieved by modifying the associated label values for the source and destination.

Figure 2.7(d) shows the source code, using indirect post-increment addressing. The memory locations at addresses 10h and 11h store the base addresses for the source and destination regions. Using MVI instructions, a data value is loaded from the source region into the A register, and then moved to the destination region. No increment instructions are needed, as in the third case, because the pointer values are automatically incremented. As a result only ten instructions are required, which occupy 22 bytes in memory, as in the indexed addressing case. The execution time is 96 clock cycles, the longest time of the four cases. The flexibility is similar to the third case except that the labels for the two base addresses have to be changed if different memory regions are involved in the data transfer.

Clearly, if flexibility and the ease of changing the source code is the main concern, indexed addressing should be used. If minimal execution time is the focus, then direct addressing provides fast data transfer while retaining flexibility.

In general, the execution time, in terms of clock cycles required for transferring a data vector of size K from address A to address B, can be expressed by:

$$Number\ of\ clock\ cycles \; = \; K \times \sum_{i=1}^{n} cycles_i \qquad (2.1)$$

where n is the number of instructions required to transfer one data value, i is the index for the set of instructions required for the transfer of one data value and $cycles_i$ represents the number of cycles required for the i–th instruction.

The execution time depends on the number of data items, K, to be transferred and the number of clock cycles required to transfer each. To reduce the execution time, direct addressing should be used when small, fixed-size amounts of data need to be accessed, and indexed addressing should be employed if large datasets, or sets of unknown size, are involved. In addition, the relative positioning of the allocated memory for the source and target also influences the number of clock cycles required. For example, manipulating compact memory regions requires the execution of fewer instructions and hence fewer clock cycles than if data are not in contiguous memory regions.

Table 2.3: Characteristics of the four specification styles.

	Immediate	Direct	Indexed	Indirect
Execution time (clock cycles)	32	40	68	96
flash memory (number of bytes)	12	12	22	22
Number of instructions	4	4	13	10
Flexibility	very low	low	high	high

A.2 SWAP Instructions

Swap instructions exchange the contents of the A, X, SP registers and SRAM, respectively. Table 2.4 lists the four kinds of SWAP instructions and is a summary of the semantics of each instruction, its opcode, the number of bytes occupied by the instruction in ROM, and the number of clock cycles required for its execution.

The semantics of the SWAP instructions is as follows:

- *SWAP A,X:* This instruction swaps the contents of the A and X registers.

- *SWAP A,[expr]:* The contents of the A register is swapped with the contents of the SRAM cell at address expr (source-direct addressing mode).

- *SWAP X,[expr]:* This instruction swaps the contents of the X register with the contents of the SRAM cell at address expr (source-direct addressing).

- *SWAP A,SP:* The contents of the A and SP registers are exchanged.

While SWAP instructions do not modify the CF flag, the ZF flag is set if a value of zero is loaded into the A register.

Table 2.4: **SWAP** instructions [2].

Instruction	Semantics	Opcode	Bytes	Cycles
SWAP A, X	aux ← X; X ← A A ← aux	0x4B	1	5
SWAP A, [expr]	aux ← SRAM[expr]; SRAM[expr] ← A A ← aux	0x4C	2	7
SWAP X, [expr]	aux ← SRAM[expr]; SRAM[expr] ← X X ← aux	0x4D	2	7
SWAP A, SP	aux ← SP; SP ← A A ← aux	0x4E	1	5

A.3 POP and PUSH Instructions

POP and PUSH instructions are used to implement stack operations. PUSH places a value on the stack, at the location pointed to by the stack pointer, and then increments the stack pointer register. POP returns a value from the stack, and decrements the stack pointer register. Table 2.5 lists the various types of POP and PUSH instructions. This table is a summary of the semantics of the instructions, their respective opcodes, the number of bytes they occupy in SRAM, and the number of clock cycles required for execution.

The semantics for POP and PUSH instructions is as follows:

- *POP A:* The SRAM cell value at address SP − 1 is loaded into the A register. The stack pointer, stored in the SP register, is decremented.

- *POP X:* The SRAM cell value at address SP − 1 is loaded into the X register. The stack pointer is decremented.

- *PUSH A* - The content of the A register is stored on the stack at the location pointed to by SP register. The stack pointer is incremented.

- *PUSH X* - The value in register X is stored on the stack. The stack pointer is incremented.

Table 2.5: **POP** and **PUSH** instructions [2].

Instruction	Semantics	Opcode	Bytes	Cycles
POP A	**A ← SRAM[SP-1]** **SP ← SP-1**	0x18	1	5
POP X	**X ← SRAM[SP-1]** **SP ← SP-1**	0x20	1	5
PUSH A	**SRAM[SP] ← A** **SP ← SP+1**	0x08	1	4
PUSH X	**SRAM[SP] ← X** **SP ← SP+1**	0x10	1	4

POP and PUSH instructions do not modify the CF flag and PUSH instructions do not affect the ZF flag. If a value of zero is loaded by a POP instruction into the Aregister, the ZF flag is set.

The SRAM page, which can be located in any of the eight SRAM pages, is selected by the three least significant bits of the STD_PP (STD_PP[2:0]) register located at address 0,D1H. Because the stack is located in a single SRAM page, after reaching address FFH, the stack pointer, SP, returns to address 00H. Upon reset, the default value of the STD_PP register is set to 00H, and stack operations then refer to SRAM page zero, unless the STD_PP register is modified. Figure 2.8(a) shows the stack implementation for the M8C.

Example. Figures 2.8(b) and (c) show the source code for "safe POP" and "safe PUSH" instructions. The safe PUSH operation places a value on the stack only if the stack is not full. Similarly, the safe POP operation pops a value from the stack only if the stack is not empty. Location 60h in SRAM stores the number of values on the stack and the value at location 61h indicates the size of the stack. If the number of values is equal to the stack size, then the "zero flag" is set after the SUB instruction in line 2 of the Figure 2.8(b). In this case, execution jumps to the label END_PUSH. Otherwise, if the stack is not full, the value is pushed onto the stack (instruction 5) and the variable holding the number of values is incremented (instruction 6). The pushed value is found in the SRAM at address 80h (instruction 4).

The safe POP operation first checks whether there are values in the stack. If the number of values is zero (i.e., the stack is empty) and the ZF flag is set after executing instruction 1 in Figure 2.8(c) execution jumps to the "end" label. Otherwise, the value on top of the stack is first POP-ped into the A register (instruction 3), and then copied into SRAM at address 81h (instruction 4). Finally, the variable defining the number of values in the stack is decremented (instruction 5).

A.4 Data transfer from nonvolatile memory: ROMX and INDEX instructions

ROMX instructions transfer data, for ceample global constants, from the nonvolatile memory (i.e. flash memory) to the A register. Table 2.6 is a summary of the semantics of the instruction, its opcode, the number of bytes required to store the instruction, and the number of clock cycles required for execution. The current value of the PC register is stored in two temporary locations, t_1 and t_2 with the less significant byte in the former and the more significant byte in the latter. The address of the memory location to be accessed is determined by the contents of the X and A registers. The X register contains the less significant byte of the address, and the A register the more significant byte. After loading the PC register with the address, the value is copied into the A register. Finally, the PC register is restored to the value saved in the temporary locations, t_1 and t_2.

Table 2.6: **ROMX** instructions [2].

Instruction	Semantics	Opcode	Bytes	Cycles
ROMX	$t_1 \leftarrow PC[7:0]$; $PC[7:0] \leftarrow X$; $t_2 \leftarrow PC[15:8]$; $PC[15:8] \leftarrow A$; $A \leftarrow ROM[PC]$; $PC[7:0] \leftarrow t_1$; $PC[15:8] \leftarrow t_2$	0x28	1	11

If the A register is loaded with a value of zero, then the zero flag is set. However, the Carry Flag is not changed.

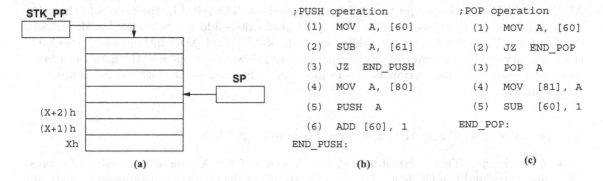

Figure 2.8: Stack description and stack-based operations.

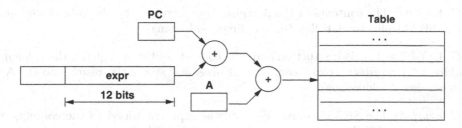

Figure 2.9: Index instruction.

Table 2.7: **INDEX** instructions [2].

Instruction	Semantics	Opcode	Bytes	Cycles
INDEX expr	**A ← ROM[A+expr+PC]** **-2048 ≤ expr ≤ 2047**	0xFx	2	13

INDEX instructions are useful for transferring data stored in tables in the flash memory. The A register is loaded with the contents of the flash memory address obtained by adding the values stored in A, expr, and PC. Table 2.7 is a summary of the main attributes of this instruction. The index value that points to a table entry is computed as shown in Figure 2.9. expr is a 12-bit offset relative to the program counter, and is expressed as its two's complement. Hence, its most significant bit is the sign bit. The remaining 11 bits represent offsets in the range -2048 to 2047. The sum of PC and expr gives the starting address of the table and the A register contains the table offset of the entry to be accessed.

Using INDEX instructions to access tables is illustrated by an example at the end of this chapter.

B. Instructions for Arithmetic Operations

Instructions for arithmetic operations include the following instructions: ADC, ADD, SBB, SUB, INC, DEC, CMP, ASL, ASR, RLC, and RRC instructions. The instructions are presented next.

B.1 Instruction for Addition: ADC and ADD Instructions

ADC and ADD instructions perform addition operations. The ADC instructions add the value of the CF flag and the two operands. The ADD instructions add the operands but do not add the CF flag. Table 2.8 presents the different kinds of ADC and ADD instructions. This table is a summary of the semantics of the instructions, and shows their opcode, the number of bytes required to store the instruction in the memory, and the number of clock cycles required for execution.

The semantics of ADC and ADD instructions is as follows:

- *ADC A,expr:* This instruciton adds the contents of the A register, the value of expression expr, and the CF flag. The result is stored in the A register (source-immediate addressing).

- *ADC A,[expr]:* The contents of the A register are increased by the value found at address expr in SRAM plus the CF flag (source-direct addressing).

- *ADC A,[X+expr]:* This instruction adds the contents of the A register, the value in SRAM at address "X register + expr", and the CF flag, and stores the result into the A register (source-indexed addressing).

- *ADC [expr],A:* The SRAM contents at address expr are added to the contents of the A register and the CF flag, and the result is stored in SRAM at address expr (destination-direct addressing).

Table 2.8: **ADC** and **ADD** instructions [2].

Instruction	Semantics	Opcode	Bytes	Cycles
ADC A,expr	$A \leftarrow A + expr + CF$	0x09	2	4
ADC A,[expr]	$A \leftarrow A + SRAM[expr] + CF$	0x0A	2	6
ADC A,[X+expr]	$A \leftarrow A + SRAM[X+expr] + CF$	0x0B	2	7
ADC [expr],A	$SRAM[expr] \leftarrow SRAM[expr] + A + CF$	0x0C	2	7
ADC [X+expr],A	$SRAM[X+expr] \leftarrow SRAM[X+expr] + A + CF$	0x0D	2	8
ADC [expr$_1$],expr$_2$	$SRAM[expr_1] \leftarrow SRAM[expr_1] + expr_2 + CF$	0x0E	3	9
ADC [X+expr$_1$],expr$_2$	$SRAM[X+expr_1] \leftarrow SRAM[X+expr_1] + expr_2 + CF$	0x0F	3	10
ADD A,expr	$A \leftarrow A + expr$	0x01	2	4
ADD A,[expr]	$A \leftarrow A + SRAM[expr]$	0x02	2	6
ADD A,[X+expr]	$A \leftarrow A + SRAM[X+expr]$	0x03	2	7
ADD [expr],A	$SRAM[expr] \leftarrow SRAM[expr] + A$	0x04	2	7
ADD [X+expr],A	$SRAM[X+expr] \leftarrow SRAM[X+expr] + A$	0x05	2	8
ADD [expr$_1$],expr$_2$	$SRAM[expr_1] \leftarrow SRAM[expr_1] + expr_2$	0x06	3	9
ADD [X+expr$_1$],expr$_2$	$SRAM[X+expr_1] \leftarrow SRAM[X+expr_1] + expr_2$	0x07	3	10
ADD SP,expr	$SP \leftarrow SP + expr$	0x38	2	5

- *ADC [X+expr],A:* The SRAM value at address "X + expr" is added to the value in the A register and the CF flag, and the result is stored in SRAM at address register "X + expr" (destination indexed addressing).

- *ADC [expr$_1$],expr$_2$:* The SRAM contents at address expr$_1$ are added to the value of expr$_2$ and the CF flag, and the result is saved at SRAM address expr$_1$. This instruction uses immediate addressing for referring to the source, and direct addressing for referring to the destination.

- *ADC [X+expr$_1$],expr$_2$:* This instruction adds the value of expr$_2$ to the value found in SRAM at address "X + expr$_1$" and the CF flag, and stores the result at the latter address (source-immediate and destination indexed addressing).

- *ADD A,expr:* The value of the A register is increased by the value of expr (source-immediate addressing).

- *ADD A,[expr]:* The contents of the A register are added to the value found in the SRAM at address expr, and the result is stored in the A register (source-direct addressing).

- *ADD A,[X+expr]:* The value in the A register is increased by the memory value pointed to by X register indexed with the value expr (source-indexed addressing).

- *ADD [expr],A:* The SRAM value at address expr is increased by the value in the A register (destination-direct addressing).

- *ADD [X+expr],A:* The memory value at address "X register + expr" is increased by the value stored in the A register (destination indexed addressing).

- *ADD [expr$_1$],expr$_2$:* The memory contents at address expr$_1$ are increased by the value expr$_2$.

- *ADD [X+expr$_1$],expr$_2$* - Value expr$_2$ is added to the value in memory at address "register X + expr$_1$", and the result is stored at the same address.

- *ADD SP,expr* - The contents of the SP register are increased by the value of expr.

ADC and ADD instructions set the ZF flag if the result is zero, and clear the flag otherwise. If the result is larger than 255, i.e., the maximum value that can be represented by eight bits, the CF flag is set to one. If not, CF is set to zero.

B.2 Instruction for Subtraction: SBB and SUB Instructions

SBB and *SUB* instructions execute subtraction operations. The SBB instructions subtract the value of the CF flag and the two operands. The SUB instructions subtract the operands but not the CF flag. Table 2.9 presents the different kinds of SBB and SUB instructions. This table is a summary of the semantics of the instructions, and shows their opcode, the number of bytes required to store the instruction in the memory, and the number of clock cycles needed for execution.

The semantics of SBB and SUB instructions is as follows.

- *SBB A,expr:* subtracts the values of expr and CF flag from the A register. The result is stored in the A register (source-immediate addressing).

- *SBB A,[expr]* - The value found at address expr in SRAM plus the CF flag are subtracted from the A register (source-direct addressing).

Table 2.9: *SBB* and *SUB* instructions [2].

Instruction	Semantics	Opcode	Bytes	Cycles
SBB A, expr	A ← A - (expr + CF)	0x19	2	4
SBB A, [expr]	A ← A - (SRAM[expr] + CF)	0x1A	2	6
SBB A, [X+expr]	A ← A - (SRAM[X+expr] + CF)	0x1B	2	7
SBB [expr], A	SRAM[expr] ← SRAM[expr] - (A + CF)	0x1C	2	7
SBB [X+expr], A	SRAM[X+expr] ← SRAM[X+expr] - (A + CF)	0x1D	2	8
SBB [expr₁], expr₂	SRAM[expr₁] ← SRAM[expr₁] - (expr₂ + CF)	0x1E	3	9
SBB [X+expr₁], expr₂	SRAM[X+expr₁] ← SRAM[X+expr₁] - (expr₂ + CF)	0x1F	3	10
SUB A, expr	A ← A - expr	0x11	2	4
SUB A, [expr]	A ← A - SRAM[expr]	0x12	2	6
SUB A, [X+expr]	A ← A - SRAM[X+expr]	0x13	2	7
SUB [expr], A	SRAM[expr] ← SRAM[expr] - A	0x14	2	7
SUB [X+expr], A	SRAM[X+expr] ← SRAM[X+expr] - A	0x15	2	8
SUB [expr₁], expr₂	SRAM[expr₁] ← SRAM[expr₁] - expr₂	0x16	3	9
SUB [X+expr₁], expr₂	SRAM[X+expr₁] ← SRAM[X+expr₁] - expr₂	0x17	3	10

- _SBB A,[X+expr]_: - The value at address "register X + expr" in memory and the Carry Flag are subtracted from the A register. The result is stored in the A register (source-indexed addressing).

- _SBB [expr],A_: - The A register and the CF flag are subtracted from the value in SRAM at address expr (destination-direct addressing).

- _SBB [X+expr],A_: Similar to the previous instruction, the A register and CF flag values are subtracted from the SRAM value at address "register X + expr" (destination indexed addressing).

- _SBB [expr$_1$],expr$_2$_: This instruction subtracts the value expr$_2$ and the CF flag from the value at address expr$_1$ in the memory. The result is stored in the memory at address expr$_1$ (source-immediate addressing and destination-direct addressing).

- _SBB [X+expr$_1$],expr$_2$_: The memory contents at the address pointed to by the X register, indexed with the value expr$_1$ is decreased by the value expr$_2$ plus the CF flag (source-immediate and destination indexed addressing),

- _SUB A,expr_: The value represented by expr is subtracted from the A register, and the result is stored in the A register (source-immediate addressing).

- _SUB A,[expr]_: The value found at address expr in SRAM is subtracted from the A register, and the result is stored in the A register (source-direct addressing).

- _SUB A,[X+expr]_: The value in the A register is decreased by the value of memory entry at the address pointed to by the X register plus the value expr (source-indexed addressing).

- _SUB [expr],A_: subtracts the contents of the A register from the value at address expr in the memory, and the result is stored in memory at the same address (destination-direct addressing mode).

- _SUB [X+expr],A_: The A register is subtracted from the memory value at address "register X + expr", and the result is stored in memory at the same address (destination indexed addressing).

- _SUB [expr$_1$],expr$_2$_: This instruction subtracts the value expr$_2$ from the value at address expr$_1$ in memory. The result is stored in the memory at address expr$_1$ (source-immediate addressing and destination-direct addressing).

- _SUB [X+expr$_1$],expr$_2$_: This instruction subtracts the value expr$_2$ from the value at address given by "register X + expr$_1$": (source-immediate and destination indexed addressing).

B.3 Instruction for Incrementing and Decrementing: INC and DEC Instructions

The _INC_ instructions increment the values in the A and X registers, and SRAM. Table 2.10 lists the four kinds of increment instructions and shows the semantics of the instructions, their opcodes, the number of memory bytes, and their respective execution times.

The _DEC_ instructions decrement the values in the A and X registers, and SRAM. Table 2.11 lists the four kinds of decrement instructions and shows the semantics of the instructions, their opcodes, the number of memory bytes, and their respective execution times.

Table 2.10: **INC** instructions [2].

Instruction	Semantics	Opcode	Bytes	Cycles
INC A	$\mathbf{A \leftarrow A + 1}$	0x74	1	4
INC X	$\mathbf{X \leftarrow X + 1}$	0x75	1	4
INC [expr]	**SRAM [expr] ← SRAM[expr] + 1**	0x76	2	7
INC [X+expr]	**SRAM [X+expr] ← SRAM[X+expr] + 1**	0x77	2	8

Table 2.11: **DEC** instructions [2].

Instruction	Semantics	Opcode	Bytes	Cycles
DEC A	$\mathbf{A \leftarrow A - 1}$	0x78	1	4
DEC X	$\mathbf{X \leftarrow X - 1}$	0x79	1	4
DEC [expr]	**SRAM[expr] ← SRAM[expr] - 1**	0x7A	2	7
DEC [X+expr]	**SRAM[X+expr] ← SRAM[X+expr] - 1**	0x7B	2	8

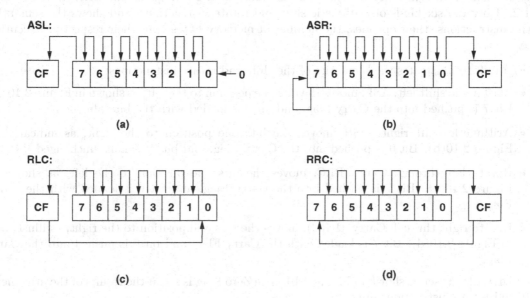

Figure 2.10: Shift and rotate instruction.

Table 2.12: **CMP** instructions [2].

Instruction	Semantics	Opcode	Bytes	Cycles
CMP A, expr	**A - expr**	0x39	2	5
CMP A, [expr]	**A - SRAM[expr]**	0x3A	2	7
CMP A, [X+expr]	**A - SRAM[X+expr]**	0x3B	2	8
CMP [expr$_1$], expr$_2$	**SRAM[expr$_1$] - expr$_2$**	0x3C	3	8
CMP [X+expr$_1$], expr$_2$	**SRAM[X+expr$_1$] - expr$_2$**	0x3D	3	9

B.4 Instruction for Comparison: CMP Instructions

The goal of *CMP* instructions is to set the CF and ZF flags. This instruction subtracts the second operand of the instruction from the first operand, but the value is not stored. If the result of the subtraction is negative, the CF flag is set to one, otherwise the flag is cleared. If the result of the subtraction is zero the ZF flag is set to one. If not, the ZF flag is reset to zero. Table 2.12 is a summary of the main attributes of the *CMP* instruction.

B.5 Instruction for Shifting and Rotation: ASL, ASR, RLC, and RRC Instructions

The shift and rotate instructions modify the values in the A register and SRAM. Tables 2.13 and 2.14 list the six kinds of arithmetic shift and rotate instructions, and shows the semantics of the instructions, their opcodes, the number of memory bytes, and their respective execution times.

Figure 2.10 illustrates the semantics of the shift and rotate instructions:

- Arithmetic shift left, ASL, moves the bits one position to the left, as shown in Figure 2.10(a). Bit 7 is pushed into the Carry Flag, and bit 0 is loaded with the bit value 0.

- Arithmetic shift right, ASR, moves the bits one position to the right, as indicated in Figure 2.10(b). Bit 0 is pushed into the Carry Flag, and bit 7 remains unchanged.

- Rotate left through Carry, RLC, moves the bits one position to the left, as shown in Figure 2.10(c). Bit 7 is pushed into the Carry Flag, and bit 0 is loaded with the Carry Flag.

- Rotate right through Carry, (RRC), moves the bits one position to the right, as illustrated in Figure 2.10(d). Bit 7 is loaded with the Carry Flag, and bit 0 is pushed into the Carry Flag.

The Carry Flag is set as shown in Figure 2.10. The Zero Flag is set, if the results of the arithmetic shift and rotate instructions are zero.

Table 2.13: **ASL** and **ASR** instructions [2].

Instruction	Semantics	Opcode	Bytes	Cycles
ASL A	see in Figure 2.10	0x64	1	4
ASL [expr]	see in Figure 2.10	0x65	2	7
ASL [X+expr]	see in Figure 2.10	0x66	2	8
ASR A	see in Figure 2.10	0x67	1	4
ASR [expr]	see in Figure 2.10	0x68	2	7
ASR [X+expr]	see in Figure 2.10	0x69	2	8

Example (Multiplication of two four-bit unsigned values). Figure 2.11(a) illustrates the algorithm for multiplying the decimal value 14 (represented by the 4-bit long, binary bitstring 1110) with the decimal value 10 (represented by the 4-bit long, binary bitstring 1010). The result is 8-bits long and has the value 10001100 (i.e., the decimal value 140). In general, if two M-bit unsigned numbers are multiplied, the result is 2 M bits long.

This algorithm uses an auxiliary variable to store the final product and examines the second operand bitwise, starting from the least significant bit. If the analyzed bit is 0 then the first operand is shifted to the left by one position, which is equivalent to multiplying it by 2. If the analyzed bit is '1' then the value of the first operand is added to the auxiliary variable holding the partial result, and the first operand is shifted by one position. After all bits of the second operand have been examined, the auxiliary variable holds the result of the product.

Figure 2.11(b) shows the assembly language source code for the corresponding multiplication routine. The two operands for the multiplication routine _mul8 are passed to the A and X registers. Because the contents of these registers change during execution, the routine begins by saving the two operands on the stack. Then, the X register is set to point to the top of the stack. The A register, initialized to zero, stores the partial results of the multiplication algorithm. The CMP instruction, at address 03CC, is used to determine if the multiplication algorithm has finished, which is indicated by the value of the second operand being zero. Note that the second operand always becomes zero after performing a number of rotate right shifts equal to the position of its most significant nonzero bit. In the case, where the algorithm did not finish, it resets the carry bit, that is bit 2 of the CPU_F register, by performing a bitwise AND of the flags register, CPU_F, with the value 251 representing the bitvector 11111011 (address 03D1). Then, the second operand is shifted right with a carry. This pushes the next bit to be examined into the carry bit of the flags register,CPU_F. If there is no carry, then the currently analyzed bit of the second operand is 0, and the first operand is shifted to the left by one position (address 03D9). This instruction is equivalent to multiplying the second operand by 2. If there is a carry (i.e., the currently analyzed bit) if the second operand is 1, then the content of the first operand is added to the A register holding the partial results. The algorithm iterates until the stop condition at address 03CC is met.

```
1110 x
    1010
   00000 (initial) +
                          shift left operand 1
   11100                  add operand 1 and then shift left operand 1
                          shift left operand 1
 1110000                  add operand 1 and then shift left operand 1
10001100
```

(a)

```
__mul8:
03C7:    PUSH   X
03C8:    PUSH   A
03C9:    MOV    X, SP
03CA:    MOV    A, 0
03CC:    CMP    [X-1], 0
03CF:    JZ     0x03DD
03D1:    AND    F, 251
03D3:    RRC    [X-1]
03D5:    JNC    0x03D9
03D7:    ADD    A, [X-2]
03D9:    ASL    [X-2]
03DB:    JMP    0x03CC
03DD:    ADD    SP, 254
03DF:    RET
```

(b)

Figure 2.11: Assembly code for multiplication.

Table 2.14: **RLC** and **RRC** instructions [2].

Instruction	Semantics	Opcode	Bytes	Cycles
RLC A	**See Figure 2.10**	0x6A	1	4
RLC [expr]	**See Figure 2.10**	0x6B	2	7
RLC [X+expr]	**See Figure 2.10**	0x6C	2	8
RRC A	**See Figure 2.10**	0x6D	1	4
RRC [expr]	**See Figure 2.10**	0x6E	2	7
RRC [X+expr]	**See Figure 2.10**	0x6F	2	8

C. Instructions for Logical Operations

Logical operations are performed by instructions, for example a bitwise *AND*, bitwise *OR*, bitwise exclusive-*OR* (XOR), and bitwise complement (CPL). The *TST* instruction performs a bitwise *AND* operation, but its purpose is only to set the ZF flag, in as much as the result is not stored. Logical operations can be performed on the flags register, CPU_F, to set the CF and *ZF* flags to the desired values.

The *AND* instructions perform a bitwise *AND* on the corresponding bits of their two operands, for example a bitwise *AND* of bits zero of the two operands, bitwise *AND* of bits one of the operands, and so on. Table 2.15 lists the different kinds of bitwise AND instructions and is a summary of the semantics of the instructions, their opcode, the number of bytes used to represent the instruction in the memory, and the number of clock cycles required for execution. Rows 1-7 in the table show *AND* instructions utilizing different methods of addressing their source and destination, such as source-immediate, source-direct, source-indexed addressing, destination-direct and destination indexed. Row 8 shows the instruction *AND REG[expr₁], expr₂* which has, as an operand, the value in register address expr$_1$. Similarly, the instruction in row 9 uses an operand stored at address "X + expr$_1$" in the register space. Finally, row 10 shows a bitwise AND instruction applied to the flags register (register CPU_F) and the bit value represented by expr. This instruction is used to set, or clear, the M8C flags.

The bitwise *OR* instructions compute the bitwise logical *OR* of the corresponding bits of the two operands. For example, bit 0 of the result is the logical OR of the bits 0 (least significant bits) of the two operands. Bit 1 of the result is the logical *OR* of the bit 1 of the two operands, etc. Table 2.16 shows the different kinds of *OR* instructions, their semantics, opcode, number of bytes occupied in memory, and number of clock cycles required. Rows 1–7 in the table show *OR* instructions using a different source and destination addressing modes. Rows 8 and 9 contain *OR* instructions with one of the operands located in the register space. The two instructions utilize direct and indexed addressing, respectively. Row 10 presents bitwise *OR* instruction involving the flags register, CPU_F. This instruction is the only bitwise *OR* instruction that can modify the CF flag, that is bit 2 in register CPU_F. This instruction can also change the ZF flag which is bit 1 of the CPU_F register. In addition, the ZF is set whenever a bitwise *OR* instruction produces a result with all its bits being zero. Otherwise, the ZF flag is cleared.

The *XOR* instructions compute, bitwise, the exclusive-*OR* (*XOR*) X of the corresponding bits of the *XOR* instruction. For example, bit 0 of the result is the logical X*OR* of bit 0 of operand one and bit 0 of the second operand. Table 2.17 is a summary of the different types of *XOR* instructions. As in the case of *AND* and *OR* instructions, *XOR* instructions employ source-immediate, direct, and indexed addressing, and destination-direct and indexed addressing, as shown by the instructions in rows 1–7 of the table. Rows 8 and 9 show *XOR* instructions operating on operands stored in the register space. Finally, row 10 presents the *XOR* instruction with the flags register, CPU_F, as one of the operands. *XOR F* affects the CF flag, but the instructions in rows 1–9 do not change the flag. The ZF flag is set if the result is zero, otherwise the flag is cleared. In addition, *XOR F, expr* can also modify the ZF flag.

The *CPL* instruction stores the bitwise complement of the value initially found in the A register, in the A register. Table 2.18 is a summary of the characteristics of this instruction. Although

Table 2.15: Bitwise AND instructions [2].

Instruction	Semantics	Opcode	Bytes	Cycles
AND A, expr	$A \leftarrow A$ & expr	0x21	2	4
AND A, [expr]	$A \leftarrow A$ & SRAM[expr]	0x22	2	6
AND A, [X+expr]	$A \leftarrow A$ & SRAM[X+expr]	0x23	2	7
AND [expr], A	SRAM[expr] $\leftarrow$ SRAM[expr] & A	0x24	2	7
AND [X+expr], A	SRAM[X+expr] $\leftarrow$ SRAM[X+expr] & A	0x25	2	8
AND [expr$_1$], expr$_2$	SRAM[expr$_1$] $\leftarrow$ SRAM[expr$_1$] & expr$_2$	0x26	3	9
AND [X+expr$_1$], expr$_2$	SRAM[X+expr$_1$] $\leftarrow$ SRAM[X+expr$_1$] & expr$_2$	0x27	3	10
AND REG[expr$_1$], expr$_2$	REG[expr$_1$] $\leftarrow$ REG[expr$_1$] & expr$_2$	0x41	3	9
AND REG[X+expr$_1$], expr$_2$	REG[X+expr$_1$] $\leftarrow$ REG[X+expr$_1$] & expr$_2$	0x42	3	10
AND F, expr	$F \leftarrow F$ & expr	0x70	2	4

Table 2.16: Bitwise **OR** instructions [2].

Instruction	Semantics	Opcode	Bytes	Cycles
OR A, expr	$A \leftarrow A - expr$	0x29	2	4
OR A, [expr]	$A \leftarrow A - SRAM[expr]$	0x2A	2	6
OR A, [X+expr]	$A \leftarrow A - SRAM[X + expr]$	0x2B	2	7
OR [expr], A	$SRAM[expr] \leftarrow SRAM[expr] - A$	0x2C	2	7
OR [X+expr], A	$SRAM[X+expr] \leftarrow SRAM[X+expr] - A$	0x2D	2	8
OR [expr$_1$], expr$_2$	$SRAM[expr_1] \leftarrow SRAM[expr_1] - expr_2$	0x2E	3	9
OR [X+expr$_1$], expr$_2$	$SRAM[X+expr_1] \leftarrow SRAM[X+expr_1] - expr_2$	0x2F	3	10
OR REG[expr$_1$], expr$_2$	$REG[expr_1] \leftarrow REG[expr_1] - expr_2$	0x43	3	9
OR REG[X+expr$_1$], expr$_2$	$REG[X+expr_1] \leftarrow REG[X+expr_1] - expr_2$	0x44	3	10
OR F, expr	$CPU\_F \leftarrow CPU\_F - expr$	0x71	2	4

Table 2.17: Bitwise **XOR** instructions [2].

Instruction	Semantics	Opcode	Bytes	Cycles
XOR A, expr	$A \leftarrow A \otimes expr$	0x31	2	4
XOR A, [expr]	$A \leftarrow A \otimes \textbf{SRAM}[expr]$	0x32	2	6
XOR A, [X+expr]	$A \leftarrow A \otimes \textbf{SRAM}[X+expr]$	0x33	2	7
XOR [expr], A	$\textbf{SRAM}[expr] \leftarrow \textbf{SRAM}[expr] \otimes A$	0x34	2	7
XOR [X+expr], A	$\textbf{SRAM}[X+expr] \leftarrow \textbf{SRAM}[X+expr] \otimes A$	0x35	2	8
XOR [expr$_1$], expr$_2$	$\textbf{SRAM}[expr_1] \leftarrow \textbf{SRAM}[expr_1] \otimes expr_2$	0x36	3	9
XOR [X+expr$_1$], expr$_2$	$\textbf{SRAM}[X+expr_1] \leftarrow \textbf{SRAM}[X+expr_1] \otimes expr_2$	0x37	3	10
XOR REG[expr], expr$_2$	$\textbf{REG}[expr_1] \leftarrow \textbf{REG}[expr_1] \otimes expr_2$	0x45	3	9
XOR REG[X+expr$_1$], expr$_2$	$\textbf{REG}[X+expr_1] \leftarrow \textbf{REG}[X+expr_1] \otimes expr_2$	0x46	3	10
XOR F, expr	$\textbf{CPU\_F} \leftarrow \textbf{CPU\_F} \otimes expr$	0x72	2	4

Table 2.18: **CPL** instructions [2].

Instruction	Semantics	Opcode	Bytes	Cycles
CPL A	**A ← $\bar{A}$**	0x73	1	4

Table 2.19: **TST** instructions [2].

Instruction	Semantics	Opcode	Bytes	Cycles
TST [expr$_1$],expr$_2$	**SRAM[expr$_1$] & expr$_2$**	0x47	3	8
TST [X+expr$_1$],expr$_2$	**SRAM[X+expr$_1$] & expr$_2$**	0x48	3	9
TST REG[expr$_1$],expr$_2$	**reg[expr$_1$] & expr$_2$**	0x49	3	9
TST REG[X+expr$_1$],expr$_2$	**reg[X+expr$_1$] & expr$_2$**	0x4A	3	10

CF is not modified by this instruction, the ZF flag is set if the resulting value, after complement, is zero.

The *TST* instructions (test with mask) calculate a bitwise *AND* of their operands, but do not store the result. Only the ZF flag is modified as a result of this instruction, but it does not modify the CF flag. Table 2.19 summarizes the *TST* instructions.

Example (Different Bit Manipulations). Figures 2.12(a), (b), and (c) present code that sets, resets, and inverts a single bit in a data byte, respectively. Examples (d), (e), and (f) set, reset, and invert a group of bits, respectively.

Example (a) sets bit 2 of the A register to 1, leaving all other bits unchanged. Mask *MASK_1* has 0 bits for all positions that remain unchanged, and a 1 for the position that is set. A bitwise *OR* instructions sets the desired bit to one.

Example (b) resets bit 2 of the A register leaving all other bits unmodified. The used mask has a bit 1 for all the positions that are not affected, and a bit 0 for the position that is reset. A bitwise *AND* instruction resets the desired bit to zero.

Example (c) complements the third bit of the A register, but other bits are unmodified. The used mask is similar to the mask for example (a). A bitwise *XOR* instruction complements the third bit.

Example (d) sets a group of four bits, bits 2-5, to 1111. The mask has a bit value of 1 for all the positions that are set, and a bit 0 for each unchanged positions. Similar to case (a), a bitwise *OR* instruction is used.

Example (e) resets bits 2 and 4-5, but the other bits are unchanged. The mask used has a bit value of 1 corresponding to the positions that are not modified, and a bit value of 0 for the positions that are reset. The bitwise *AND* instruction resets the wanted bits.

Example (f) complements bits 0-1 and bits 5-6 of the A register. An *XOR* instruction is used with a mask that has a bit value of 1 for all positions that are inverted, and the remaining positions have each a bit value of 0.

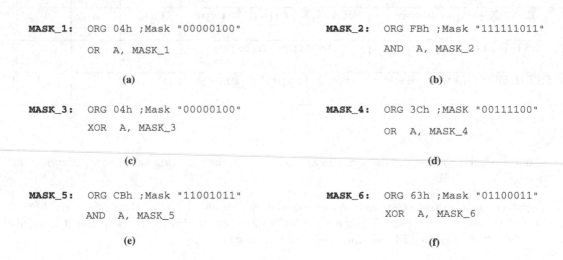

Figure 2.12: Bit manipulations.

D. Instructions for Flow Control

The instructions in this group define the execution flow of programs beyond the sequential execution of instructions. This group of instructions includes *JACC*, *JC*, *JMP*, *JNC*, *JNZ*, *JZ*, *LJMP*, *CALL*, *LCALL*, *RET*, *RETI*, and *SSC* instructions. Table 2.20 is a summary of the M8C's *JMP* instructions.

- *JACC expr:* This is an unconditional jump instruction to the address represented by (PC + 1) plus the contents of the A register plus the value of expr expressed as a 12-bit, two's complement value. The accumulator is unaffected by this instruction.

- *JC expr:* If the Carry Flag (CF) is set, it causes program execution to jump to the address represented by (PC + 1) plus the value of expr expressed as a 12-bit, two's complement value. The current value of the PC register points to the first byte of the *JC* instruction.

Table 2.20: Jump instructions [2].

Instruction	Semantics	Opcode	Bytes	Cycles
JACC expr	$\mathbf{PC \leftarrow PC + 1 + A + expr}$	0xEx	2	7
JC expr	$\mathbf{PC \leftarrow PC + 1 + expr}$ **-2048 $\leq$ expr $\leq$ 2047**	0xCx	2	5
JMP expr	$\mathbf{PC \leftarrow PC + 1 + expr}$ **-2048 $\leq$ expr $\leq$ 2047**	0x8x	2	5
JNC expr	$\mathbf{PC \leftarrow PC + 1 + expr}$ **-2048 $\leq$ expr $\leq$ 2047**	0xDx	2	5
JNZ expr	$\mathbf{PC \leftarrow PC + 1 + expr}$ **-2048 $\leq$ expr $\leq$ 2047**	0xBx	2	5
JZ expr	$\mathbf{PC \leftarrow PC + 1 + expr}$ **-2048 $\leq$ expr $\leq$ 2047**	0xAx	2	5
LJMP expr	$\mathbf{PC \leftarrow expr}$ **0 $\leq$ expr $\leq$ 65535**	0x7D	3	7

- *JMP expr:* This instruction causes an unconditional jump to the address represented by (PC + 1) plus the value of expr expressed as a 12-bit, two's complement value. The current value of the PC register points to the first byte of the *JC* instruction.

- *JNC expr:* If the CF flag is not set, program execution jumps to the address represented by (PC + 1) plus the value of expr expressed as a 12-bit, two's complement value. The current value of the PC register points to the first byte of the *JNC* instruction.

- *JNZ expr:* If the zero flag (ZF flag) is not set, the instruction causes program execution to jump to the address represented by (PC + 1) plus the value of expr expressed as a 12-bit, two's complement value. The current value of the PC register points to the first byte of the *JNZ* instruction.

- *JZ expr:* If the ZF flag is set, program execution jumps to the address represented by (PC + 1) plus the value of expr expressed as a 12-bit, two's complement value. The current value of the PC register points to the first byte of the *JZ* instruction.

- *LJMP expr:* This instruction causes an unconditional long jump, that is to the 16 bit address represented by *expr*.

The *CALL* instruction causes the PC register to be loaded with the sum of PC+2 and expr. The current value of PC points to the first byte of the *CALL* instruction and expr is a 12–bit, signed number expressed in two's complement form. The current value of the Program Counter is pushed onto the stack, by first pushing the more significant byte followed by the less significant byte. CF and ZF are not changed by this instruction. Table 2.21 shows the main characteristics of the *CALL* instruction.

The *LCALL* instruction loads the value of expr into the PC register expressed as a 16-bit, unsigned value of the physical address. The address of the instruction following the *LCALL* instruction is pushed onto the stack. This instruction is at an address pointed to by the current PC + 2. The more significant byte of the address is saved on the stack, followed by the less

Table 2.21: **CALL** and **LCALL** instructions [2].

Instruction	Semantics	Opcode	Bytes	Cycles
CALL expr	**PC ← PC + 2 + expr** **-2048 ≤ expr ≤ 2047**	0x9x	2	11
LCALL expr	**PC ← expr** **0 ≤ expr ≤ 65535**	0x7C	3	13

significant byte. The CF and ZF flags are not affected by the execution of this instruction. The *LCALL* instruction is summarized in Table 2.21.

Table 2.22: **RET** and **RETI** instructions [2].

Instruction	Semantics	Opcode	Bytes	Cycles
RET	**SP ← SP - 1; PC[7:0] ← SRAM[SP];** **SP ← SP - 1; PC[15:8] ← SRAM[SP];**	0x7F	1	8
RETI	**SP ← SP - 1; CPU_F ← SRAM[SP];** **SP ← SP - 1; PC[7:0] ← SRAM[SP];** **SP ← SP - 1; PC[15:8] ← SRAM[SP];**	0x7E	1	10

RET instructions are used to return from a routine called by the instructions *CALL* and LCALL. The Program Counter is restored from the stack, and program execution resumes at the next instruction following the call instruction. This instruction does not modify the CF and ZF flags. Table 2.22 is a summary of the RET instructions.

RETI instructions are used to return from interrupt service routines (ISRs) and system supervisory calls (SSCs). After the CPU_F register is restored from the stack, the CF and ZF flags are updated to the new value of the flags register. Then, the Program Counter is restored. Table 2.22 is a summary of the characteristics of *RETI* instructions.

Example (Calling assembly code routines from C programs). This example discusses the assembly code that corresponds to a function call in a high-level programming language, such as C language. This example is important in that it illustrates the structure required for assembly code routines that are called from programs in high-level languages.

Figure 2.13(a) shows an arbitrary function *f*, with three parameters *a*, *b*, and *c*. Each of these parameters is of type char and *f* returns a value of type char. The type char was selected for simplicity, so that each parameter can be stored as one byte. In addition, the function has two local variables *loc1* and *loc2*. Function *f* is called using the parameters *v1*, *v2*, and *v3*, and the returned value is assigned to variable *d*. The example illustrates the use of *PUSH* and *POP* instructions to transfer data between the calling environment and the called function. The transferred data includes the parameters of the function, the returned value of the function, and the return address.

Figure 2.13(c) presents the stack structure, called the *activation record*, AR that is created when function f is called. The bottom of AR is pointed to by the X register, so that its different fields can be easily accessed by source-indexed addressing. The bottom entry contains the return value of the function. The old value of the X register is stored on top of it, so that it can be retrieved after returning from the function call. Next, AR contains entries for the three parameters of the function. Then, the return address (the address of the instruction following the function call) is saved on the stack, with the less significant byte sitting on top of the more significant byte. Finally, the two top entries store the values of the local variables *loc1* and *loc2*.

Figure 2.13(b) shows the assembly language code for the function call and the called function f. The calling environment first saves the value of the X register. Then, the X register is loaded with the value of the stack pointer SP, resulting in the X register pointing to the bottom of AR. Instruction 3 increases the stack pointer value thereby allocating a stack entry for the returned value of function f. Because the returned values are of type char, one stack entry (one byte) is sufficient. Instructions 4-9 push the values of the three parameters on the stack. Instruction 10 calls function f and as a result, the address of the instruction following the call is saved on the stack.

The assembly language source code for function f starts at instruction 11. The first two *PUSH* instructions reserve stack entries for the two local variables. Instruction 13 stores a value to the local variable *loc2*, and instruction 14 to variable *loc1*. Instruction 15 assigns a value to the returned value of function f. Before returning from the function call, instructions 16-17 remove the two entries for the local variables from the stack. Instruction RET finds the return address on top of the stack, and the PC register is loaded with the address of instruction 19.

After returning from the function call, the POP instructions, in lines 19–21, remove the entries for the three function parameters from the stack. Instruction 22 pops the returned value of the function, and stores the value in the memory location for variable *d*. The last instruction restores the X register to its original value prior to the function call.

Table 2.23: **SSC** instructions [2].

Instruction	Semantics	Opcode	Bytes	Cycles
SSC	SRAM[SP] ← PC[15:8]; SP ← SP + 1; SRAM[SP] ← PC[7:0]; SP ← SP + 1; SRAM[SP] ← CPU_F; PC ← 0x0000; CPU_F ← 0x00	0x00	1	15

System supervisory calls (SSCs) are used to call predefined routines that are stored in ROM and used to perform system functions. Initially, the more significant byte of the *PC* register is saved on the stack, followed by the less significant *PC* byte. Next, the flags register is pushed

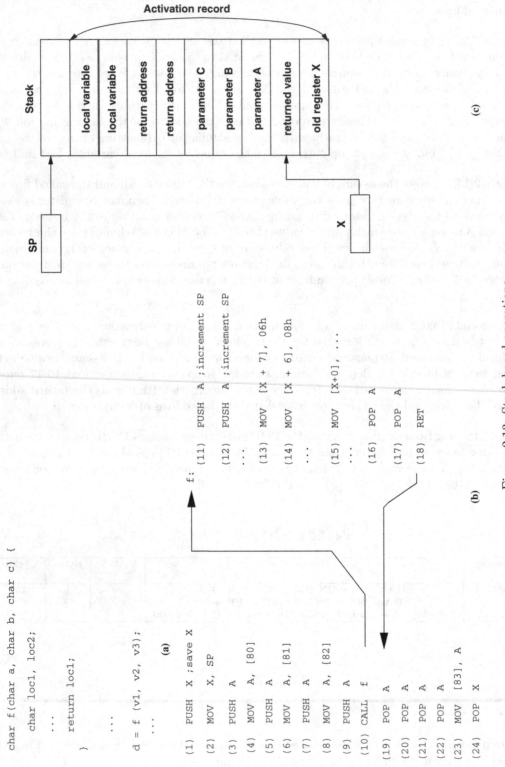

Figure 2.13: Stack-based operations.

onto the stack. After being saved, the flags register is reset. Program execution returns from an *SSC* by executing an *RETI* instruction. Table 2.23 is a summary of the *SSC* instructions.

E. Other Instructions

The *HALT* instruction halts program execution by the microcontroller, and suspends further microcontroller activity pending a hardware reset, for example Power-On, Watchdog Timer, or External Reset. Table 2.24 is a summary of the main characteristics of the *HALT* instruction.

The *NOP* instruction is executed in four clock cycles but has no effect other than to introduce a quantifiable delay in terms of program execution time. It does increment the program counter but does not effect anything else, for example: the CF and ZF flags are unaffected. Table 2.25 summarizes *NOP* instructions.

Example (Design of a sequence detector). Sequence detectors recognize predefined patterns in a stream of bits and are useful in various data communication protocols. The goal of this example is to design a sequence detector that outputs a bit value of 1 whenever an odd number of 0 bits and an odd number of 1 bits are encountered in the input bit stream.

Figure 2.14(a) presents the interface signals of the sequence detector which has two one-bit inputs and two one-bit outputs. Input *In* is the port for the input stream. Input *Ready* is an external signal, which is reset each time a new bit is available at input *In*. After a new input is read by the sequence detector, the *Ack* signal is generated by the detector and is used to remove the *Ready* signal. After the new input has been processed, the corresponding value of the output, *Out*, is produced by the sequence detector. Figure 2.14(b) illustrates the sequencing of the four interface signals.

The behavior of the sequence detector can be described in terms of a finite state machine (FSM) with four states:

- *State A*: This state is entered if an even number of 0s and an even number of 1s occur in the input stream, or upon receiving a reset signal.

- *State B*: This state is entered if an odd number of 0s and an even number of 1s was present in the input stream.

- *State C*: This state is entered if an even number of 0s and an odd number of 1s occur in the input stream.

- *State D*: This state is entered if an odd number of 0s and an odd number of 1s occur in the input stream.

Figure 2.14(c), the state transition diagram, shows all transitions to new states for each possible state of the detector and each possible input value. If the system is in state *A* and the input bit is 0, then the new state is *B*. If the input bit is 1 the new state is *C*. The state diagram also shows the output bits generated for each transition. Output 0 is produced for the transition from state *A* to state *B*. Output 1 is generated for the transition from state *C* to state *D*. The graphical description in Figure 2.14(c) is characterized by the tables provided in Figure 2.14(d).

The design of the sequence detector is given in Figure 2.15 with the data structures organized as illustrated in Figure 2.15(a). The current state of the detector is stored in the variable

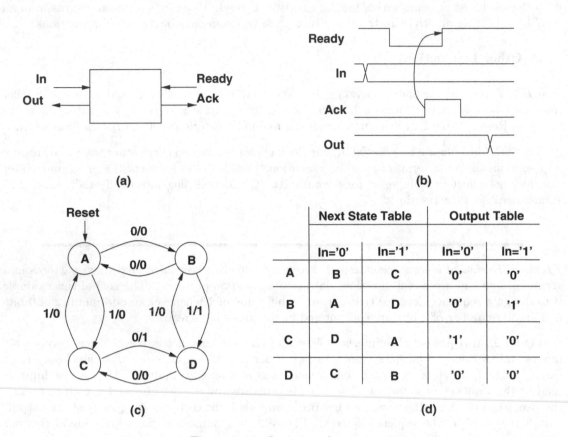

Figure 2.14: Sequence detector.

Table 2.24: **HALT** instruction [2].

Instruction	Semantics	Opcode	Bytes	Cycles
HALT	**REG[CPU_SCR] ← REG[CPU_SCR] + 1**	0x30	1	∞

Table 2.25: **NOP** instruction [2].

Instruction	Semantics	Opcode	Bytes	Cycles
NOP	-	0x40	1	4

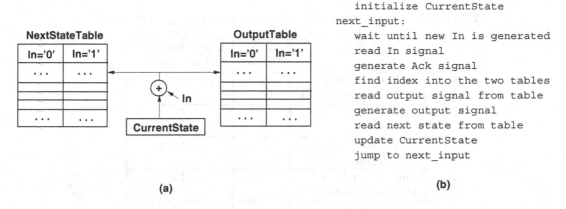

(a)

(b)

Figure 2.15: Sequence detector design.

CurrentState. The next state and output information are organized as two separate tables. Access is based on the values of the *CurrentState* variable and signal *In*. Figure 2.15(b) shows the pseudocode of the detector. The first step is to initialize the *CurrentState* variable. The detector then waits until a new input signal, *In*, is produced. After the detector reads the In bit, the *Ack* output is generated as shown in Figure 2.14(b). The next step calculates the index used to access tables *NextStateTable* and *OutputTable*. Using the computed index, Table *OutputTable* is first accessed to find the signal value *Out* corresponding to the current state and input signal. Then, the next state is retrieved from Table *NextStateTable*, and used to update variable *CurrentState*. The last step moves execution to the label called next_input for processing the following input bit.

Figure 2.16(a) shows the interface implementation. Pin 1, pin 2, pin 3, and pin 4 of a PSoC chip are used to connect the four I/O signals of the detector to the external environment. The four pins are connected to the bits of the Port 0 register, that is the *PRT0DR* register at address 0,00h in the register space, as shown in the figure. For example, pin 1 is connected to bit 1, and is used for implementing the input signal *In*. Similarly, port 4 is linked to bit 7 of the port register, and used for signal *Ready*. This figure also shows the configuration of the four control registers for port 0 and the *PRT0DM0*, *PRT0DM1*, *PRT0DM2*, and *PRT0GS* registers. (Additional information about the programming of these registers is provided in Chapter 3.)

Implementation using INDEX instructions. Figure 2.16(b) shows the assembly language source code for a sequence detector. This code corresponds to the pseudocode in Figure 2.15(b). *PRT0DR* is the address of the port 0 register, and *READY_MSK* is the mask used to access bit *READY*, that is bit 7 of the port register. *IN_MSK* is the mask to access bit *In* (bit 1). *ACK_1* is the mask used to assign a 1 for output *ACK*. *ACK_0* is the mask for producing a 0 for output ACK. *CRT_STATE* is the SRAM address that stores the current state of the sequence detector. *TEMP1* and *TEMP2* are the addresses in SRAM of two locations used to store temporary values.

Instruction 1 initializes the current state variable with the value 00h representing state *A*. Instructions 2-4 implement the busy waiting for the next input bit *In*. The *PRT0DR* register is first read into the A register. Then using mask *READY_MSK* (10000000), bit 7 of the port is selected. If the bit is nonzero then the program jumps back to label *BACK*. If a new input is detected, then a copy of the *PRT0DR* register is saved in SRAM location *TEMP1* (instructions 5-7). Instruction 6 resets bit Out. Instructions 8–9 produce the output signal *Ack*, first the

Table 2.26: Performance of the sequence detector implementation.

FSM Step	ROM (bytes)	Execution time (clock cycles)
wait until new input In is generated	6	$15 \times \#_{BW}$
read In signal	6	15
generate Ack signal	6	18
find index into the two tables	7	19
read output signal from table	2	13
generate output signal	4	11
read next state from table	4	18
update CurrentState	2	5
wait for Ready Signal to go High	6	$15 + \#_{BW}$
jump to next_input	2	5
Total	45	149 (without busy waiting)

signal value is 1 and then 0. Next, instructions 10–13 compute the index used to access the two tables. Instruction 10 selects the input bit using mask *IN_MSK* (00000010). The A register is shifted one position to the right, so that it can be used for the index. The index value is stored in the SRAM location *TEMP2*. Instruction 14 accesses Table *OutputTable* to find the bit value that has to be output by the detector. Instructions 15–16 set bit 5 of the port register to the required value. Instructions 17–18 access Table *NextStateTable* to get the encoding of the next state. The encoding is stored in the variable *CurrentState* in instruction 19. Instructions 20–22 delay execution until the *READY* signal is one (inactive). Finally, program execution jumps to label FSM to start the processing of the next input bit.

Performance analysis. Table 2.26 is a summary of the memory and timing performance of the implementation. Rows correspond to the steps shown in Figure 2.15(b). The first row corresponds to "busy-waiting" until a new input bit is found at the sequence detector. The timing of the step depends on the number of executed iterations ($\#_{BW}$). The last row shows the amount of memory and the number of clock cycles required to execute the code. Assuming that the $\#_{BW} = 1$, new input data are available each time a new iteration starts. Because one iteration of the code is executed in 149 clock cycles, thus for a clock frequency of 24 MHz, this results in an execution time of 149×0.041 μsec, or 6.109 μsec per iteration. Thus, input bit streams of

$$\frac{1}{6.109\ 10^{-6}} \text{bits/sec} \approx 163 \text{ kbits/sec} \tag{2.2}$$

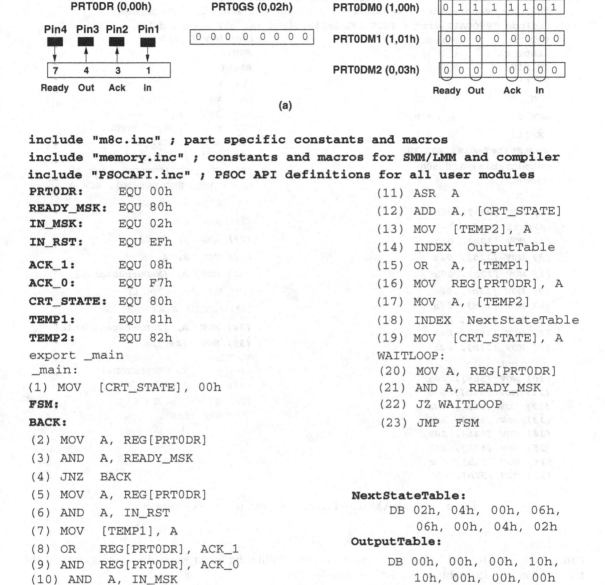

(a)

```
include "m8c.inc" ; part specific constants and macros
include "memory.inc" ; constants and macros for SMM/LMM and compiler
include "PSOCAPI.inc" ; PSOC API definitions for all user modules
PRT0DR:        EQU 00h                    (11) ASR   A
READY_MSK:     EQU 80h                    (12) ADD   A, [CRT_STATE]
IN_MSK:        EQU 02h                    (13) MOV   [TEMP2], A
IN_RST:        EQU EFh                    (14) INDEX OutputTable
                                          (15) OR    A, [TEMP1]
ACK_1:         EQU 08h                    (16) MOV   REG[PRT0DR], A
ACK_0:         EQU F7h                    (17) MOV   A, [TEMP2]
CRT_STATE:     EQU 80h                    (18) INDEX NextStateTable
TEMP1:         EQU 81h                    (19) MOV   [CRT_STATE], A
TEMP2:         EQU 82h               WAITLOOP:
export _main                              (20) MOV A, REG[PRT0DR]
_main:                                    (21) AND A, READY_MSK
(1) MOV   [CRT_STATE], 00h                (22) JZ WAITLOOP
FSM:                                      (23) JMP   FSM
BACK:
(2) MOV   A, REG[PRT0DR]
(3) AND   A, READY_MSK
(4) JNZ   BACK
(5) MOV   A, REG[PRT0DR]
(6) AND   A, IN_RST                   NextStateTable:
(7) MOV   [TEMP1], A                      DB 02h, 04h, 00h, 06h,
(8) OR    REG[PRT0DR], ACK_1                 06h, 00h, 04h, 02h
(9) AND   REG[PRT0DR], ACK_0          OutputTable:
(10) AND  A, IN_MSK                       DB 00h, 00h, 00h, 10h,
                                             10h, 00h, 00h, 00h
```

(b)

Figure 2.16: Sequence detector implementation using INDEX instructions.

can be processed without losing a bit. For higher input frequencies, some of the data might be lost, because the sequence detector then becomes too slow to process all the inputs.

This analysis also shows that about 33% of the total execution time, excluding busy waiting time, is consumed accessing the two tables, 10% of the time is required to read data from the input port, 12% of the time is required to generate the *Ack* signal, 7% of the time is required to produce the *Out* signal, 3% to update the state variable, and 3% of the time is required to jump to the beginning of the detection code. Thus, speedingup execution for processing faster inputs

```
include "m8c.inc" ; part specific constants and macros
include "memory.inc" ; constants and macros for SMM/LMM and compiler
include "PSOCAPI.inc" ; PSOC API definitions for all user modules
```

PRT0DR: EQU 00h   *(18) MOV X, 00h*
READY_MSK: EQU 80h   FSM:
IN_MSK: EQU 02h   BACK:
IN_RST: EQU EFh   (19) MOV A, REG[PRT0DR]
ACK_1:   EQU 08h   (20) AND A, READY_MSK
ACK_0:   EQU F7h   (21) JNZ BACK
                                              (22) MOV A, REG[PRT0DR]
TEMP1:   EQU 81h   (23) AND A, INT_RST
NextStateTable: EQU 10h   (24) MOV [TEMP1],A
OutputTable: EQU 50h   (25) OR   REG[PRT0DR], ACK_1
export _main   (26) AND REG[PRT0DR], ACK_0
_main: (27) AND A, IN_MSK
(1) MOV [00h], 00h   (28) ASR A
; NextStateTable   *(29) ADD A, [X+0]*
(2) MOV [10h], 02h   *(30) MOV X, A*
(3) MOV [11h], 04h   *(31) MOV A, [X+OutputTable]*
(4) MOV [12h], 00h   (32) OR   A, [TEMP1]
(5) MOV [13h], 06h   (33) MOV REG[PRT0DR],A
(6) MOV [14h], 06h   *(34) MOV A, [X+NextStateTable]*
(7) MOV [15h], 00h   *(35) MOV [X+0],A*
(8) MOV [16h], 04h   WAITLOOP:
(9) MOV [17h], 02h   (36) MOV A, REG[PRT0DR]
; OutputTable   (37) MOV A, READY_MSK
(10) MOV [50h], 00h   (38) JZ WAITLOOP
(11) MOV [51h], 00h   (39) JMP FSM
(12) MOV [52h], 00h
(13) MOV [53h], 10h
(14) MOV [54h], 10h
(15) MOV [55h], 00h
(16) MOV [56h], 00h
(17) MOV [57h], 00h

Figure 2.17: RAM-based sequence detector implementation.

can be achieved by two orthogonal improvements: (1) finding faster ways of accessing data in the two tables, and (2) using faster I/O interfaces.

RAM-based implementation. Figure 2.17 presents an alternative implementation, in which the FSM data are first moved from the flash memory to the RAM. Data accessing is simpler and faster than for the example in Figure 2.16(b). Instead of using *INDEX* instructions, the program uses indexed addressing to access the FSM tables. The modifications to the code are shown in bold.

2.2   Memory Space

This section discusses PSoC's SRAM and ROM space.

A. SRAM Space

The PSoC architecture can have as many as eight, 256 -byte, memory pages. The 256 bytes of any given page are addressed by an eight bit address. Different memory pages can be used for storing the data variables of a program, the stack, and the variables accessed by the two indexed addressing modes. Having multiple memory pages also simplifies the process of simultaneous execution of multiple tasks in multitasking applications. The variables of a task can be stored in a separate page, and reducing the amount of data transfer required when switching from the execution of one task to another. On the negative side, the assembly code programming becomes slightly more difficult, because programmers must then manage data stored over multiple SRAM pages.

Figure 2.18 presents the paging mechanism of the SRAM space. Bit 7 of the *CPU_F* register is used to enable and disable the SRAM paging. If the bit is zero then all memory accesses are to page zero. If the bit is set, then the last three bits of the *CUR_PP* register at address 0, D0H (CUR_PP[2:0]) select the current SRAM page. Then, with the exception of *PUSH*, *POP*, *CALL*, *LCALL*, *RETI*, *RET*, and *MVI* instructions, all other SRAM accesses refer to this page.

In addition to the current SRAM page, the PSoC architecture also gives the option to setup dedicated pages for the stack and for the indexed addressing mode. The stack page is selected

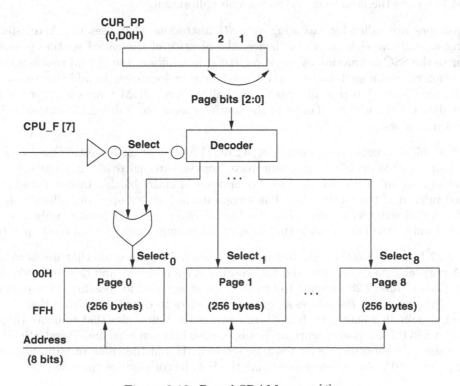

Figure 2.18: Paged SRAM space [4].

by the bits 2–0 of the *STK_PP* register at address 0, D1H. After a reset operation, the three bits are set to 000, and hence all stack operations refer to page 0.

The indexed addressing mode uses one of three possible SRAM pages depending on the value of the two bits 7–6 of the *CPU_F* register at address x,F7H:

- If the bits are 0 then page 0 is used for the indexed addressing mode.

- If the bits are 01 or 11 then the used page is that pointed by the bits 2–0 of the STK_PP register at address 0,D1H. Among other situations, this mode is useful to access the elements of a routine's activation record in the memory stack.

- If the bits are 10 then indexed addressing uses the page pointed by the bits 2–0 of the IDX_PP register at address 0,D3H. This mode can be used for transferring data located in different memory pages, for example when data need to be sent from one task to another, assuming that each task has its own memory page.

Four temporary data registers, *TMP_DRx*, are available to help in accessing the data transfer between pages. The registers are always accessible independent of the current SRAM page. The *TMP_DR0* register is at address x, 60H, the *TMP_DR1* register at address x, 61H, the *TMP_DR2* register at address x, 62H, and the TMP_DR3 register is located at address x, 63H.

B. SROM Space

SROM stores the code of eight routines used in booting the PSoC chip, reading from and writing blocks (64 bytes) to the flash memory, and circuit calibration.

The routines are called by executing an *SSC* instruction that uses the A register for distinguishing among the eight routines (hence, the identity of the called routine is passed as a parameter to the SSC instruction by the A register). In addition, the SROM routines use a set of extra parameters, which are located at predefined memory locations. In addition to the returned values that are returned as part of their functionality, each SROM function also returns a code value that signals the success or failure of the routine execution. Table 2.29 enumerated the four possible return values.

Each flash block is protected against illegal SROM function calls by setting its accessing mode to one of four possible modes: in the unprotected mode, external read and write operations are allowed, as well as internal writes and read operations of entire blocks. Internal read operations are allowed in each of the four modes. The second mode (factory upgrade) allows only external write and internal write operations. The third mode (field upgrade) permits only internal write operations. Finally, the fourth mode (full protection) permits only internal read operations.

Table 2.27 lists the SROM eight functions in column one, the value that needs to be loaded into the A register before executing the SSC routine in column two, and the required stack space in column three. Table 2.28 presents the variables that are used for passing extra parameters to the SROM functions, and the addresses of these variables in the SRAM. Note that all variables are located in page 0. Finally, Table 2.29 enumerates the four codes that indicate the status at the end of an SROM function execution. To distinguish between legal and illegal SROM function calls, the value of parameter *KEY1* must be set to 3AH and the value of parameter *KEY2* to the stack pointer, SP, plus 3 before executing the SSC instruction for the call.

The behavior of the SROM functions is as follows:

Table 2.27: SROM functions [1].

Function name	Function code	Stack space
SWBootReset	**00H**	**0**
ReadBlock	**01H**	**7**
WriteBlock	**02H**	**10**
EraseBlock	**03H**	**9**
TableRead	**00H**	**3**
CheckSum	**00H**	**3**
Calibrate0	**00H**	**4**
Calibrate1	**00H**	**3**

- SROM functions called at system reset.

  - *CheckSum*: The function computes the checksum over a number of blocks in the flash memory. The number of blocks is defined by the parameter *BLOCKID* (see Table 2.28). The checksum value is 16 bits long, and is returned using the following parameters: *KEY1* which holds the least significant byte, and *KEY2* which holds the more significant byte of the checksum value.

  - *SWBootReset*: This function is automatically executed upon hardware reset. It initializes some of the CPU registers and page 0 of SRAM. It first verifies the checksum of the calibration data. For valid data, it loads the following registers with the value 00H: A register, X register, CPU_F register, SP register, and the Program Counter, PC. Also, it sets the SRAM page 0 locations to predefined values, and starts execution of the user code at address 0000H in the flash memory.

  The SRAM page 0 memory locations are initialized, depending on their addresses, to the value 0x00, some predefined hexadecimal values, or to values programmed by the bit IRAMDIS (bit 0) of the control register, *CPU_SCR1*, at address x,FEH. If the bit is set to 0 then the corresponding SRAM cells must be initialized to 0x00H after a watchdog reset, otherwise not, therefore preserving the SRAM value before reset. More details about the specific values that are loaded into each memory location in the SRAM page 0 can be found in [1].

Table 2.28: SROM function variables [1].

Variable name	SRAM
KEY1 / RETURN CODE	**0, F8H**
KEY2	**0, F9H**
BLOCKID	**0, FAH**
POINTER	**0, FBH**
CLOCK	**0, FCH**
Reserved	**0, FDH**
DELAY	**0, FEH**
Reserved	**0, FFH**

Table 2.29: SROM return codes [1].

Return code value	Description
00H	successful completion
01H	function is not allowed because of the protection level
02H	software reset without hardware reset
03H	fatal error

- The following are the SROM functions that read and write to the flash memory:

  - *ReadBlock*: This function transfers a data block consisting of 64 bytes from the flash memory to the SRAM. Figure 2.19 presents the pseudocode for this function. First, the function verifies if the block pointed by the parameter *BLOCKID* is readable. If the block is not readable then the A register and the parameters *KEY1* and *KEY2* are loaded with values which indicate that reading was not possible due to protection restrictions.

    If the block protection allows reading, then 64 bytes are read from the flash memory using ROMX instructions, and stored in the SRAM by MVI instructions. The SRAM area is pointed to by the parameter POINTER. The SRAM page is determined by the MVW_PP register at address 0, D5H, for example for any MVI instruction. The successful completion of the operation is indicated by loading parameters *KEY1* and *KEY2* with the value 00H.

    The flash bank is selected by the FLS_PR1 register at address 1, FAH, if the architecture has multiple flash banks. Bits 1–0 of the register select one of the four possible flash banks.

  - *TableRead*: This function accesses part-specific data stored in the flash memory. For example, these data are needed to erase or to write blocks in the flash memory. The table to be accessed is selected by programming the parameter *BLOCKID*. Table 0 provides the silicon ID of the chip, tables 1 and 2 the calibration data for different power supplies and room temperatures, and table 3 the calibration data for correctly erasing or writing to the flash memory.

  - *EraseBlock*: This function erases a block in the flash memory. The block is indicated by the parameter *BLOCKID*. The function first checks the protection of the block, and, if writing is not enabled, then the value 01H, indicating failure due to protection, is loaded into the parameter *KEY1*.

    In addition to the parameter *BLOCKID*, two other parameters, *DELAY* and *CLOCK*, must also be set before calling the *EraseBlock* function. The parameters are introduced in Table 2.28. If the CPU clock is in the range 3–12 MHz, then the value of the parameter DELAY is set by the following[1],

$$DELAY \ = \ \frac{10^{-4} \times CPU_{speed} \text{ (Hz)} - 84}{13} \tag{2.3}$$

For a higher clock speed, the *DELAY* value is computed as [1]:

$$DELAY \ = \ \frac{10^2 \times 12 - 84}{13} \tag{2.4}$$

```
          SROM function ReadBlock:
       (1) if the block pointed by BLOCKID
               is NOT readable then
                   (2) load 00H into the register A;
                   (3) load 00H into KEY2;
                   (4) load 01H to KEY1;
                   (5) exit;
          else
                   (6) read 64 bytes from flash using the
                       ROMX instruction and store the
                       bytes in SRAM using MVI instructions.
                       the SRAM area is pointed by POINTER;
                       the flash bank is selected by
                       register FLS_PR1.
                   (7) load 00H to KEY1;
                   (8) load 00H to KEY2;
                   (9) exit;
       end SROM function ReadBlock.


            SROM function WriteBlock:
         (1) if the block pointed by BLOCKID
                 is NOT writeable then
                     (2) load 00H into the register A;
                     (3) load 00H into KEY2;
                     (4) load 01H to KEY1;
                     (5) exit;
             else
                     (6) read 64 bytes from SRAM using the
                         MVI instruction and store the
                         bytes in the flash memory.
                         the SRAM area is pointed by POINTER;
                         the flash bank is selected by
                         register FLS_PR1.
                     (7) load 00H to KEY1;
                     (8) load 00H to KEY2;
                     (9) exit;
             end SROM function WriteBlock.
```

Figure 2.19: ReadBlock and WriteBlock SROM functions [1].

The parameter CLOCK is defined as:

$$CLOCK = B - \frac{2 \times M \times T}{256} \tag{2.5}$$

The values for B and M are device specific, and are accessed using the *TableRead* SROM function.

Different values are used for temperatures, T below and above, 0°C. For example, for temperatures below 0°C, the value of M is stored in the flash bank 0 at address F8H, and B at address F9H. For temperatures higher than 0°C, the value of M is found at address FBH and B at address FCH.

– *WriteBlock*: This function writes a data block in the SRAM to the flash memory. Figure 2.19 shows the pseudocode for this function. The parameter *POINTER* points to the address of the block in SRAM, and the parameter *BLOCKID* points to the location in flash memory region where the data are to be copied. The flash bank is determined by bits 1-0 of the FLS_PR1 register at address 1, FAH.

This function first determines if the flash memory block pointed to by the parameter *BLOCKID* is writable, or not. If it is writable, then 64 bytes are accessed by MVI instructions, starting at the SRAM address given by the parameter *PARAMETER*.

Parameters *DELAY* and *CLOCK* must also be programmed for correct writing to the flash memory. The value of *DELAY* is set according to Equation (2.3). The parameter *CLOCK* is defined as [1]:

$$CLOCK = \frac{CLOCK_E \times MULT}{64} \tag{2.6}$$

The value $CLOCK_E$ is the clock parameter for the erase operation, and is computed according to the equation (2.5). The value of the parameter *MULT* depends on the temperature, is stored in table 3 of the flash bank 0 (at address FAH for cold temperatures, and at address FDH for hot temperatures), and can be accessed using the *TableRead* SROM function.

• The functions used in calibration are the following:

– *Calibrate0*: This function transfers the calibration values from the flash memory to the required registers.

– *Calibrate1*: This function executes the same functionality as the function *Calibrate0*, but in addition also computes the checksum of the calibration data. If the checksum is incorrect, then a hardware reset in generated, and bit IRESS (bit 7) of the register CPU_SCR1 is set. The CPU_SCR1 register is at address x, FEH.

The parameter POINTER points to a 30–byte buffer that is used for computing the checksum. Also, as MVI instructions are used for the actual data transfer, the MVR_PP and MVW_PP registers must point to the same SRAM page.

2.3 Conclusions

This chapter has presented the important characteristics of a microcontroller's instruction set and discussed programming techniques, in assembly language, for six popular applications and routines: data block transfer, stack operation, unsigned data multiplication, calling assembly routines from programs in high-level programming languages, bit manipulations, and sequence detectors. It has also discussed the memory space of embedded architectures, including the microcontroller registers, the SRAM and nonvolatile memory spaces, and the status and control registers.

PSoC's M8C microcontroller instruction set was discussed in some detail as an illustrative example. PSoC instructions can use ten addressing modes that result from combining the four basic addressing modes: immediate addressing, direct addressing mode, indexed addressing, and indirect addressing. These addressing modes allow a tradeoff between execution speed and code size and provide flexibility in mapping data to the microcontroller memory. Using immediate- and direct-addressing modes results in faster, smaller, but also less flexible code.

M8C's instruction set includes five categories of instructions: instructions for (1) data transfer, (2) arithmetic operations, (3) logic operations, (4) execution flow control, and (5) miscellaneous.

- The instructions for data transfer involve M8C's general-purpose A and X registers, the SRAM space (instructions MOV, MVI, and SWAP), the stack (instructions POP and PUSH), and the nonvolatile memory (instructions ROMX and INDEX).

- The instructions for arithmetic operations perform additions (instructions ADD and ADC), subtractions (instructions SBB and SUB), increment and decrement (instructions INC and DEC), comparison (instruction CMP), arithmetic shift (instructions ASL and ASR), and rotate (instructions RLC and RRC).

- The instructions for logic operations execute bit-level logic AND (instructions AND), OR (instructions OR), XOR (instructions XOR), and complement (instructions CPL).

- The instructions for flow control include: jumps (instructions JACC, JC, JMP, JNC, JNZ, JZ, and LJMP), subroutine calls (instructions CALL, LCALL, RET, and RETI), and system supervisory calls (instructions SSC).

- Other instructions include HALT and NOP instructions.

PSoC's memory space consists of the microcontroller registers (A, X, CPU_F, SP, and PC registers), the paged SRAM space, the nonvolatile memory (flash and ROM), and the register space for programming the mixed-signal architecture. The paged memory includes up to eight SRAM pages, inclusive, and is managed by the CUR_PP (the current memory page), STK_PP (stack page), and IDX_PP (indexed addressing) registers. The temporary registers, TMP_DR, are available for speeding-up the data transfer between pages. Eight system routines are stored in ROM that are called by the instructions SSC. The register space is organized as two register banks, which help fast reconfiguration of the architecture.

Assembly code routines for six different applications have also been provided in this chapter, for example the sequence detector design provides the most comprehensive discussion. The goal of sequence detection is to recognize a predefined bitstring in a sequence of input bits. This example presented the pseudocode of the detector specified as a finite state machine (FSM), two implementation alternatives, and the performance analysis of the implementation. The design used a handshaking mechanism to read input and output results at the general-purpose ports of the architecture. The two implementation alternatives stored the next state and output tables either in the flash memory, or in SRAM. Performance analysis estimated the execution time and the memory required. It also estimated the highest data rate of the input that can be processed without data loss. The example also accounted for the various contributions to the total execution time of the implementation.

2.4   Recommended Exercises

1. Develop Assembly routines for adding and subtracting two unsigned operands of lengths two bytes - in the first case, and then four bytes – in the second case. Many C language compilers represent `short unsigned int` variables using 2 bytes, and `unsigned int` variables using 4 bytes.

2. Write a routine in Assembly language for dividing two unsigned bytes. Extend the routine for division of numbers of arbitrary size (an arbitrary number of bytes is used for their representation).

3. Write assembly language routines for multiplication and division of signed operands of lengths 2, 4, and 8 bytes. Optimize the routines for a minimum number of execution clock cycles.

4. For a bitstring of length M, write Assembly code routines that set and reset a bit at position P. Also, develop routines that set and reset a field of length L starting at position P.

5. Write a set of routines that implement the abstract data type a set of positive integers up to 512 (the elements of the set can be the integers 0, 1, ... , 512). Propose routines for initializing a set as the empty set, adding the value v to the set, eliminating the value v from the set, and verifying if the value v is in the set. Use these routines to compute the reunion, intersection, and difference between two sets.

6. Write a routine in the M8C microcontroller Assembly language for sorting in increasing order a vector of unsigned bytes. Optimize the flexibility of your code, so that it can run for different vector sizes and for different memory regions (memory pages) storing the vector.

7. Develop an Assembly code program that searches a value in a vector of sorted values. The algorithm should implement a binary search method: at the first step, the algorithm compares the searched value with the value at the middle position of the vector. If the searched value is less then the algorithm repeats the process for the lower half, and if the value is larger then the algorithm considers the upper half. The algorithm returns the position of the searched element (if the value was found), or –1, otherwise. The starting address and the length of the vector are inputs to the procedure. Assume initially that the value is one byte long. Then, discuss the changes that must be operated so that data values of size B bytes are handled by the procedure.

8. Write Assembly code routines for multiplying two bidimensional matrices of unsigned integers. The routines should be flexible in handling matrices with different numbers of rows and columns.

9. Develop the Assembly code routines for a "software" lock. One routine must read serially eight bits at a rate of one bit per second. Then, the routine checks if the bitstring is equal to the preset code of the lock. If the two are equal than the signal *Unlock* is set to 1, otherwise it is set to 0. The second routine sets the code of the lock by reading serially eight bits at a rate of ten bits per second.

10. Develop an Assembly language routine that implements a palindrome recognizer. A palindrome is a bitstring which is equal to its reversed form. For example, the bitstring 111000111 is

a palindrom, but the bitstring 10100011 is not. Upon receiving the pulse *START*, the routine reads eight bits that are sent serially at the bit rate T. The value T is programmable. The eight bits indicate the length L of the bitstring that is checked. Then, the L bits of the bitstring are serially input at the same rate T. The system outputs bit 1 if the bitstring is a palindrom, and bit 0, otherwise.

11. Develop Assembly routines that implement a traffic-light controller. The traffic light controller controls the traffic moving along four directions: south-to-north, north-to-south, east-to-west, and west-to-east. The traffic light should stay in a state (e.g., green light) for a specified time T, and then move to the next state. If no cars are passing on a direction (e.g., north-to-south and south-to-north, or east-to-west and west-to-east) for B consecutive states then the value of the time interval T for the state "red light" should be shortened to $T/2$, and then to $T/4$, and so on, until it falls below the threshold T_H, and when it is set to 0. Once cars move again in that direction the value of the state "red light" should be set back to the value T. The moving of cars in a direction is signaled to the controller by proximity sensors, which generate an asynchronous pulse of width T_{pulse}.

12. The task is to develop Assembly code for a bottled-water vending machine. A bottle costs $0.45. The vending machine accepts nickels, dimes, and quarters. The customer begins entering coins, one at a time. If exact change has been entered then the output signal *Unlatch* is generated, so that the customer can get a bottle. If the amount of deposited money exceeds the price of a bottle, then change is given. If there is not enough change in the repository then the coins are refunded by issuing the output signal *Refund*, and the signal *Unlatch* is not generated. If there is enough change in the repository then it should be paid using the minimum amount of coins. The repository indicates to the controller the number of coins of each type available in the repository: bits Nx show the number of nickels, bits Dx the number of dimes, and bits Qx the number of quarters. Change is given to the customer by generating pulse signals for the repository: the signal *NR* releases one nickel, the signal *DR* releases one dime, and the signal *QR* releases one quarter. One coin is released at a time. (*Acknowledgment*: The exercise is based on a homework problem for Prof. R. Vemuri's VLSI design course at the University of Cincinnati).

13. Write Assembly routines that implement the basic operators of a circular FIFO (first input first output) buffer. The buffer is described by two pointers. The pointer head indicates the first element of the buffer, and the pointer tail points to the last element of the buffer. A new element is inserted at the tail of the buffer, and only the value that is pointed by the pointer head can be removed from the buffer. The FIFO buffer is of size MAX - where the value of MAX is programmable. Once the pointer *head* reaches the value MAX then it is set back to the first element of the buffer. Similarly, once the pointer *tail* indicates to the first buffer element, the pointer is set to the last element in the FIFO buffer. The assembly routines should offer the following functionality: create a new FIFO buffer starting at address xxH, reset the buffer, introduce an element into the buffer, and remove a value from the buffer.

14. Write an Assembly code routine that calculates the expression $((4 + 5) * 3) * (2 + 3)$ using only the stack for storing the operand values and the intermediate results of the expression.

15. Develop Assembly code routines for implementing the abstract data type, sorted linked lists of unsigned bytes. Define the routines for setting up an empty list, inserting a new element to the list, and removing a value from the list.

16. For the routines in Figure 2.8, identify all situations in which the "safe " pop and push routines are not safe anymore. Extend the two routines so that they are safe in all situations. Explain your procedure for identifying the conditions that lead to unsafe operation.

17. Write an Assembly routine for realizing the behavior of a pulse width modulation block. The designer should have the option of programming the pulse width α of the PWM routine through calling a C function with two parameters: the new value of the pulse width and a pointer to the PWM block data.

18. Develop the following routines and programs in Assembly code:

- A routine that displays the message "Hello world" on the LCD.

- A program that controls the blinking of an LED once every two seconds with a 50% duty cycle (the LED is on for one second, and off for one second).

- Two assembly routines that convert a binary number to BCD, and BCD to binary number, respectively.

- A routine that implements a frequency calculator.

- A routine for the cyclic redundancy check (CRC) function.

19. For the example shown in Figure 2.13, modify the structure of the activation record and show the Assembly code that must be inserted in the calling and called routines, so that input parameters of any byte size can be passed, and results of any size are returned.

20. For the C compiler that you are using for your embedded applications, analyze the Assembly code that is generated for different C statements, for example assignment, if, case, for, while statements, and different data types, such as arrays and structures. Understand what code optimizations are carried out by the compiler.

21. Develop Assembly code that reads the values of the constants B, M, and $MULT$ specific to your PSoC chip.

22. Using SSC instructions and the routines stored in the ROM memory, implement a program that writes N blocks to the flash memory, then reads M blocks of data from the flash memory to the SRAM memory, and finally, erases the M blocks of flash memory,

23. For the sequence detector design in Figures 2.15 and 2.16, estimate the execution time overhead due to busy waiting for different values of the input bit rates, such as 0.1 kbits/sec, 1 kbits/sec, 10 kbits/sec, 50 kbits/sec, and 100 kbits/sec. Discuss the results and propose a solution to decrease the busy-waiting overhead.

24. For Exercise 14, estimate the execution time in clock cycles if the data values are represented on one, two, and four bytes. Discuss the increase in execution time depending on the size of the values that are processed.

25. Find an analytical formula that predicts the execution time of an algorithm depending on the memory size of the processed values.

26. For exercise 8, assume that each of the matrices occupies multiple SRAM pages. Modify the algorithm in this case, and propose a way of distributing the matrix elements to the SRAM pages, so that the execution time of the program is minimized.

Bibliography

[1] PSoC Mixed Signal Array, Technical Reference Manual, *Document No. PSoC TRM 1.21*, Cypress Semiconductor Corporation, 2005.

[2] PSoC Designer: Assembly Language User Guide, *Spec. #38-12004*, Cypress Microsystems, December 8 2003.

[3] PSoC Designer: C Language Compiler User Guide, *Document #38-12001 Rev.*E*, Cypress Semiconductor, 2005.

[4] K. Boulos, Large Memory Model Programming for PSoC, *Application Note AN2218*, Cypress Microsystems, September 13 2004.

[5] J. Byrd, Interfacing Assembly and C Source Files, *Application Note AN2129*, Cypress Microsystems, May 29 2003.

[6] J. Holmes, Multithreading on the PSoC, *Application Note AN2132*, Cypress Microsystems, February 24 2004.

[7] D. Lewis, *Fundamentals of Embedded Software. Where C and Assembly Meet*, Saddle River, NJ: Prentice Hall, 2002.

[8] M. Raaja, Binary to BCD Conversion, *Application Note AN2112*, Cypress Microsystems, February 12 2003.

[9] W. Snyder, J. Perrin, Flash APIs, *Cypress Microsystems*, August 4 2004.

[10] S. Sukittanon, S. Dame, Embedded State Machine Design for PSoC using C Programming (Part I of III), *Application Note AN2329*, Cypress Semiconductor, February 8 2006.

[11] S. Sukittanon, S. Dame, Embedded State Machine Design for PSoC using an Automatic Code Generator (Part II of III), *Application Note AN2332*, Cypress Semiconductor, December 5 2005.

[12] S. Sukittanon, S. Dame, Embedded State Machine Design for PSoC with Interfacing to a Windows Application (Part III of III), *Application Note AN2333*, Cypress Semiconductor, January 5 2006.

[13] D. Van Ess, Unsigned Multiplication, *Application Note AN2032*, Cypress Semiconductor, June 21 2002.

[14] D. Van Ess, Signed Multi-Byte Multiplication, *Application Note AN2038*, Cypress Semiconductor, August 29 2002.

[15] D. Van Ess, A Circular FIFO, PSoC Style, *Application Note AN2036*, Cypress Microsystems, July 21 2002.

Chapter 3

Hardware and Software Subsystems of Mixed-Signal Architectures

This chapter focuses on the main hardware and software subsystems for sensing, processing and control, data communication, and actuation in embedded applications. The PSoC architecture is used to illustrate the subsystems.

The PSoC architecture incorporates the following subsystems: the analog subsystem (programmable continuous-time and switched capacitor analog blocks, analog input/output ports, analog bus, and programmable interconnect), the digital subsystem (including the CPU, volatile and nonvolatile memory, array of programmable digital blocks, and customized digital blocks), the interrupt system, the general input/output ports, the system buses, and the system clocks. The chapter also introduces the software components of the architecture, for example the boot program, interrupt service and firmware routines.

Because the architecture subsystems are the topics of the following chapters, this chapter constitutes the transition from the instruction-level presentation of the PSoC architecture in Chapter 2 to the rest of this textbook.

The chapter has the following structure:

- Section 1 discusses the main subsystems of a mixed-signal architecture, and the hardware and software components of the PSoC architecture, as illustrative examples.

- Section 2 provides a comprehensive presentation of the PSoC interrupt subsystem and a tachometer-handling, interrupt service routine case study.

- Section 3 gives a detailed description of the PSoC global input/output ports and their features and programming.

- Section 4 presents the PSoC bus subsystem, including the interconnect for the digital and analog blocks.

- Section 5 discusses PSoC's system clocks.

- Section 6 presents chapter conclusions.

A. Doboli, E.H. Currie, *Introduction to Mixed-Signal, Embedded Design*,
DOI 10.1007/978-1-4419-7446-4_3, © Springer Science+Business Media, LLC 2011

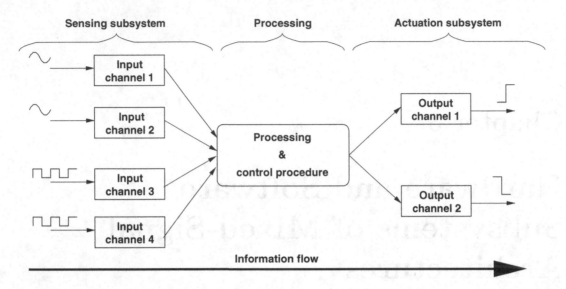

Figure 3.1: Sense-process and control-actuate flow in embedded applications.

3.1   Subsystems of the PSoC Mixed-Signal Architecture

Embedded mixed-signal architectures provide hardware and software support for implementing the data acquisition, processing, and control of an application. This textbook focuses primarily on applications involving signal sensing, embedded processing/control, actuation, and advanced interfacing, and less on applications demanding intensive data computations, such as multimedia applications. In this context, besides the digital processing capabilities, it is important that the architecture has the resources needed for interfacing to analog and digital signals, analog processing (e.g., signal conditioning, filtering and data conversion), and possesses the capability of identifying and reacting to external and internal events, including timing events.

In summary, embedded mixed-signal architectures incorporate resources for the following four kinds of activities:

- *Sensing of analog signals*: This includes analog interfacing resources, such as analog input ports, circuits for signal conditioning, analog filtering, data buffers and analog-to-digital, and digital-to-analog conversion and the related software drivers, firmware routines, and interrupt service routines for operating the circuits.

- *Sensing and communication of digital data*: The related resources are digital ports, interrupt controllers, customized digital processing blocks, modules for specific interfacing protocols (e.g., SPI, UART and I2C), and the corresponding drivers, firmware routines, and interrupt service routines.

- *Execution of the embedded processing and control algorithms*: The corresponding resources are the central processing unit (CPU), volatile and nonvolatile memory, digital blocks with customized functionality (e.g., timers, watchdog timers, multipliers), programmable digital blocks, the related drivers, firmware routines, and interrupt service routines, and the application (control) software.

- *Generating actuation signals*: This includes the support for producing the output signals used in actuation. The signals that control the operation of the supervised peripheral

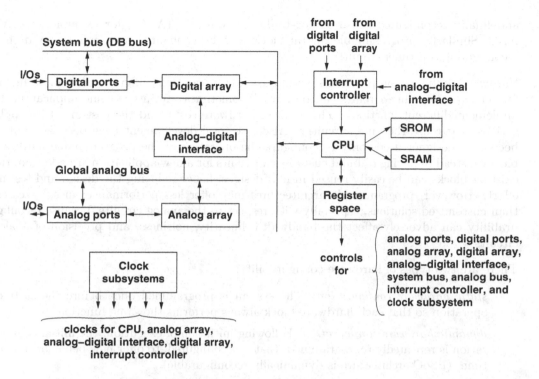

Figure 3.2: Subsystems of the PSoC architecture.

devices are both analog and digital. The architectural support includes interfacing logic, buffers, output ports, and the related driver and firmware routines.

Figure 3.1 illustrates the sense/process, and control/actuate, flow in a typical embedded control application, and the mapping of the flow to the subsystems of a mixed-signal architecture. This chapter introduces the main subsystems of an embedded, mixed-signal architecture using Cypress' Programmable System on Chip (PSoC) as an illustrative example. PSoC was selected because it is a system-on-chip (SoC) that provides broad support for analog and digital sensing, processing, actuation, and data communication. The following chapters explain the use of the subsystems for developing constraint-satisfying, and performance-optimized, implementations of embedded applications.

The PSoC mixed-signal architecture [5] incorporates hardware (i.e., analog and digital circuits) and software support in the form of libraries of firmware routines. The implementation of complex sensing, control, actuation, and data communication using PSoC is achieved by two main features of the architecture, hardware programmability and integration of the subsystems as a single chip (SoC):

- *Hardware configuration and reconfiguration*: Application developers can utilize platforms such as PSoC for a wide variety of applications because the analog and digital blocks provided by the architecture can be individually programmed, that is customized, to meet application-specific functionality and performance requirements. This allows the developer/designer to standardize on one hardware/software platform for different applications.

  The same programmable analog blocks can be dynamically reconfigured, in real time, to operate, for example, as a lowpass filter, and then be reconfigured as a bandpass filter, and finally reconfigured as an analog-to-digital converter (ADC), for example as a $\Delta\Sigma$ ADC. Similarly, programmable digital blocks can be configured as a counter, and then reconfigured as a timer circuit.

Hardware configuration provides several advantages over nonconfigurable circuits. In the former case, the same circuits can be reused, sometimes within the same application, to implement different functions. This reduces hardware costs, and the cost of future applications that reuse the same circuitry. Also, the total development time becomes shorter because a designer needs only to configure the operation of the programmable hardware blocks instead of having to build customized circuits for a new application. In addition, the existing blocks can be easily reused in new designs, hence reducing the design and testing effort. However, programmable architectures may offer less performance, in some cases, than customized solutions, especially with respect to the analog circuits, because reconfigurability can adversely affect the bandwidth, linearity, noisiness, and precision of analog circuits.

There are two types of hardware configurability:

- *static hardware configuration:* The system is programmed once before the start of operation so that each hardware block always performs the same function.
- *dynamic hardware configuration:* Following initiation of execution, the initial configuration is repeatedly reprogrammed, that is reconfigured, to offer new behavior in real-time. (PSoC architecture is dynamically reconfigurable).

- *Integration of all architectural subsystems on a single chip:* The PSoC architecture is a complete SoC, namely all of the hardware subsystems required in an embedded application are integrated and packaged in a single silicon chip. The CPU, the nonvolatile and volatile memories, the programmable analog and digital blocks, I/O ports, and so on, are all integrated as a SoC. This reduces not only the size and weight of the system compared to a system built of discrete components, but also improves an implementation's performance in terms of speed, power, and reliability.

3.1.1   PSoC Hardware Components

Figure 3.2 illustrates PSoC's block-level architecture.

- *Analog signal sensing subsystem:* PSoC provides hardware support for sensing and processing analog signals at frequencies as high as 100 kHz and includes the following modules.

  - *Matrix of configurable analog blocks:* Various analog and mixed-signal circuits and subsystems, for example different amplifier types (i.e., differential amplifiers, instrumentation amplifiers, variable gain amplifiers, etc.), comparator circuits with/without hysteresis, analog filters, analog -to-digital converters and digital-to-analog converters.

    Figure 3.3 illustrates the structure of the analog matrix which is organized as a two-dimensional array that includes blocks organized in columns. The functionality, inputs, outputs, and performance of the analog blocks are determined by programming dedicated control registers in PSoC's register space. This matrix includes two types of configurable analog blocks, that is continuous-time blocks and switched-capacitor blocks. The continuous-time blocks are denoted as ACBxx, and the switched capacitor blocks as either *ASCxx* or *ASDxx*. There are two types of switched-capacitor topology type C, and type D (i.e., blocks *ASCxx*, and *ASDxx*, respectively).

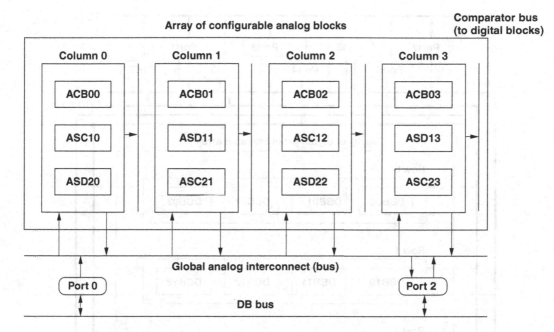

Figure 3.3: Matrix of programmable analog blocks [5].

— *Analog bus and programmable interconnect*: There are three interconnect structures that link the matrix of programmable analog blocks, and these blocks to the chip's I/O pins.

  * *The local programmable interconnect* is used to connect the outputs of analog blocks to the negative, or positive, inputs of another analog block to form larger analog structures. The local interconnect is also used to link block inputs to specific reference voltages, or to the ground.

  * The *input* of the *global analog bus* (cf. Figures 3.2 and 3.3) links the chip pins to the inputs of the programmable analog blocks.

  * The *output* of the *global analog bus*, also called the *comparator bus* [5], connects each analog column to the programmable digital blocks, and, via the global digital buses, to a fixed set of output pins.

(Chapters 6 and 7 discuss the programmable continuous-time and switched capacitor analog blocks in PSoC.)

• *Digital sensing, processing, and actuation subsystem:* Includes the modules that implement all the interfacing, processing, and communication in the digital domain, including:

  — *CPU*: An M8C microcontroller serves as the architecture's general-purpose processor. (The M8C's instruction set is discussed in Chapter 2.)

  — *Memory subsystem*: The memory system includes paged RAM used to store data, flash memory used to store the application programs, and SROM which contains eight of the more frequently used system functions. (The memory subsystem is presented in Chapter 2.)

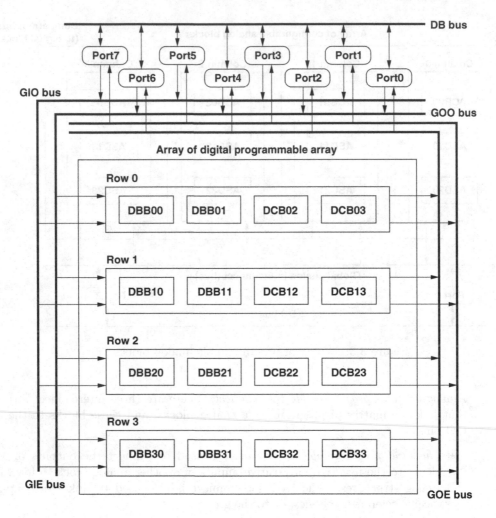

Figure 3.4: Matrix of programmable digital blocks [5].

— *Digital programmable and customized blocks*: Both kinds of blocks can be used to provide faster application execution times than a software-only implementation executed by the CPU. These blocks also support data communication functions.

The PSoC architecture incorporates up to 16 programmable digital blocks that can operate as one of the following: timers, counters, deadband generators, and cyclic redundant checkers (CRC). In addition, four of the programmable blocks (i.e., the communication blocks) can also implement serial data communication protocols, for example the protocols *SPI* and *UART* [5]. Figure 3.4 illustrates the matrix of programmable digital blocks.

The customized blocks are digital circuits with specific functionality, for example pulse width modulators (PWM), multiply accumulate (MAC) modules, decimator modules, and I2C interfacing logic.

(The programmable and customized blocks are detailed in Chapters 4 and 5.)

- *Interrupt subsystem*: Interrupt signals immediately notify the CPU of the occurrence of predefined events. These events are predefined by conditions that may occur during the *normal* operation of the system, such as exceeding a preset threshold value, elapsing of a certain time interval, availability of data at the input ports, ending of a data transmission, system reset, etc. In contrast, exceptions describe abnormal situations that occur during the system functioning, such as division by zero, overflow of memory buffers, and power loss, among others. Interrupts are widely used for interfacing to sensors, actuators, and other IO devices and have an advantage over device polling by demanding less execution time overhead.

PSoC provides hardware and software support for fast interrupt servicing [5]. The hardware support includes the following.

- – *Priority encoder*: The priority encoder is a hardware circuit that selects the pending interrupt of highest priority. This interrupt-handling capability is needed because interrupts from various devices, and architectural blocks, can arrive at the same time, and should be serviced by the system in the order of their relative importance (priority).

- – *Interrupt table*: The interrupt subsystem provides support for fast identification of the interrupt source. This is achieved by associating a unique identifier with each interrupt source, and using that identifier as an index to an interrupt table to find the physical address of the software routine corresponding to that interrupt. The software routine handling an interrupt is called an *interrupt service routine* (ISR).

An example of interrupt table indexing is shown in Figure 3.7. The interrupt table is located in the SRAM (e.g., starting at the physical address 00H), and set up by the boot program during system initialization.

- – *Support for enabling and disabling interrupts*: The instruction sets of some architectures include dedicated instructions for enabling and disabling interrupts. Other architectures, such as PSoC, do not offer such dedicated instructions, but achieve the same task by setting and resetting the associated interrupt control bits.

- *Interrupt handling*: Prior to executing each instruction, the CPU determines whether an interrupt has occurred. If so, and the interrupts are enabled, then the current program execution is suspended, the interrupt of the highest priority is identified by the priority encoder, and the associated ISR is found by indexing the interrupt table. After executing the related ISR, the execution of the interrupted program is resumed.

Section 3.2 provides more details on the PSoC interrupt subsystem and the related ISRs.

- *General input/output ports*: The PSoC architecture includes general-purpose programmable ports. The ports can be programmed as input or output ports. Also, the ports can be connected either to the CPU, to the digital resources, or to the programmable analog blocks. Designers can select the driving capabilities of a port depending on the attributes of the external device. Finally, the ports can be programmed to originate interrupt signals to the microcontroller.

- *System buses*: The system buses interconnect the architecture's input/output ports, CPU, programmable and customized digital blocks, and programmable analog blocks. The buses are organized to offer maximum data communication rates in the system, including transferring data in parallel between the blocks, and small propagation delays by using short interconnect.

The bus subsystem includes the system bus (DB) that connects the microcontroller to the other PSoC subsystems, the global digital interconnect between the ports and the programmable digital blocks, the local interconnect for the digital blocks, and the analog interconnect for the programmable analog blocks.

- *System clocks*: The PSoC architecture includes three possible clocking sources: the internal main oscillator, the internal low speed oscillator, and the external clocking sources. The internal main oscillator can operate in two precision modes, the moderate precision mode and the high precision mode. Depending on the specifics of the application, designers can select lower clock frequencies to reduce the power and energy consumption. Also, they can use clocks of higher accuracy (e.g., for the switched capacitor analog circuits), but at the penalty of increasing the implementation cost due to the external crystal oscillator that is required and the power consumption of the implementation.

3.1.2 PSoC Software Components

The software running on the mixed-signal architecture includes routines for managing the analog and digital blocks of the architecture (the routines are called firmware routines), for implementing the behavior executed in response to the external and internal events (the routines are called Interrupt Service Routines), for booting the system at initialization, and implementing the embedded algorithm of the application.

- *Boot program* initializes the general-purpose hardware resources of the architecture, configures the application-specific programmable analog and digital blocks, the system interconnect, system clocks, and supply voltage, initializes the software runtime environment, including the data structures and global variables, and then calls the main program of the embedded application. The steps are summarized in Figure 3.5(a).

- *Embedded application algorithms* implement the main control and communication functionality of the application. Figure 3.1 illustrates the principle of the algorithms. Input data are sensed (read) from multiple sources. Data are read either periodically at fixed intervals of time, or in response to the occurrence of an event. The first case represents *discrete-time systems*, and the second case corresponds to *reactive systems*. For example, a discrete system might sample continuous-time input signals (e.g., a sinusoidal voltage signal) after each one millisecond. In contrast, a reactive system will read its input only if an event occurs (e.g., one of the inputs exceeding a specific threshold value).

Some inputs correspond to analog signals and others to digital signals. In addition, inputs are sensed by hardware modules with different characteristics. Conceptually, each of the input sources defines an *input channel* from where data are read by the embedded application. Similarly, the actuation data computed and output by the application introduced several *output channels*, which are mostly for digital signals. Each channel provides a set of *methods* that are called by the main algorithm to interact with the channels. These methods instantiate a channel, initialize a channel, modify and retrieve the parameters of a channel, read data from a channel, and put data into a channel. The channel methods help abstracting away all physical (hardware) details of a channel. For example, when sensing the voltage produced by a temperature sensor, the read method of the corresponding channel returns a floating point number representing the voltage value, or an integer number for the temperature. Channel methods use firmware routines to interact with the hardware.

```
execute boot program:
    - initialize general-purpose resources;
    - configure application specific modules;
    - initialize run time environment;
    - disable interrupts;
call main application routine;
```

(a)

```
void main() {
    start system timers;
    initialize application-specific modules;
    initialize global variables;
    instantiate application-specific channels;
    enable interrupts;
    while (1) {
        wait for events (enabled interrupts);
        read values from input channels;
        execute control procedure and
                compute actuation data;
        output actuation data to output channels;
    }
}
```

(b)

Figure 3.5: (a) Boot program and (b) application routine.

Figure 3.5(b) presents the generic structure of the *main* routine implementing the embedded control algorithm. The routine starts by initializing the system timers followed by initializing all hardware modules used by the application. Note that before executing the application routine, the boot program configures the control registers of the needed hardware modules, thereby defining the functionality of the modules. However, all the application-dependent parameters of a module (e.g., pulse width, period of a pulse etc.) are set in the initialization step of the application routine. Then, the global variables of the application are defined. The next step instantiates all signal channels required by the application. At this point, the application routine is ready to sense, process, and actuate. Therefore, the next step enables all related interrupts giving the system the capability of identifying external and internal events.

The control algorithm is specified as a *while (1)* loop (execute "while true" loop), which terminates only if special events occur, such as the system being reset or shut down. The loop body comprises the following actions. The system first waits for the occurrence of an event. For discrete-time systems, events are generated by the system timers (e.g., an event is produced after each one millisecond), or for reactive systems, by hardware modules that signal the occurrence of specific conditions (e.g., exceeding of threshold values).

Upon identifying an event, the algorithm reads the input channel, and then performs the control algorithm to compute the corresponding actuation values. The last step outputs the computed values to the output channels.

- *Interrupt service routines* are executed upon the occurring of an event of interest to the system. If an event occurs, then a specific *interrupt* signal is generated for the system. The events of interest have their corresponding interrupt signals *enabled*. Before executing the next instruction of the embedded code, the CPU verifies if an enabled interrupt was produced. In the case where there is an interrupt, the CPU suspends the execution of the program, and jumps to the corresponding ISR. The interrupt service routine executes all actions defined for the interrupt (e.g., sampling the input signal and storing the value in a reserved area). After the ISR ends, execution resumes back to the point of the embedded code at which execution was suspended.

- *Firmware routines* configure and operate the hardware modules of the system, including input and output ports, programmable hardware blocks, system clocks, and supply voltages. Typical firmware routines are for starting and stopping a hardware module, configuring the parameters of the module, enabling and disabling interrupts, sending and retrieving data to and from the module, and accessing the status of the hardware module.

Firmware routines offer a higher-level interfacing mechanism between the application algorithms and the hardware resources of the architecture. Specifically, the firmware routines can be grouped into two categories:

- *Low-level firmware routines* implement all the hardware-related operations, for example setting the control registers of the hardware modules to the values for the required functionality, accessing the data registers of the modules, enabling and disabling the circuits, activating and deactivating the interrupts, and so on. The user does not have to specify the hardware-related operations, but must specify some low-level attributes, such as the physical addresses of the related hardware modules.

- *High-level firmware routines* offer to application programs an abstract interfacing mechanism to hardware. In addition to the abstractions offered by the low-level firmware routines, the high-level routines eliminate all low-level attributes: the hardware modules are identified by abstract names, and the configuration values are denoted by abstract symbols.

To distinguish between the two, low-level firmware routines are referred to simply as firmware routines, and the high-level firmware routines as channels methods (or APIs). Chapters 4 and 5 offer examples of firmware routines for interacting with specific hardware blocks, including timer modules, pulse width modulators, and *SPI* and *UART* communication blocks.

The operation of a PSoC-based, mixed-signal system can be summarized as follows. Upon resetting or powering up the system, the boot program is executed. The program first initializes the general-purpose resources of the system, and sets the value for voltage stabilization, enables the watchdog timer module, selects the desired clock signal, and sets up the stack program. Next, the application initializes specific modules and their interconnects. This is achieved by loading the control registers with configurations that define amplifiers, analog filters, digital counters, *I2C* and *UART* interfacing blocks, and so on. The runtime environment is then initialized, and

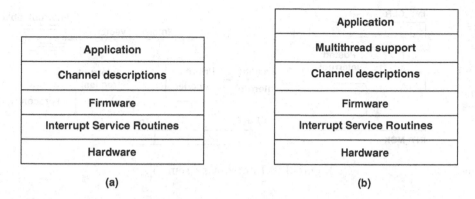

Figure 3.6: Hierarchy of abstraction levels in a mixed-signal architecture.

the interrupts are disabled. Final execution control is passed to the routines implementing the embedded control algorithms.

Execution flow includes parts for the application code (accounting also for the channel methods) interleaved with parts for ISRs. When switching from the application code to an ISR and then back to the application code, the system performs what is termed *context-switching*. Context switching involves saving the state information of the application on the stack, and then retrieving this information so that the execution of the application code can be resumed without being affected by the ISR execution. Context-switching constitutes overhead, in terms of execution time and consumed energy, and should be minimized as much as possible.

In summary, the components of mixed-signal architectures are organized to form successive levels of abstractions, as shown in Figure 3.6(a). The bottom level corresponds to the *mixed-signal hardware*, including analog circuits, digital CPU and processing blocks, memory, interfacing logic, ports, buses and clocks. *Firmware routines* are software routines that provide a more abstract interface to the hardware modules. The routines implement all the low-level steps needed for interacting with hardware, including synchronization, data formatting, and reset. Firmware might not provide a totally abstract interface: some physical details about hardware (e.g., physical addresses and module parameters) have to be provided to a routine.

Also, some firmware functionality relies on *interrupt service routines*. ISRs might access the hardware directly. Therefore, firmware routines can be seen as being on top of ISRs, even though both operate directly with the hardware. *Channels* offer an abstract interface (referred to as an application interface or API) to manipulate input and output data. The user of a channel does not have to know any details about the underlying hardware, such as physical addresses, and configuration parameters, among others. Channels relate the high-level entities in the control algorithm (e.g., channel identifiers, constants, channel types) to the low-level data (e.g., physical addresses of registers, ports, and memory areas and configuration data). Finally, the top level corresponds to the control algorithm of the application.

If an application requires having multiple execution threads, such as in multitasking applications, then the architecture must include software support for multitasking. The corresponding

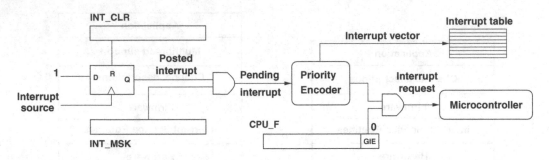

Figure 3.7: Interrupt system [5].

hierarchy is shown in Figure 3.6(b). Multitasking support includes capabilities to create and terminate tasks, task synchronization, and intertask data communication.

The following section introduces the concept of interrupts, and discusses the developing of interrupt service routines, and the PSoC hardware of the interrupt subsystem.

3.2   The PSoC Interrupt Subsystem

Figure 3.7 shows PSoC's interrupt system[5]. Interrupt signals are generated to signal the occurrence of predefined events. In response, the microcontroller suspends execution of the current program, saves the present state (contained in the A, F, X, and PC registers) on the stack, and executes the interrupt service routine corresponding to the signaled event. After the ISR ends, the microcontroller resumes execution of the program suspended due to the interrupt.

Figure 3.7 illustrates the signal flow that occurs in PSoC hardware for handling an interrupt. Interrupts are generated by different conditions and sources, including reset, supply voltage, general-purpose I/O ports (GPIOs), digital blocks, analog blocks, variable clocks, sleep timer, and the I2C interface. Table 3.1 lists all the possible interrupt sources.

Interrupts are globally enabled (or disabled) by setting (or resetting) bit *GIE* (bit 0) of the register *CPU_F* at address *x,F7H*. If interrupts are enabled, then the interrupt vector (eight bits) of the interrupt selected by the *Priority Encoder* is used as an index in the *Interrupt Table*. The selected table entry contains the address of the ISR routine corresponding to the interrupt. The priority encoder circuit selects the interrupt of highest priority among the (multiple) pending interrupts. An interrupts is called *pending*, if it was posted (i.e., that the corresponding event occurred), and if it was individually enabled by setting the content of the interrupt mask register *INT_MSK*. Also, an interrupt generated by an interrupt source is considered to be *posted* as long it is not cleared by the *INT_CLR* register.

Table 3.1 enumerates the 25 different kinds of interrupt signals, their priority, and the address of the corresponding ISRs.

Interrupt Mask Registers INT_MSKx control the enabling and disabling of the individual interrupts. There are four different mask registers, each controlling different kinds of interrupts. If a bit is set to zero, then the corresponding interrupt is disabled. If the bit is set to one, then a posted interrupt becomes pending. Table 3.2 shows how the bits of the *INT_MSK* registers relate to the different interrupts. The register addresses are also given in the table. For example, if bit 0 of register *INT_MSK3* is set, then posted I2C interrupts become pending. Bit 5 of register *INT_MSK2* controls the interrupts posted by the digital PSoC basic block *DBB31* the block

Table 3.1: List of interrupt types [5].

Interrupt name	Semantics	Interrupt priority	Interrupt address
Reset	generated upon system reset	0 (highest)	0000H
Supply voltage monitor	supply voltage	1	0004H
Analog column 0	generated by analog column 0	2	0008H
Analog column 1	generated by analog column 1	3	000CH
Analog column 2	generated by analog column 2	4	0010H
Analog column 3	generated by analog column 3	5	0014H
VC3	variable clock VC3	6	0018H
GPIO (in Section 3.3)	produced by the GPIO ports	7	001CH
PSoC block DBB00	produced by digital block DBB00	8	0020H
PSoC block DBB01	produced by digital block DBB01	9	0024H
PSoC block DCB02	produced by digital block DBB02	10	0028H
PSoC block DCB03	produced by digital block DBB03	11	002CH
PSoC block DBB10	produced by digital block DBB10	12	0030H
PSoC block DBB11	produced by digital block DBB11	13	0034H
PSoC block DCB12	produced by digital block DBB12	14	0038H
PSoC block DCB13	produced by digital block DBB13	15	003CH
PSoC block DBB20	produced by digital block DBB20	16	0040H
PSoC block DBB21	produced by digital block DBB21	17	0044H
PSoC block DCB22	produced by digital block DBB22	18	0048H
PSoC block DCB23	produced by digital block DBB23	19	004CH
PSoC block DBB30	produced by digital block DBB30	20	0050H
PSoC block DBB31	produced by digital block DBB31	21	0054H
PSoC block DCB32	produced by digital block DBB32	22	0058H
PSoC block DCB33	produced by digital block DBB33	23	005CH
I2C	I2C circuit	24	0060H
Sleep timer	sleep timer circuit	25	0064H

situated in row 3 and column 1. Similarly, bit 2 of register INT_MSK2 enables the interrupts due to the digital PSoC communication block $DCB22$ - located in row 2 and column 2. The bits of register INT_MSK0 correspond to interrupts produced by variable clocks, sleep timer, $GPIO$, analog PSoC blocks, and supply voltage monitor.

The bit content of *Interrupt Clear Registers* (INT_CLRx) indicates whether the corresponding interrupts were posted. If a bit is one, then the interrupt was posted. There are four INT_CLR registers, each of their bits controlling a different interrupt. Table 3.3 presents the interrupts for each of the bits in the registers. The register addresses are also given in the table. For example, if bit 2 of the INT_CLR1 register is set, then there exists a posted interrupt coming from the digital block $DCB02$.

The bit $ENSWINT$ in register INT_MSK3 (bit 7 of the register) defines the way the INT_CLR bits are handled. If bit $ENSWINT$ is zero, then a posted interrupt can be canceled by setting the corresponding INT_CLR bit to zero. Setting the INT_CLR bit does not have any effect. If bit

Table 3.2: Structure of **INT_MSK** registers [5].

Register (address)	Bit 7	Bit 6	Bit 5	Bit 4	Bit 3	Bit 2	Bit 1	Bit 0
INT_MSK3 (0,DEH)	ENSWINT	-	-	-	-	-	-	I2C
INT_MSK2 (0,DFH)	DCB33	DCB32	DBB31	DBB30	DCB23	DCB22	DBB21	DBB20
INT_MSK1 (0,E1H)	DCB13	DCB12	DBB11	DBB10	DCB03	DCB02	DBB01	DBB00
INT_MSK0 (0,E0H)	VC3	Sleep	GPIO	Analog3	Analog2	Analog1	Analog0	VMonitor

Table 3.3: Structure of **INT_CLR** registers [5].

Register (address)	Bit 7	Bit 6	Bit 5	Bit 4	Bit 3	Bit 2	Bit 1	Bit 0
INT_CLR3 (0,DAH)	-	-	-	-	-	-	-	I2C
INT_CLR2 (0,DBH)	DCB33	DCB32	DBB31	DBB30	DCB23	DCB22	DBB21	DBB20
INT_CLR1 (0,DDH)	DCB13	DCB12	DBB11	DBB10	DCB03	DCB02	DBB01	DBB00
INT_CLR0 (0,DCH)	VC3	Sleep	GPIO	Analog3	Analog2	Analog1	Analog0	VMonitor

ENSWINT is one, then bit *INT_CLR* being set to one determines the corresponding interrupt to be posted. An *INT_CLR* bit set to zero does not have any effect. Bit *ENSWINT* allows producing *software enabled interrupts*, in which the ISRs are executed as a result of the program executing on the CPU, and not by ports, or the digital and analog hardware blocks of the architecture.

The *Interrupt Vector Clear* register (*INT_VC* at address *0, E2H*) contains the highest priority pending interrupt. If the register is read, then the interrupt address of the highest priority pending interrupts, or the value 00H if there are no pending interrupts, is returned. A write to the register clears all pending interrupts.

The PSoC microcontroller executes the following steps once an interrupt has been issued [5]:

1. The execution of the current instruction finishes.

2. Thirteen clock cycles are needed to execute the internal interrupt routine. (i) *PC* (first, the more significant byte followed by the less significant byte), and (ii) *CPU_F* registers are pushed onto the stack. Then, (iii) the *CPU_F* register is set to zero, hence disabling any future interrupts. (iv) The more significant byte of the *PC* (*PC[15:8]*) register is set to zero. (v) The interrupt vector from the interrupt system is loaded into the lower byte of the *PC* register. For example, I2C interrupt causes the value 60H (see Table 3.1) to be loaded into the *PC* register.

3. The execution control jumps to the ISR of the received interrupt.

4. The corresponding ISR is performed.

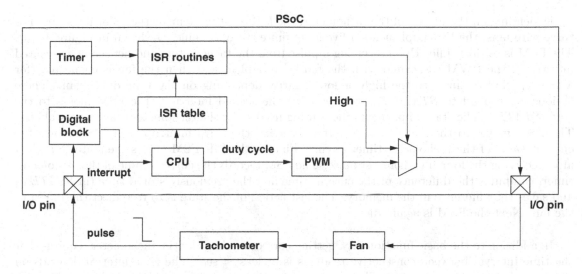

Figure 3.8: Tachometer interfacing to PSoC.

5. A *RETI* instruction is executed at the end of an ISR. The *PC* and *CPU_F* registers are restored to their values saved in the stack.

6. The program resumes execution at the instruction following the last instruction performed before the interrupt handling.

3.2.1 Case Study: Tachometer Interrupt Service Routines

This subsection details the tachometer servicing ISR. The discussion is based on the ISR used in the PSoC Express design environment [3] to connect a tachometer to PSoC. This example illustrates the principal elements involved in interrupt handling by ISRs. It is interesting to note that the ISR implementation includes both hardware and software. Two programmable digital blocks are used to produce interrupts for the microcontroller upon receiving tachometer pulses, and to handle the abnormal functioning of the tachometer, respectively. Correct operation implies that three tachometer pulses are received within a specified time interval. This case study also defines the data structures needed for handling multiple tachometers.

Figure 3.8 shows the structure of an embedded system including a fan and a tachometer for adjusting the speed of the fan. This example was introduced as a case study in Chapter 1. The tachometer generates pulses that are applied to one of PSoC's input pins. Consecutive pulses correspond to one rotation of the fan, and are used by the control algorithm to measure the time interval between pulses, and thereby compute the rotational speed of the fan.

The tachometer pulses are used to produce interrupt signals internally for the microcontroller. This avoids long microcontroller idle times, which result if the microcontroller polls the tachometer inputs. Interrupts are generated by a PSoC programmable digital block upon receiving a pulse from the tachometer. The interrupts are serviced by the corresponding ISR routines that have their starting address stored in the ISR table. The interrupt vector produced by the priority encoder module for the specific interrupt becomes the index in the ISR table. The fan speed is controlled by setting the duty cycle of the pulse width modulation circuit. The PWM is connected at one output pin of the circuit as shown in Figure 3.8.

Depending on the nature of the tachometer, the ISR must implement the following steps. For three-wire fans, the ISR implements a five-state finite state machine, as shown in Figure 3.9(a). The FSM is initially idle. Upon receiving a pulse from the tachometer (and hence, an associated interrupt), the PWM associated with the fan is decoupled, and then the fan is connected (for a short period of time) to the high or low polarity depending on the type of the fan. These actions correspond to $STATE_1$. After receiving the second interrupt, the FSM moves to the state $STATE_2$. The state implements the waiting needed for the tachometer reading to stabilize. The FSM moves to the state $STATE_3$ after receiving the third interrupt. The ISR stores the current value of the tachometer timer circuit. Finally, the ISR FSM enters the state $STATE_4$ after receiving the fourth tachometer pulse. The FSM records the current value of the tachometer timer, computes the difference of the current time and the previously stored time (in $STATE_3$), and saves the difference in the memory. The last action of the state is to reconnect the PWM to the fan. Next the FSM is again idle.

In addition to the basic functionality, getting correct readings by the tachometer requires that the time interval between consecutive readings is at least 4 sec. The time interval depends on the type of the tachometer used. This constraint can be expressed in the ISR FSM by adding an additional state, called state $STATE_5$. The purpose of the state is to introduce a delay of 4 sec before the FSM returns to the state $STATE_1$. The resulting ISR FSM is shown in Figure 3.9(b).

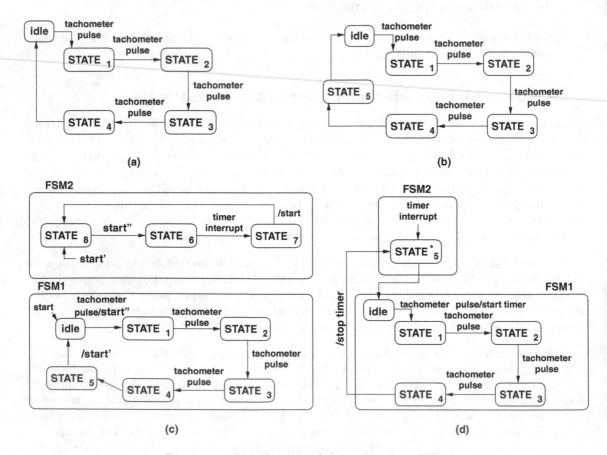

Figure 3.9: Specification of the tachometer ISR.

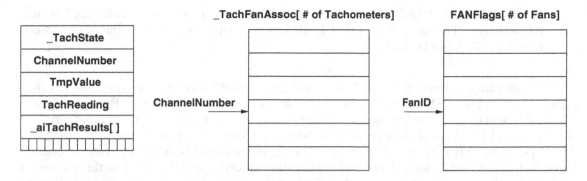

Figure 3.10: Data structure for handling multiple tachometers.

Although the formulated specification captures the main defining characteristics of the tachometer ISR, the routine can be improved to handle certain additional aspects: (1) provide the capability for handling multiple tachometers connected at different input pins, and (2) introduce the functionality for handling cases in which the tachometer does not generate, within a specified time interval, the three consecutive interrupts required to bring the FSM back to the idle state (and therefore the FSM is stuck in an intermediate state). The two issues are discussed next.

Handling multiple tachometers by the same ISR can be implemented by defining the data structures that store the characteristics and state information for each separate tachometer. In addition, the state $STATE_5$ must switch the ISR to the next tachometer to be serviced in addition to delaying the FSM for the required time interval. Note that the tachometer handling procedure must offer fair treatment of all the connected tachometers. A simple policy is *round-robin*, in which tachometers are considered successively, one after another, and starting from the tachometer with the lowest identifier. After the last tachometer has been considered, the procedure reiterates, starting again from the first tachometer.

Figure 3.10 introduces the data structures for multiple tachometers. Following are the required data structures:

- Variable _TachState stores the current state of the ISR FSM.

- Variable ChannelNumber is a unique index associated with each tachometer. The channel for the tachometer has an abstract name in the *main* program, for example *Tach*. However, at the ISR level, it is identified by the unique *ChannelNumber*.

- The variable *ChannelNumber* is also used as an index in the table _TachFanAssoc that indicates the fan associated with each tachometer. The size of the table is equal to the number of tachometers in the application.

- Variable *TmpValue* is an auxiliary variable for the ISR.

- Variable *TachReading* stores two bytes read from the tachometer timer.

- Table _aiTachResults has a two byte entry for each existing tachometer. The entry is used to store the MSB and LSB found by executing the ISR.

- Finally, the table *FANFlags* stores the control information associated with each fan (e.g., its polarity). The index *FanID* that is used to access the table is the content of an entry in Table _*TachFanAssoc*.

One exceptional situation occurs if the three pulses, and hence the corresponding interrupt signals, for the ISR *FSM* are not generated in due time by the tachometer. For example, this situation would occur if, for any reason, the tachometer were not connected to an input pin. This situation can be addressed by introducing a supervising routine that resets the tachometer ISR. The supervising ISR is activated by an interrupt signal produced by a timer circuit, in case the three tachometer pulses are not received within the required time interval. The timer circuit is loaded with the predefined time interval, and generates an interrupt after decrementing the value to zero.

Figure 3.9(c) presents the resulting FSM structure. *FSM1* is the ISR handling the tachometer, and *FSM2* is the supervising FSM. *FSM2* is initiated by the state $STATE_1$ of *FSM1*, hence after the system received the first pulse from the tachometer *FSM1* generates the start signal *start'* for *FSM2*. After the predefined time interval has expired (expressed through state $STATE_6$ in *FSM2*, the timer circuit generates an interrupt, which brings *FSM2* into state $STATE_7$. In this state, *FSM2* resets *FSM1*, and brings it back to the initial state. If *FSM1* reaches the state $STATE_5$ before the time limit ends then the state $STATE_5$ enables the transition of *FSM2* to state $STATE_8$. Note that this solution requires two communicating FSMs (each FSM handling different interrupt signals). The FSMs interact through the signals indicated by dashed line arrows.

The number of FSM states can be reduced by restructuring the two FSMs as shown in Figure 3.9(d). The state $STATE_5$ of *FSM1* is merged with the states $STATE_6$, $STATE_7$, and $STATE_8$ of *FSM2* to produce the new state called $STATE_5*$. After *FSM1* receives the fourth tachometer pulse, the state $STATE_4$ starts the execution of *FSM2*. The single state of *FSM2* handles the abnormal cases for the current tachometer, switches to the next tachometer to be handled, implements the required delay of 4 sec between consecutive readings using the same tachometer, and initiates the execution of *FSM1* for the selected tachometer.

Figure 3.11 defines the pseudocode of the *FSM1* ISR:

- First, the routine pushes the values of the registers A and X on the stack.

- If the ISR is in the state $State_1$, then it uses the *ChannelNumber* variable as an index into the table _*TachFanAssoc* to find the identifier of the PWM of the associated fan. Then, the information about the PWM is retrieved from Table *FANFlags*. If the fan does not control the tachometer (e.g., if the tachometer control pin of the fan is not set), then the value of the FSM state variable is changed to $State_2$, the registers A and X are restored to the values stored on the stack, and the ISR execution ends. Note that the RPM reading is incorrect in this case. If the fan controls the tachometer, then the ISR next finds the polarity of the fan. If the fan has positive polarity, then the PWM is decoupled from the fan, and the fan is controlled using the high signal. If the fan has negative polarity, then it is driven with a low signal, after decoupling its PWM block. The registers A and X are restored to their saved values, and the execution returns to the interrupted program.

- In the state $State_2$, the ISR only changes the FSM state to $State_3$ and then restores the registers A and X to their values previous to the ISR execution.

```
routine TACH_Timer_LSB_ISR is
    push registers X and A on the stack;
    switch (current_state){
    case STATE1:
        find the fan associated to the tachometer;
        find the PWM driving the fan;
        get the state information of the fan;
        if the PWM is currently driving the fan then
            find the driving polarity of the fan;
            unconnect the PWM from the fan AND
            drive the fan with the necessary polarity;
        current_state = STATE2; break;
    case STATE2;
        current_state = STATE3; break;

    case STATE3:
        record the timer value;
        current_state = STATE4; break;
    case STATE4:
        stop the timer;
        disable capture interrupts for timer;
        record the timer value;
        compute the difference between the two timer recordings
        reconnect the PWM to the fan;
        save the difference in the entry of the tachometer;
        start the tachometer timer;
        current_state = STATE5*; break;
    }
    pop registers X and A from the stack;
end routine
```

Figure 3.11: Pseudocode of the FSM1 ISR.

- In the state $State_3$, the FSM reads the MSB and LSB of the tachometer timer, and stores the two values as the two bytes of the SRAM variable *TachReading*. Then, registers A and X are restored to their saved values, and the ISR execution ends.

- $State_4$ first stops the tachometer timer and disables its interrupts. Then, it reads the value of the tachometer timer into the registers A and X. Then, it computes the difference of the value stored in variable *TachReading* and the current reading present in registers A and X. The result is stored in variable *TachReading*. The fan is reconnected next (by reconnecting the PWM block to the fan). The current state of the FSM is set to state $STATE_5*$ (which actually starts the execution of *FSM2*). Then, the result in variable *TachReading* is stored in Table *_aiTACH_Results*, in the entry corresponding to the tachometer channel. The tachometer timer is restarted, registers A and X are restored, and the ISR execution ends.

Figure 3.12 shows the pseudocode for the *FSM2* ISR. The routine corresponds to the FSM structure in Figure 3.9(d). The purpose of the routine is to restore the correct operation of *FSM1* in the case of incorrect operation of the tachometer, and switching to the next monitored tachometer:

```
routine TACH_Timer_MSB_ISR is
    push registers A and X on the stack;
    disable the capture interrupt of the timer;
    if current_state <> STATE5* then
        find the results entry for the tachometer;
        store value FFFFh in the tachometer entry;
        current_state = STATE5*;
        reconnect the PWM to the corresponding fan;
    stop the timer of the tachometer;
    move to the next monitored tachometer;
    find the time of the last tachometer reading;
    compute the difference of the two time readings;
    if difference < 4 sec then
        start the timer;
    else
        get the current system time;
        store the current time into the tachometer entry
        start the timer;
        enable the capture interrupts of the timer;
        current_state = STATE1;
    pop registers A and X from the stack;
end routine
```

Figure 3.12: Pseudocode of the FSM2 ISR.

- After saving registers A and X values on the stack and disabling the timer interrupts, the routine verifies if the tachometer ISR received three interrupts within the predefined time limit. In the case where the *FSM1* ISR is stuck in an intermediate state (different from state $STATE_5*$) then the routine attempts to recover the *FSM1* from the erroneous situation: (i) the value *FFFFH* is loaded into the results entry of the tachometer to indicate an incorrect reading, (ii) the state of the tachometer FSM is reset to state $STATE_5$, and (iii) the PWM is re-connected to the fan.

- The routine switches to the next monitored tachometer by following the round-robin policy. The routine computes the time interval elapsed from the last reading of the tachometer. If the time interval is less than the specified limit (4 sec), then the timer circuit is restarted, and the ISR busy waits until the time constraint is met. Once the time constraint is met, the state information for the tachometer is updated (e.g., the time of the last reading), the capture interrupts of the timer are enabled, and the state of *FSM1* is modified to the state $STATE_1$. This starts a new reading process for the tachometer.

The global structure of the hardware and software modules implementing the two tachometer-related ISRs is given in Figure 3.14. Figure 3.13 illustrates the interconnecting of multiple tachometers to the same PSoC chip. Each tachometer is connected to a different input pin of the chip, but the connection of a pin to the interrupt handling resources can be programmed by PSoC's programmable interconnect.

The ISR implementation in Figure 3.14 uses two programmable hardware blocks, blocks *DCB02* and *DCB03*, in addition to the software routines. The hardware blocks produce interrupt signals for the microcontroller upon receiving inputs from the tachometer, and also reduce the timing

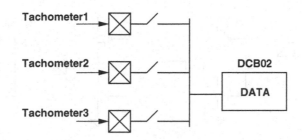

Figure 3.13: Connecting multiple tachometers to PSoC.

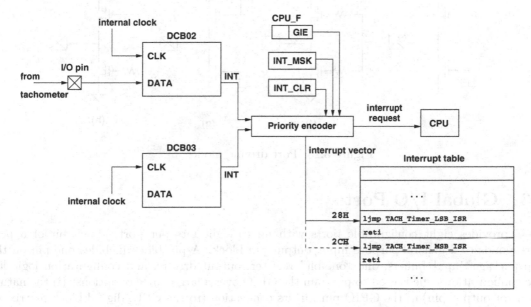

Figure 3.14: Hardware and software components implementing the tachometer ISR.

overhead due the routines that measure the time intervals related to the tachometer ISRs (e.g., measuring the time interval between two consecutive readings of the same tachometer). The programmable digital block *DCB02* is configured as a timer circuit with counter functionality. (Chapter 4 provides more details about the programmable blocks.) Each input signal from the tachometer generates an interrupt signal for the microcontroller. The interrupt vector for the block *DCB02* is the index 28H in the interrupt table. After saving the registers of the microcontroller, the execution flow of the microcontroller jumps to the ISR *TACH_Timer_LSB_ISR* that implements the functionality of *FSM1*.

Similarly, the programmable block *DCB03* is configured as a timer circuit with terminal count functionality. The timer count is set to the value of the predefined time interval (e.g., 4 sec in this case). An interrupt is generated by the digital block after the time interval elapses. The interrupt vector 2CH is produced for this interrupt, and therefore the execution jumps to the ISR *TACH_Timer_MSB_ISR*. The ISR implements *FSM2*.

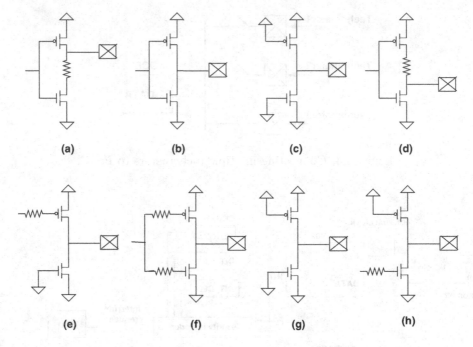

Figure 3.15: Port driving modes [5].

3.3   Global I/O Ports

PSoC provides eight configurable ports with up to eight bits per port. Each bit of a port corresponds to a general-purpose input/output pin block. A pin block includes one pin on the chip package, input buffers, one "one-bit" register, output drivers, and configuration logic [5]. The configuration logic is used to program the GPIO over a large range of options: (i) the nature (input or output pin) of the GPIO pin, (ii) its connection (to the CPU, digital block matrix, or analog blocks matrix), (iii) driving capabilities, and (iv) the associated interrupts. These aspects are discussed next.

Figure 3.2 introduces the concepts of input ports, and that of output ports, for which data coming from different PSoC internal sources (e.g., CPU, digital blocks, and analog blocks) are routed to the output pin to drive external peripheric devices. The selection of the external or internal source signals for a pin is achieved by configuration logic, which incorporates tri-state buffers, multiplexer circuits, and control bits programmed by the control registers of the PSoC GPIO pin block. Each pin block includes a programmable block to select the driving mode of the pin block depending on the interfacing required for the pin block.

Figure 3.15 illustrates the circuits for the different driving modes of a PSoC pin.

- Figure 3.15(a) shows the *resistive pull down mode*:

  If data are low, then the pin is connected to the positive power supply through the on PMOS transistor, therefore offering strong driving capabilities. If data are high, then the pin is connected to ground by the resistor and the conducting NMOS transistor. The fall time of the signal at the GPIO pin is large due to the increased resistance in series with the on NMOS transistor.

- Figure 3.15(b) shows the circuit for the *strong drive* mode. The circuit operates as an inverter that provides strong driving capabilities independent of the data value.

- Figures 3.15(c) indicates the *high impedance* mode.The data are not connected to the circuit inputs. The NMOS and PMOS transistors are both off, and hence the circuit is not connected to the pin. This mode is used when the GPIO pin is implemented as an input pin.

- Figure 3.15(d) depicts the circuit for the *resistive pull up* mode. Its functioning is similar to the resistive pull down mode: if the input is high, then the NMOS transistor is on offering a low resistance connection to ground. If the input to the circuit is low, then the PMOS transistor is on, but the rising time of the output signal at the pin is larger due to the resistance in series with the conducting PMOS transistor.

- Figure 3.15(e) presents the circuit for the *open drain, drives high* mode. For a high input, both MOSFET transistors are off, and hence the circuit output is in high impedance state. For a low input, the PMOS transistors turn on raising the output to high, and giving strong driving capabilities at the output. The fall time of the input is longer due to the resistor in series with the PMOS gate.

- Figure 3.15(f) illustrates the circuit for the *slow strong drive* mode. The circuit operates as an inverter, and provides strong output driving capabilities independent of the input value (by the conducting NMOS and PMOS transistors, respectively). The two resistances in series with the MOSFET transistors increase the transition time of the input signal, therefore offering a slow turn on and off of the transistors.

- Figures 3.15(g) indicates the *high impedance* mode. The operation is similar to that of the circuit in Figure 3.15(c), but in this case the digital input is also disabled, as shown in Figure 3.16. This mode is set at system reset, and also used if analog signals are routed to the pin.

- Figure 3.15(h) shows the *open drain, drives low* mode. If the input is low, then both transistors are turned-off, hence the output is in high impedance state. For a high input, the output is low and has a strong driving capability by the conducting NMOS transistor. The rise time of the input signal is lengthened by the resistor in series with the NMOS gate. This mode provides compatibility with I2C interfaces.

The driving capabilities of each port pin are controlled by the *PRTxDM2*, *PRTxDM1*, and *PRTxDM0* registers. Table 3.4 presents the values for programming the port modes. Each pin of a port is controlled by the corresponding bits of the three control registers, for example pin 3 is controlled by bits 3 of the registers *PRTxDM2*, *PRTxDM1*, and *PRTxDM0*.

There are in total 24 registers *PRTxDMx*, which control the PSoC I/O ports as follows:

- Port *0* is controlled by registers *PRT0DM2* at address *0,03H*, *PRT0DM1* at address *1,01H*, and *PRT0DM0* at address *1,00H*.

- Port *1* by registers *PRT1DM2* at address *0,07H*, *PRT1DM1* at address *1,05H*, and *PRT1DM0* at address *1,04H*.

- Port *2* by registers *PRT2DM2* at address *0,0BH*, *PRT2DM1* at address *1,09H*, and *PRT2DM0* at address *1,08H*.

- Port *3* by registers *PRT3DM2* at address *0,0FH*, *PRT3DM1* at address *1,0DH*, and *PRT3DM0* at address *1,0CH*.

Table 3.4: Programming of the port driving modes [5].

Driving mode	DM2	DM1	DM0
Resistive pull down	0	0	0
Strong drive	0	0	1
High impedance	0	1	0
Resistive pull up	0	1	1
Open drain high	1	0	0
Slow strong drive	1	0	1
High impedance analog	1	1	0
Open drain low	1	1	1

- Port *4* by registers *PRT4DM2* at address *0,13H*, *PRT4DM1* at address *1,11H*, and *PRT4DM0* at address *1,10H*.

- Port *5* by registers *PRT5DM2* at address *0,17H*, *PRT5DM1* at address *1,15H*, and *PRT5DM0* at address *1,14H*.

- Port *6* by registers *PRT6DM2* at address *0,1BH*, *PRT6DM1* at address *1,19H*, and *PRT6DM0* at address *1,18H*;

- Port *7* by registers *PRT7DM2* at address *0,1FH*, *PRT7DM1* at address *1,1DH*, and *PRT7DM0* at address *1,1CH*.

The connections to and from any GPIO port can be programmed to operate as in one of the following three cases:

- *Digital IOs* connect the chip pins to the microcontroller of the architecture. This includes both sending data to the CPU and getting data from the CPU. There are the following programming options for an I/O pin used as a digital I/O:

    - An IO pin is configured as a digital IO by setting to 0 its corresponding bit (called bit *BYP*) in register *PRTxGS*. Bit *BYP* is used in Figure 3.16 to control the multiplexer circuit of the output path, and the tri-state buffer going to the global interconnect in the input path.

      There are eight *PRTxGS* registers, one for each I/O port: *PRT0GS* at address *0,02H* corresponds to port *0*, *PRT1GS* at address *0,06H* is for port *1*, *PRT2GS* at address *0,0AH* refers to port *2*, *PRT3GS* at address *0,0EH* corresponds to port *3*, *PRT4GS* at address *0,12H* is for port *4*, *PRT5GS* at address *0,16H* relates to port *5*, *PRT6GS* at address *0,1AH* is for port *6*, and *PRT7GS* at address *0,1EH* corresponds to port *7*. Each bit of a *PRTxGS* register corresponds to a pin of the corresponding I/O port (bit *0* to pin *0*, bit *1* to pin *1*, etc.).

    - Data communication to/from the CPU is achieved by the *PRTxDR* registers. The microcontroller reads input data from the I/O port by accessing the *PRTxDR* register for the port. Note that the accessed data are not the value currently stored in the *PRTxDR* register but instead the values corresponding to the signals at the port pins.

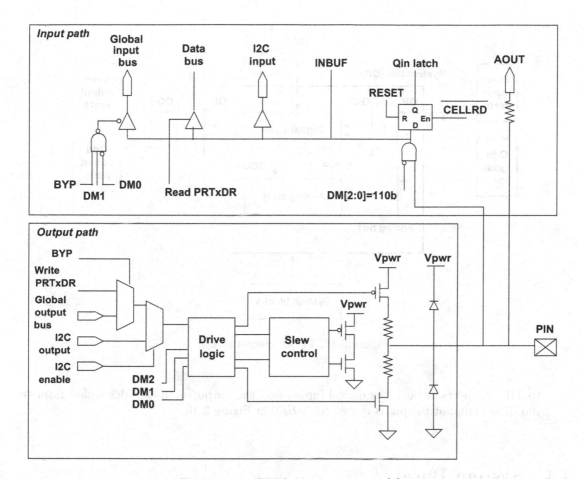

Figure 3.16: GPIO block structure [5].

Similarly, the CPU transfers output data to a port by writing to the corresponding *PRTxDR* register.

The *PRTxDR* registers are associated with the eight I/O ports: *PRT0DR* at address *0,00H* corresponds to the port *0*, *PRT1DR* at address *0,04H* relates to the port *1*, *PRT2DR* at address *0,08H* is for port *2*, *PRT3DR* at address *0,0CH* relates to the port *3*, *PRT4DR* at address *0,10H* is for port *4*, *PRT5DR* at address *0,14H* corresponds to port *5*, *PRT6DR* at address *0,18H* corresponds to port *6*, and *PRT7DR* at address *0,1CH* relates to port *7*. Each bit of a *PRTxDR* register corresponds to a pin of the corresponding IO port.

- *Global IOs* connect the GPIO ports to PSoC's digital configurable blocks. A pin of a GPIO port is configured as global IO by setting to 1 its corresponding bit *BYP* in the *PRTxGS* register. In Figure 3.16, bit *BYP* controls the multiplexer of the output path, threby allowing therefore data on the *Global output bus* to be routed to the pin. For the input path, the bit controls the tristate buffer going to the *Global input bus*.

- *Analog IOs* link the GPIO ports to PSoC's configurable analog blocks. This is programmed by setting the corresponding bits of the *PRTxDM2*, *PRTxDM1*, and *PRTxDM0* registers

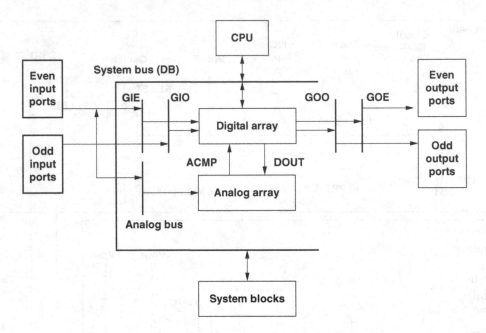

Figure 3.17: PSoC system buses.

to 110. As a result both the digital inputs and the output paths are decoupled from the pin. The signal at the pin is routed to $AOUT$ in Figure 3.16.

3.4   System Buses

System buses connect the input/output ports and the different PSoC subsystems, such as the microcontroller, the arrays of programmable analog and digital blocks, clocking circuits, and memory [5]. The bus characteristics influence the performance of the system including the total system latency, power consumption, and processing accuracy. For example, if the CPU executes a task followed by transmitting data to a digital block, the total latency of the processing (including the processing on the CPU and the digital block) also includes the time required to communicate the data between the CPU and digital block. Also, the interconnect structure between the data memory and the data processing subsystems determine the time delay involved in transferring data to, and from, the subsystems. Finally, the energy (power) consumption of the entire application includes the energy (power) consumption of data processing plus the energy (power) consumption of the data transfers between the architecture blocks.

The data communication delay on a bus depends on two factors:

$$T_{\text{data com}} \quad = \quad T_{\text{delay}} + T_{\text{busy waiting}} \tag{3.1}$$

The value $T_{data\ com}$ is the time delay required to send one information data value on the bus, and the value $T_{busy\ waiting}$ is the waiting time for starting the communication due to the ongoing communications allocated to the same bus. For simplicity, it was assumed that the bit-width

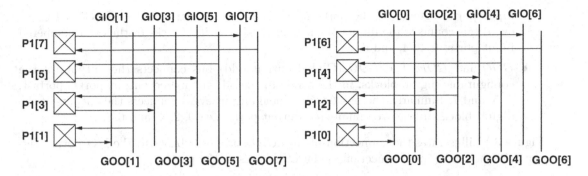

Figure 3.18: Bus connections to the I/O ports [5].

for one data value of information is equal to the bus width, so that the entire data value can be transmitted simultaneously on the bus.

The time delay for communicating a data value depends on the bus characteristics, and is approximated reasonably well by the following (Elmore delay) [10],

$$T_{\text{delay}} \approx \frac{r\,c\,L^2}{2} \tag{3.2}$$

where r is the resistance per unit length of the bus, c is the capacitance per unit length, and L is the length of the bus. This expression shows that the bus delay increases rapidly with the length of the bus, because there is a quadratic dependency between the delay time and the length of the bus. More accurate bus delay models, and the impact of the bus delay on the architecture design, are discussed in the literature [7, 8, 11]. The busy waiting time $T_{busy\ waiting}$ of a communication depends on communication scheduling.

Depending on the kind of signals they interconnect, buses are grouped into (i) *data buses*, (ii) *address buses*, and (iii) *control buses* (e.g., for control, chaining, and interrupt signals). This subsection discusses only data buses. Address buses were presented in Chapter 2, and the control signals are discussed as part of the subsystems to which they pertain.

Figure 3.17 illustrates PSoC's data bus architecture. These buses can be grouped into several categories, depending on the blocks they interconnect: (1) the global system bus connects the microcontroller to the other PSoC subsystems, (2) the global digital interconnect links the programmable digital blocks of the PSoC architecture, and (3) the analog interconnect is used for the programmable analog blocks. The three bus structures are detailed next.

1. *Global system bus* (*DB*): The global system bus is an eight-bit shared bus that interconnects all the main subsystems of the architecture: input and output ports, CPU, memories, digital blocks, and system resources (timers, watchdogs, etc.). The *DB* bus is used to transfer data to and from the CPU, I/O ports, digital blocks, and other system resources. The data transfer is under the control of the CPU, which executes the corresponding data transfer instructions.

2. *Global digital interconnect bus* (*GDI*): The global digital interconnect links the input and output ports to the digital array of blocks. There are four *GDI* buses:

   - *GIO* and *GIE*: These are each four-bit input buses. The *GIO* (Global input odd) bus connects the inputs of the configurable digital blocks in the array to the odd number input ports. These are the ports *1*, *3*, *5*, and *7*. Similarly, the *GIE* (Global Input even) bus connects the inputs of the configurable digital blocks to the even numbered input ports: 0, 2, 4, and 6.

- *GOO* and *GOE*: The *GOO* (Global output odd) bus connects the outputs of the configurable digital blocks, in the array, to the odd numbered output ports: ports 1, 3, 5, and 7. Similarly, the bus *GOE* (Global output even) connects the outputs of the digital blocks to the even numbered output ports, i.e., 0, 2, 4, and 6.

Figure 3.18 illustrates the connection of the *GIO* and *GOO* buses to the I/O ports. Figure 3.19 shows the interconnection of the *GIE* buses to the ports.

The global digital interconnect buses are configured by programming the *GDI_O_IN*, *GDI_E_IN*, *GDI_O_OU*, and *GDI_E_OU* registers. The architecture allows direct data transfer between the *GIE/GOE*, and *GIO/ GOO* buses. Figure 3.20 shows the possible connects, and indicates the PSoC registers that control the transfer:

- *Register GDI_O_IN*: The register *GDI_O_IN* at address *1,D0H* controls the connection of the input bus, *GIO*, to the output bus *GOO*. Table 3.5 describes the meaning of each register bit. For example, if the bit 5 is set to 1, then there is a connection between the bit 5 of the bus *GIO* and the same bit of the bus *GOO*. Otherwise, the bit 5 of bus *GIO* does not drive the corresponding bit of *GOO*.

- *Register GDI_E_IN*: The register *GDI_E_IN* at address *1,D1H* controls the connection of the input bus *GIE* to the output bus *GOE*. Table 3.5 describes the meaning of each register bit. For example, if the bit 2 is set to 1, then there is a connection between the bit 2 of the bus *GIE* and the bit 2 of the bus *GOE*. Otherwise, the two bits are not connected to each other.

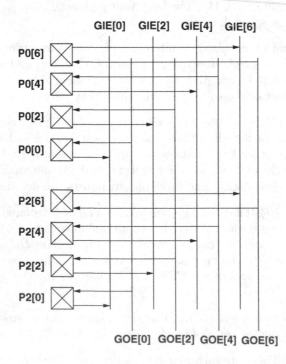

Figure 3.19: Bus connections to the I/O ports [5].

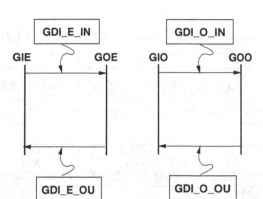

Figure 3.20: Direct connection of buses GIx and buses GOx [5].

Table 3.5: Structure of the registers *GDI_O_IN* and *GDI_E_IN* [5].

Register	0/1	Bit 7	Bit 6	Bit 5	Bit 4
GDI_O_IN	'0'	NOC	NOC	NOC	NOC
GDI_O_IN	'1'	GIO[7]→GOO[7]	GIO[6]→GOO[6]	GIO[5]→GOO[5]	GIO[4]→GOO[4]
GDI_E_IN	'0'	NOC	NOC	NOC	NOC
GDI_E_IN	'1'	GIE[7]→GOE[7]	GIE[6]→GOE[6]	GIE[5]→GOE[5]	GIE[4]→GOE[4]

Register	0/1	Bit 3	Bit 2	Bit 1	Bit 0
GDI_O_IN	'0'	NOC	NOC	NOC	NOC
GDI_O_IN	'1'	GIO[3]→GOO[3]	GIO[2]→GOO[2]	GIO[1]→GOO[1]	GIO[0]→GOO[0]
GDI_E_IN	'0'	NOC	NOC	NOC	NOC
GDI_E_IN	'1'	GIE[3]→GOE[3]	GIE[2]→GOE[2]	GIE[1]→GOE[1]	GIE[0]→GOE[0]

- *Register GDI_O_OU*: As shown in Table 3.6, the register *GDI_O_OU* at address *1,D2H* programs the connection between the output bus *GOO* and the input bus *GIO*. Setting a bit of the register to 1 establishes a connection from the corresponding bit of the output bus to the same bit of the input bus.

- *Register GDI_E_OU*: The bits of the output bus *GOE* drive the corresponding bits of the input bus *GIE*, if the same bits of the register *GDI_E_OU* at address *1,D3H* are set to the value 1. Table 3.6 details the bit structure of the register, and the connections programmed by each of the register bits.

The Row digital interconnect (*RDI*) includes the buses that offer data interconnection at the level of the digital array rows. It was already mentioned and illustrated in Figure 3.4 that the programmable digital blocks are organized as identical rows. Figure 3.21 details the interconnect structure between *GDI* and *RDI*. The *BCR* buses are the broadcast signals produced by one row for the other rows of the digital block array. The *ACMP* signals are generated by the comparator circuits of the programmable analog blocks. More details about these signals are in the following subsections of the chapter.

Figure 3.22 provides a more detailed presentation of the *RDI* structure [5]. The figure shows the generation of the signals needed for the digital programmable blocks, for example the input signals *DATA* and *AUX*, the output signal *RO*, the clock *CLK*, the interrupt signals *INT*, and the chaining signals *FPB*, *TPB*, *TNB*, and *FNB*. Chapters 4 and 5 present

Table 3.6: Structure of the registers *GDI_O_OU* and *GDI_E_OU* [5]

Register	0/1	Bit 7	Bit 6	Bit 5	Bit 4
GDI_O_OU	'0'	NOC	NOC	NOC	NOC
GDI_O_OU	'1'	GOO[7]→GIO[7]	GOO[6]→GIO[6]	GOO[5]→GIO[5]	GOO[4]→GIO[4]
GDI_E_OU	'0'	NOC	NOC	NOC	NOC
GDI_E_OU	'1'	GOE[7]→GIE[7]	GOE[6]→GIE[6]	GOE[5]→GIE[5]	GOE[4]→GIE[4]

Register	0/1	Bit 3	Bit 2	Bit 1	Bit 0
GDI_O_OU	'0'	NOC	NOC	NOC	NOC
GDI_O_OU	'1'	GOO[3]→GIO[3]	GOO[2]→GIO[2]	GOO[1]→GIO[1]	GOO[0]→GIO[0]
GDI_E_OU	'0'	NOC	NOC	NOC	NOC
GDI_E_OU	'1'	GOE[3]→GIE[3]	GOE[2]→GIE[2]	GOE[1]→GIE[1]	GOE[0]→GIE[0]

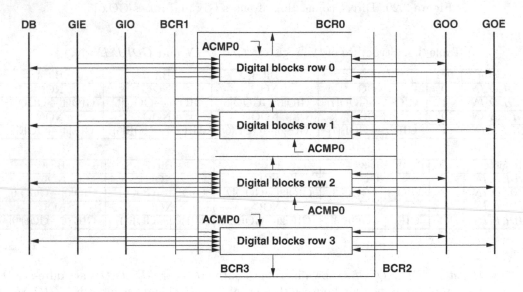

Figure 3.21: Array digital interconnect [5].

in detail how these signals are used to realize the functionality of each programmable digital block. The block signals are routed as follows.

- *DATA*: The data signals for the digital blocks are selected among the following sources: the row input interconnect *RI* connected to the input buses, the interconnect *RO* connected to the output buses, the data from the previous digital block, the *ACMP* signals from the analog comparators, the broadcast signals from other rows, and the low and high DC signals:

  - *RI interconnect*: The local bus is four bits, and is connected to the global input buses *GIE* and *GIO*, as shown in Figure 3.22. For example, the bit 0 of the bus *RI* is selected among the bits 0 and 4 of the two global input buses. The figure presents the source for the other *RI* bus bits.

  - *RO interconnect*: The four-bit wide bus is driven by the programmable digital block outputs. More details about the signals follow.

  - *Broadcast signals*: Each row of the digital array can generate the broadcast signal *BCROW* (BCR) for the other rows of the architecture. For example, the row 0

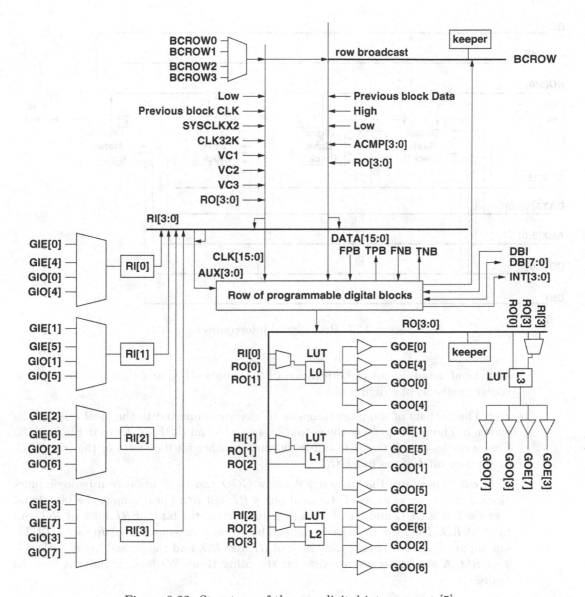

Figure 3.22: Structure of the row digital interconnect [5].

originates the signal *BCROW0*, and so on. Each *BCROW* net has a *keeper* circuit that stores the last driving value of the net in case the net becomes undriven [5]. Upon resetting the system, the keeper circuit is set to the value 1.

- *Analog comparator signals*: The *ACMP* signals are the output signals of the comparator circuits of the PSoC analog array. More details about the signals follow in this section and are also given in [5].

- *Other signals*: The block input *DATA* can be also connected to other signals, for example the data input of the previous digital block, and the constant signals high and low.

- *AUX*: These signals are the inputs to the auxiliary input of the digital blocks. The signals are selected from the bits of the local *RI* bus. Chapters 4 and 5 provide

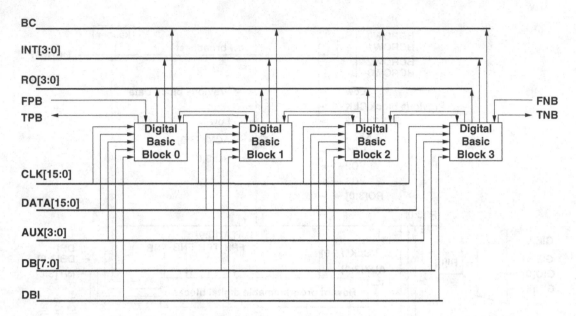

Figure 3.23: Row digital interconnect [5].

additional details about the utilization of the inputs AUX, and their selection by the control registers of a digital block.

- RO: The outputs of the programmable blocks are connected to the local bus RO, as shown in Figure 3.23, and to the output buses GOO and GOE, as shown in Figure 3.22. For example, the bit 0 of RO can be output as either bit 0 or bit 4 of the bus GOO, or bit 0 or bit 4 of the bus GOE.

  Figure 3.22 indicates that the output buses GOO and GOE are driven by logic functions defined over the bits of the local buses RI and RO. For example, the bits 0 and 4 of the buses GOO and GOE are a logic function of the bit 0 of RI, bit 0 of RO, and bit 1 of RO. The logic function is defined based on a two-input Lookup table (LUT): one input is selected among the bits 0 of RI and RO, and the second input is the bit 1 of RO. A similar structure exists for the other three RO bits, as indicated in the figure.

- *Chaining signals*: The programmable blocks of the digital array can be chained together using the FPB signals (from the previous block), TPB (to the previous block), FNB (from the next block), and TNB (to the next block). Block chaining is used for increasing the bit width of data processing, for example to increase the maximum count value of the counter circuit, if the counter circuit consists of the programmable digital blocks.

  Consecutive digital blocks on the same row can be connected with each other using the four signals. Also, the last block on a row can be linked to the first block of the next row, as shown in Figure 3.24.

- CLK: The clock signals for the digital blocks are selected among the following sources: the $SYSCLKX2$, $CLK32$, $VC1$, $VC2$, $VC3$ system clocks (detailed in Section 3.5 of this chapter), the constant low signal, the broadcast signals of the digital rows, the local buses RI and RO, and the clock signal of the previous digital block. The clock

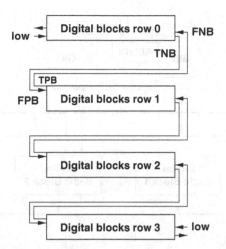

Figure 3.24: Chaining of the programmable digital blocks.

signal is selected separately for each digital block by programming the control registers of the block, as described in Chapters 4 and 5.

Figure 3.25 describes the control registers that are involved in the programming of the RDI interconnect: *RDI_RI*, *RDI_SYN*, *RDI_IS*, *RDI_LT*, and *RDI_RO* [5]. The registers have the following bit structure.

- *Register RDI_RI*: The registers select the driving signals for the row inputs: the register *RDI0RI* at address $x,B0H$ is for row 0, the register *RDI1RI* at address $x,B8H$ is for row 1, the register *RDI2RI* at address $x,C0H$ is for row 2, and the register *RDI3RI* at address $x,C8H$ is for row 3.

  The structure of the registers *RDIxRI* is the following:

    - *Bits 7-6*: The bits 7-6 select the input source for the bit 3 of the local bus *RI*: the value 00 selects *GIE[3]*, 01 chooses *GIE[7]*, the bits 10 pick the bit *GIO[3]*, and 11 uses *GIO[7]*.
    - *Bits 5-4*: The bits decide the input source for the bit 2 of *RI*: 00 selects *GIE[2]*, 01 chooses *GIE[6]*, the bits 10 pick *GIO[2]*, and 11 selects *GIO[6]*.
    - *Bits 3-2*: The bits 3-2 select the input source for the bit 1 of the local bus *RI*: the value 00 selects *GIE[1]*, 01 chooses *GIE[5]*, the bits 10 pick the bit *GIO[1]*, and 11 uses *GIO[5]*.
    - *Bits 1-0*: The bits pick the input source for the bit 0 of *RI*: 00 selects *GIE[0]*, 01 chooses *GIE[4]*, 10 picks *GIO[0]*, and 11 uses *GIO[4]*.

- *RDI_SYN*: These registers define the synchronization signals for the RDI sources. By default, the source signals are synchronized with the system clock, *SYSCLOCK*. The system clock is detailed in Section 3.5 of the chapter. One control register exists for each digital array row: the *RDI0SYN* register at address $x,B1H$ corresponds to the row 0, the *RDI1SYN* register at address $x,B9H$ is for the row 1, the *RDI2SYN* register at address $x,C1H$ is for the row 2, and the *RDI3SYN* register at address $x,C9H$ is related to the row 3.

  The *RDI_x_SYN* registers have the following bit structure:

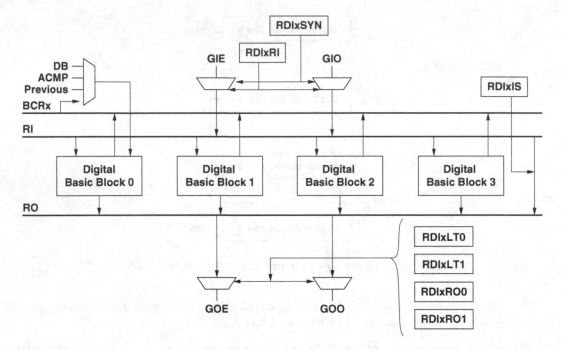

Figure 3.25: Control registers for programming row digital interconnect [5].

- *Bit 3*: If the bit is set to 1, then for the source to the bus bit 3 of *RI*, there is no synchronization of the source, otherwise the source to *RI[3]* is synchronized with the system clock *SYSCLOCK*.

- *Bit 2*: The bit being set to 1 defines that there is no synchronization of the source to the bus bit *RI[2]*. Otherwise, similar to the bit 3, the source is synchronized to the clock *SYSCLOCK*.

- *Bit 1, bit 0*: The behavior of the two bits is similar to that of the bits 3 and 2, the differences being that bit 1 controls the synchronization of the source to *RI[1]*, and bit *0* that of *RI[0]*, respectively.

- *RDI_IS*: For each row, this dedicated register defines the nature of the broadcast row signal (*BCROW*) and the inputs to the four LUTs of the RDI net. The register *RDI0IS* at address *x,B2H* corresponds to the row 0, the register *RDI1IS* at address *x,BAH* is for the row 1, the register *RDI2IS* at address *x,C2H* is for the row 2, and the register *RDI3IS* at address *x,CAH* is for the row 3.

The structure of the *RDIxIS* registers is the following:

- *Bits 5-4*: The two bits select which the *BCROW* signal drives the local broadcast net of the row corresponding to the *RDIxIS* register. The value 00 selects the signal *BCROW0*, 01 the *BCROW1* signal, 10 *the BCROW2* signal and 11 the *BCROW3* signal.

- *Bits 3-0*: These bits determine the input A to the four LUTs: bit 3 controls the LUT 3, bit 2 the LUT 2, and so on. If the bit is 0, then the input A is the corresponding bit *RO*, otherwise the corresponding bit *RI* (cf. Figure 3.22 for details).

- *Registers RDI_LT*: The logic equations of the four LUTs for each row are programmed by the *RDILT0* and *RDILT1* registers corresponding to the row. The registers *RDI0LT0* and *RDI0LT1* at addresses $x,B3H$ and $x,B4H$ are for the row *0*, the register *RDI1LT0* at address x,BBH and the register *RDI1LT1* at address x,BCH are for the row 1, the registers *RDI2LT0* and *RDI2LT1* at addresses $x,C3H$ and $x,C4H$ are for the row 2, and the registers *RDI3LT0* and *RDI3LT1* at addresses x,CBH and x,CCH are for the row 3.

The structure of the *RDIxLT0* and *RDIxLT1* registers is as follows:

  - *Bits 7-4*: The bits of the register *RDIxLT0* program the logic function for the LUT 1, and the bits of the register *RDIxLT1* decide the logic function of the LUT 3.
  - *Bits 3-0*: The bits of the register *RDIxLT0* define the logic function for the LUT 0, and the bits of the register *RDIxLT1* select the logic function of the LUT 2.

Table 3.7 shows the 16 logic functions that can be programmed into the four LUTs.

Table 3.7: The logic functions implemented in LUTs [5]

Bit Value	Logic Function	Bit Value	Logic Function
0000	*FALSE*	1000	*A NOR B*
0001	*A AND B*	1001	*A XNOR B*
0010	*A AND B*	1010	*B*
0011	*A*	1011	*A OR B*
0100	*A AND B*	1100	*A*
0101	*B*	1101	*A OR B*
0110	*A XOR B*	1110	*A NAND B*
0111	*A OR B*	1111	*TRUE*

- *Registers RDI_RO*: The outputs of the four LUTs of a row are connected to the global output buses *GOO* and *GOE* depending on the values of the control registers *RDIRO0* and *RDIRO1* of that row. Depending on the row they are controlling, the two registers are at the following addresses: for the row 0, the register *RDI0RO0* is at the address $x,B5H$ and the register *RDI0RO1* is at the address $x,B6H$, for the row 1, the register *RDI1RO0* is at the address x,BDH and the register *RDI1RO1* is at the address x,BEH, for the row 2, the register *RDI2RO0* is at the address $x,C5H$ and the register *RDI2RO1* is at the address *x,C6H*, and for the row 3, the register *RDI3RO0* is at the address *x,CDH* and the register *RDI3RO1* is at the address *x,CEH*.

The bit structure of the registers *RDIxRO0* is as follows (cf. Figure 3.22 for details):

- *Bits 7–4*: The bits control the output of the LUT 1: the bit 7 enables the tristate buffer for *GOO[5]*, the bit 6 enables the tristate buffer for *GOO[1]*, the bit 5 enables the tristate buffer for *GOE[5]*, and the bit 4 enables the tristate buffer for *GOE[1]*.
- *Bits 3–0*: The bits control the output of the LUT 0: the bit 3 enables the buffer for *GOO[4]*, the bit 2 enables the buffer for *GOO[0]*, the bit 1 enables the buffer for *GOE[4]*, and the bit 0 enables the buffer for *GOE[0]*.

The bit structure of the registers *RDIxRO1* is as follows:

- *Bits 7–4*:The bits control the output of LUT 3: the bit 7 enables the tri-state buffer for *GOO[7]*, the bit 6 enables the tri-state buffer for *GOO[3]*, the bit 5 enables the tri-state buffer for *GOE[7]*, and the bit 4 enables the tri-state buffer for *GOE[3]*.
- *Bits 3–0*:The bits control the output of LUT 2: the bit 3 enables the tri-state buffer for *GOO[6]*, the bit 2 enables the tri-state buffer for *GOO[2]*, the bit 1 enables the tri-state buffer for *GOE[6]*, and the bit 0 enables the tri-state buffer for *GOE[2]*.

3. *Analog programmable interconnect and global bus*: The programmable analog blocks are interconnected to each other, to the I/O pins, and to the digital blocks by three types of interconnect structures: (i) local programmable interconnect, (ii) the input part of the global analog bus, and (iii) the output part of the global analog bus. The three structures are detailed next.

- *Local programmable interconnect*: This interconnect is used to link several programmable analog blocks to form complex analog networks. This interconnect is either between neighboring blocks or to some predefined reference voltages and ground. The precise structure of the local interconnect is different for continuous-time blocks and for switched capacitor blocks.

  The programmable interconnect for the continuous-time analog blocks has three types of connection: (i) *NMUX connections* that connect the negative inputs of the continuous-time analog blocks, (ii) *PMUX connections* for the positive inputs of the blocks, and (iii) *RBotMux* for implementing certain analog functionalities, for example instrumentation amplifiers. The local connections are specific for each analog block. Figure 3.26 shows the interconnect patterns for the three structures. Blocks *ACBxx* are continuous-time analog blocks, and Blocks *ASCxx* and *ASDxx* are switched capacitor blocks.

  Figure 3.26(a) shows the *NMUX* connections, Figure 3.26(b) the *PMUX* connections, and Figure 3.26(c) the *RBotMux* interconnect. Note that each continuous-time block is connected to one of their continuous-time neighbors, and to the switched capacitor blocks of the second row. Also, the output of each block can be fed back to its (positive or negative) input. In addition, the negative inputs of each block can be linked to analog ground, to a high reference voltage (voltage *RefHi* in the figure), to a low reference voltage (voltage *RefLo* in the figure) and to a pin. Besides the connections to the neighbors illustrated in the figure and the feedback connections,

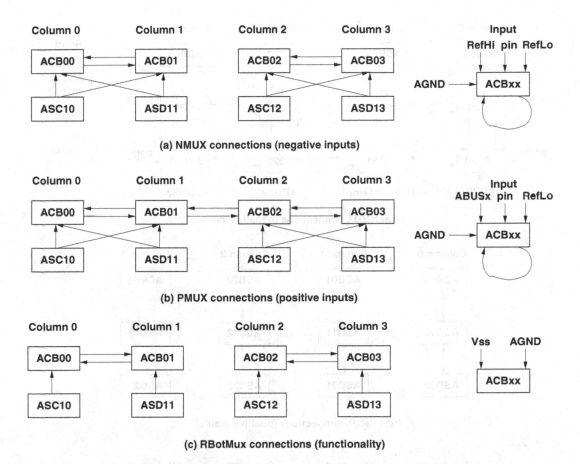

(a) NMUX connections (negative inputs)

(b) PMUX connections (positive inputs)

(c) RBotMux connections (functionality)

Figure 3.26: Local programmable interconnect for the continuous-time analog blocks [5].

the positive inputs can also be linked to the analog ground, and to the reference *RefLo* for the edge blocks, to the analog bus, and to a pin.

Figure 3.27 summarizes the possible interconnect structures for switched capacitor blocks.

- *Global analog bus input*: Figure 3.28 presents the programmable input port connections to the programmable analog blocks. The inputs are selected by the four "four-to-one" multiplexer circuits *ACMx*, one for each column. The two select bits *ACIx* for multiplexer *ACMx* are part of the control register *AMX_IN* of the architecture. In addition, columns *1* and *2* have the possibility to select their inputs either from their own multiplexer *ACMx*, or from the multiplexer of the neighboring columns. This functionality is decided by two "two-to-one" multiplexers *ACx* controlled by register *ABF_CR0*. The multiplexer *AC1* - controlled by signal *ACOL1MUX*, selects the input to Column *1* to be either the output of the multiplexer *ACM0* or that of the multiplexer *ACM1*. Similarly, the multiplexer *AC12* - controlled by signal *ACOL2MUX*, fixes the input to Column *2* to be either the multiplexer output *ACM2*, or that of the multiplexer *ACM1*.

Table 3.8 summarizes the programming of the register *AMX_IN* (Analog Input Select Register). The register is located at the physical address *0,60H*. Bits 7-6 select the

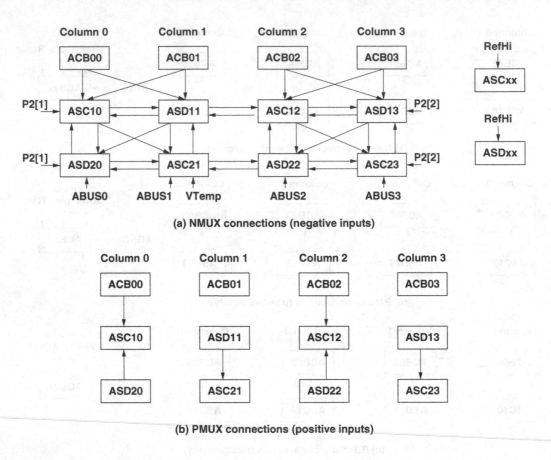

(a) NMUX connections (negative inputs)

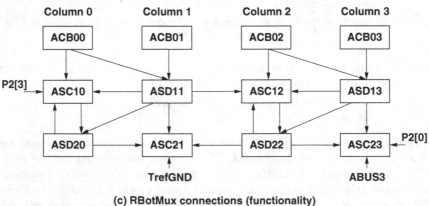

(b) PMUX connections (positive inputs)

(c) RBotMux connections (functionality)

Figure 3.27: Local programmable interconnect for the switched capacitor analog blocks [5].

input to the multiplexer *ACM3*, bits 5-4 the input to the multiplexer *ACM2*, bits 3-2 that of the multiplexer *ACM1*, and bits 1-0 the input to the multiplexer *ACM0*. For example, if bits 7-6 of the register are set to the value 00, then the pin 0 of port 0 is selected as the output of the multiplexer *ACM3*. The selected output becomes available as input to the analog blocks on the third column. If bits 7-6 are configured as 01, then pin 2 of port 0 is the selected input of the multiplexer *ACM3*. Similarly,

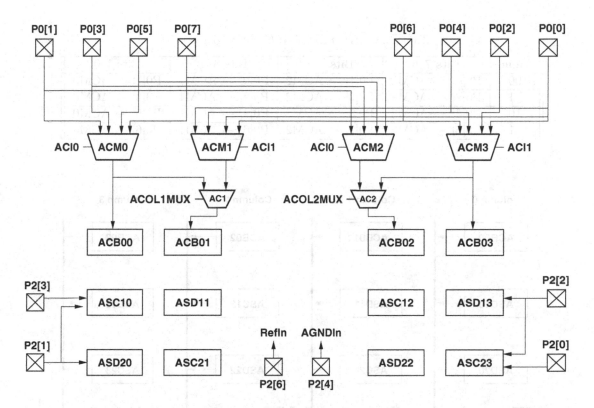

Figure 3.28: Input port connections for the programmable analog blocks [5].

the value 10 picks the pin 4 of port 0, and the value 11 chooses the pin 6 of the port 0 as output of the multiplexer.

The control signals *ACOLxMUX* of the two two-to-one multiplexer circuits going to the analog columns one and two are configured by the *ABF_CR0* register at address *1,62H*. Bit 7 of the register implements select signal *ACol1Mux*, and bit 6 corresponds to the signal *ACol2Mux*. If bit 7 is 0, then the input to column 1 is the output of the multiplexer *ACM1*. Otherwise, the input to the column is the output of the circuit *ACM0*. Similarly, if bit 6 is 0, then the output of multiplexer *ACM2* becomes the input to column 2, otherwise the *ACM3* output is the selected input.

- *Global analog bus output*: Figure 3.29 illustrates the interconnect structure to the output ports of a chip. This bus is also called *analog output bus* (*ABUS*). All analog blocks on a column are linked to a dedicated global interconnect that is linked by an output buffer to the pins of the chip. The output buffer provides the required driving capabilities. For example, Column 0 is connected to pin 3 of the GPIO port 0, Column 1 to pin 5 of port 0, Column 2 to pin 4 of port 0, and Column 3 to pin 2 of port 0.

  The interconnect of a column can be driven by any of the analog blocks on that column. However, the application developer must verify that only one circuit drives the column interconnect at any time. The analog output bus is programmed by setting the control register, *ABF_CR0* (Analog Output Buffer Control Register 0), at address *1,62H*:

  – *Bit ABUF1EN*: If bit 5 of the register is set to 1, then the analog output buffer for Column 1 is enabled. The buffer is disabled if the bit is to 0.

Table 3.8: Structure of *AMX_IN* register [5].

Value	Bits 7–6	Bits 5–4	Bits 3–2	Bits 1–0
00	P0[0] → ACM3	P0[1] → ACM2	P0[0] → ACM1	P0[1] → ACM0
01	P0[2] → ACM3	P0[3] → ACM2	P0[2] → ACM1	P0[3] → ACM0
10	P0[4] → ACM3	P0[5] → ACM2	P0[4] → ACM1	P0[5] → ACM0
11	P0[6] → ACM3	P0[7] → ACM2	P0[6] → ACM1	P0[6] → ACM0

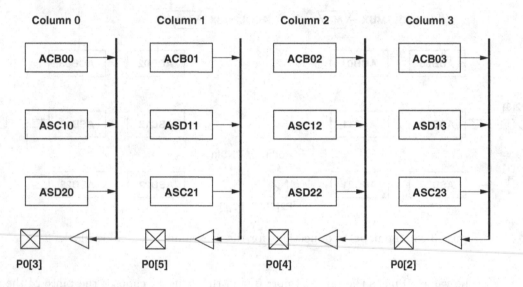

Figure 3.29: Output port connections for the programmable analog blocks [5].

- *Bit ABUF2EN*: The bit 4 set to 1 enables the analog output buffer for Column 2, and disables the buffer if the bit is to 0.

- *Bit ABUF0EN*: Setting bit 3 of the register to 1 enables the buffer for Column 0, and programming the bit as 0 disables the buffer.

- *Bit ABUF3EN*: The bit 2 set to 1 enables the analog output buffer for Column 3, and disables the buffer if the bit is set to 0.

- *Bypass*: If the bit is set to 1, then the positive input of the amplifier in a re-configurable analog block is connected directly to the output. In this case, the operational amplifier of the block must be disabled also. The bypass connection is disabled by programming the bit as 0.

- *PWR*: The bit 0 of the register controls the power level of all output buffers: if it is programmed as 0, then the power level is low, and the power level is high if the bit is configured as 1.

3.5   System Clocks

The clocking of the modules in PSoC can use nine different clocks, some of which have programmable values.

PSoC's clocking subsystem is shown in Figure 3.30. The architecture can use three different clocking sources: two internal sources, the internal main oscillator (IMO) and the internal low speed oscillator (ILO), and one external clocking source available from pin 4 of port 1.

- The IMO is disabled (if the external clock is used) by setting $IMODIS$ (bit 1) of the register OSC_CR2 to 1.

- The frequency of the external clock can be between 1 MHz and 24 MHz. The external clocking source is enabled by setting $EXTCLKGEN$ (bit 2) of the OSC_CR2 register located at address $1,E2H$ to 1.

Depending on the required precision, the internal main oscillator can operate in two modes:

- *Moderate precision IMO*: It is based on a 24 MHz internal clock, if the required precision is not extremely high. In this case, the clock frequency can vary with the temperature as much as $+/-$ 2.5%. For some applications (e.g., analog-to-digital converters) this clock frequency variation, also called clock jitter, is too high, and the second IMO mode can be used.

- *High-precision IMO*: If high accuracy is needed, then the internal frequency can be locked to the frequency of a high-precision external crystal oscillator (ECO). The ECO has a frequency of 32.768 KHz, and is connected to pins 1 and 0 of port 1. As shown in Figure 3.30,

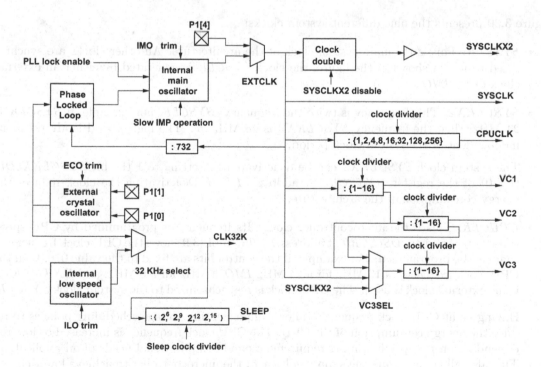

Figure 3.30: PSoC system clocks [5].

the IMO frequency divided by 732 is locked (matched) to the ECO frequency, which gives a more constant IMO frequency of 732×32.768 KHz $= 23.986176$ MHz, hence very close to 24 MHz.

The locking of the two frequencies is achieved by the phase locked loop (PLL) circuit. The enabling of the PLL circuit is controlled PLL Mode (bit 6) of the *OSC_CR0* register located at address *1,E0H*. If this bit is set to 1, then the PLL circuit is enabled. Note that in this case, *EXTCLKGEN* (bit 2) of the *OSC_CR2* register must be 0. If PLL Mode is 0, then the PLL circuit is disabled.

The *lock time* of the PLL circuit (i.e., the time required for the two frequencies to become equal) is either $\approx 10\ ms$, if the PLL circuit operates in the low gain mode, or $\approx 50\ ms$, if the circuit functions in the high gain mode[5]. However, in the high gain mode, the locked frequency varies less. The gain mode of the PLL circuit is set by *PLLGAIN* (bit 7) of the *OSC_CR2* register located at address *1,E2H*. If this bit is 0, then the circuit is in the low gain mode, and in high gain mode [5] if it is set to 1. The sequence of steps to lock the IMO frequency is the following [5]:

1. Enable the ECO by setting the bit *EXTCLKGEN* in register *OSC_CR2*.

2. Set the CPU frequency to 3 MHz or less.

3. Program the PLL gain (bit *PLLGAIN* in register *OSC_CR2*), and enable the PLL circuit by setting the bit *PLLMode* in register *OSC_CR0*.

4. Wait for the locking time.

5. Increase the CPU speed, if needed.

Figure 3.30 presents the nine different system clocks:

- *SYSCLK*: This is the main system clock of the architecture. All other clocks are synchronized to it. As shown in the figure, the clock *SYSCLK* is generated by either an external clock or by *IMO*.

- *SYSCLKX2*: This frequency is twice the frequency *SYSCLK*. For example, if *SYSCLK* is 24 MHz, then the frequency *SYSCLKX2* is 48 MHz [5]. The high period is always 21 ns independent of the system clock period.

  The system clock *SYSCLKX2* can be deactivated by setting to 1 the bit *SYSCLKX2DIS* (bit 0) of the register *OSC_CR2* at address *1,E2H*. Deactivating the clock reduces the energy consumption of the architecture.

- *CPUCLK*: This is the microcontroller clock. Its frequency is programmed by CPU speed (bits 2–0) of register *OSC_CR0* at address *1,E0H*. Table 3.9 shows the CPU clock frequencies that can be programmed. For example, if the control bits are fixed to the value 010, then the CPU clock frequency is 12 MHz for a 24 MHz *IMO*, 3 MHz for a 6 MHz, and *EXTCLK/2*, if the external clock is used. The CPU clock is resynchronized to the system clock *SYSCLK*.

  Having eight CPU clock frequencies allows implementing voltage scheduling policies to reduce the energy consumption of the CPU. The CPU clock frequency is increased or lowered depending on the specific timing requirements and computational needs of an application. This also allows changing the supply voltage of the microcontroller to achieve lower energy consumption.

Table 3.9: Programming the CPU clock [5].

Value	24 MHz IMO	6 MHz IMO	External clock
000	3 MHz	750 kHz	EXTCLK/8
001	6 MHz	1.5 MHz	EXTCLK/4
010	12 MHz	3 MHz	EXTCLK/2
011	24 MHz	6 MHz	EXTCLK
100	1.5 MHz	375 kHz	EXTCLK/16
101	750 kHz	187.5 kHz	EXTCLK/32
110	187.5 kHz	93.7 kHz	EXTCLK/128
111	93.7 kHz	46.9 kHz	EXTCLK/256

- *VC1*: The variable clock one (*VC1*) is obtained by dividing the system clock *SYSCLK*. The divider value can be programmed in the range 1–16 by setting the bits *VC1 divider* (bits 7-4) of register *OSC_CR1* (oscillator control register 1) at address *1,E1H*. Columns 2 and 3 in Table 3.10 present the clock *VC1* frequencies for different values of the bits *VC1* divider. Column 2 is for the case when the architecture uses the internal main oscillator, and column 3 is for the situation when an external clock is employed. For example, if the four bits are set to the value 1100, then the frequency of clock *VC1* is 2.18 MHz for IMO, and *EXTCLK*/11 for an external clock.

Table 3.10: Programming the *VC1* and *VC2* clocks [5].

Value	24 MHz IMO	External Clock	VC2
0000	24 MHz	EXTCLK	VC1
0001	12 MHz	EXTCLK/2	VC1/2
0010	8 MHz	EXTCLK/3	VC1/3
0011	6 MHz	EXTCLK/4	VC1/4
0100	4.8 MHz	EXTCLK/5	VC1/5
0101	4.0 kHz	EXTCLK/6	VC1/6
0110	3.43 kHz	EXTCLK/7	VC1/7
0111	3.0 kHz	EXTCLK/8	VC1/8
1000	2.67 MHz	EXTCLK/9	VC1/9
1001	2.40 MHz	EXTCLK/10	VC1/10
1010	2.18 MHz	EXTCLK/11	VC1/11
1011	2.00 MHz	EXTCLK/12	VC1/12
1100	1.85 MHz	EXTCLK/13	VC1/13
1101	1.87 kHz	EXTCLK/14	VC1/14
1110	1.6 kHz	EXTCLK/15	VC1/15
1111	1.5 kHz	EXTCLK/16	VC1/16

- *VC2*: The variable clock two (*VC2*) is obtained by dividing the clock *VC1* with a value programmable in the range 1–16. The frequency value is selected by setting the bits *VC2 divider* (bits 3–0) of the register *OSC_CR1*. Column 4 in Table 3.10 presents the *VC2* clock frequencies for the bit values shown in column 1. For example, if the bits are 1110 then, the frequency of clock $VC2$ is the frequency of clock $VC1$ divided by 15.

- $VC3$: The variable clock three ($VC3$) is the most flexible among the three variable clocks available in the PSoC architecture. It generates a clock signal by dividing a clock signal selectable among four clocks $SYSCLK$, $SYSCLKX2$, $VC1$, and $VC2$ by a value in the range 1-256. As a result, the frequency of the clock $VC3$ can span a very broad range: it can go as low as about 1 Hz (if an external clock is used) or 366 Hz (if the internal 24 MHz clock is utilized), and as high as 48 MHz.

  The source to the $VC3$ clock is selected by $VC3$ *input select* (bits 1-0) of the OSC_CR4 register located at address *1, DEH*. The source is the $SYSCLK$ clock, if the bits are set to 00, the $VC1$ clock if the bits are 01, the $VC2$ clock if the bits are 10, and the $SYSCLKX2$ clock for 11. The divider value for the clock $VC3$ is set by programming the bits $VC3$ *divider* (bits 7-0) of the register OSC_CR3 at address *1,DFH*. The divider value is equal to the decimal value programmed into register OSC_CR3 plus one.

  The clock $VC3$ can be programmed to generate an interrupt after a time interval equal to the period of $VC3$. To enable the $VC3$ interrupt, the bit GIE, in the CPU_F register, must be set (thus, the CPU has the interrupt system enabled), and the bit $VC3$ (bit 7) of the register INT_MSK0 must be also set. The register INT_MSK0 is at the address *0,E0H*. To clear the $VC3$ interrupt, bit $VC3$ (bit 7) of the register INT_CLR0 at address *0,DAH* must be set to 1.

- $CLK32k$: This clock is produced either by the internal low-speed oscillator or the external crystal oscillator.

- $CLK24M$: This clock is produced by the internal main oscillator of 24 MHz.

- $SLEEP$: The sleep interval can be programmed from 1.95 ms to 1 sec.

Clock signals can drive bus *Global Output Even*. This is controlled by the register OSC_GO_EN at address *1,DDH*. The register bits control the following clock connections to the GOE bits: Bit 7 ($SLPINT$): the connection of the sleep interrupt to bit 7 of GOE, Bit 6 ($VC3$): the connection of $VC3$ to bit 6 of GOE, bit 5 ($VC2$): $VC2$ to bit 5 of GOE, bit 4 ($VC1$): $VC1$ to bit 4 of GOE, bit 3 ($SYSCLKX2$): $SYSCLKX2$ to bit 3 of GOE, bit 2 ($SYSCLK$): the connection of $SYSCLK$ to bit 2 of GOE, bit 1 ($CLK24M$): the 24 MHz clock to bit 1 of GOE, and bit 0 ($CLK32k$): the 32k clock to bit 0 of GOE.

3.6 Conclusions

This chapter has focused on the main hardware and software components that support sensing, processing and control, data communication, and actuation activities in embedded applications. The PSoC architecture is used to illustrate the subsystems of an embedded mixed-signal architecture.

The PSoC architecture is based on two main concepts: the integration as a system-on-chip of the main subsystems in an embedded application, and programmability of the SoC hardware resources. The two capabilities lead to lower cost designs, and shorter development time, including design and testing effort. The PSoC SoC integrates, in addition to the CPU and memory (volatile and nonvolatile memory), also programmable analog and digital circuits, customized digital circuits, interrupt controller, I/O ports, and multiple clocks. Reconfiguration permits repeated programming of the hardware (analog and digital blocks, and ports) to implement new functionality and performance constraints. The PSoC analog blocks can be programmed to provide the necessary functionality, such as signal conditioning, filtering, analog-to-digital conversion and digital-to-analog converters. The PSoC digital blocks can be reconfigured as timers, counters, deadband generators, CRCs, and data communication blocks, for example SPI and UART modules.

The PSoC interrupt subsystem includes the priority encoder, interrupt table, and support for enabling and disabling interrupts. The chapter presents the programming of the registers involved in interrupt management, for exmple the registers *INT_MSK* (for masking individual interrupts), *INT_CLR* (for clearing posted interrupts), and *INT_VC* (for finding the highest priority pending interrupts). The chapter also details the general structure of interrupt service routines for interrupt handling. A case study, the tachometer handling ISR, is presented in the chapter. The ISR measures the time interval of one fan rotation based on the pulses generated by the tachometer. The ISR routine is described as two communicating FSMs. In addition, the ISR routine must receive the tachometer pulses within a predefined interval, which sets a hard timing constraint for the tachometer. The case study defines the data structures needed for handling multiple tachometers. It is interesting to note that the tachometer routine is a hardware-software design, including two programmable digital blocks that produce the interrupts for the microcontroller and implement the required timing constraint.

PSoC incorporates programmable general-purpose input output ports (GPIO). The designer can fix the nature of the ports (input or output ports), driving capabilities, connections to CPU, analog and digital blocks, and the like and associated interrupts. The GPIO attributes are controlled by the registers *PRTxDMx* (for driving capabilities), *PRTxGS* (for selecting a port as either analog or digital), and *PRTxDR* (for data communication with the CPU).

The PSoC system buses interconnect the input/output ports, the microcontroller, analog and digital blocks, and clocks. The data buses include the following types of interconnect.

- An eight-bit global data bus that is shared by all PSoC subsystems.

- The global digital interconnect (GDI) links the input ports to the four-bit buses GIO and GIE, and the four-bit buses GOO and GOE to the output ports. These buses are programmed by the registers *GDI_O_IN*, *GDI_E_IN*, *GDI_O_OU*, and *GDI_E_OU*.

- The row digital interconnect (RDI) connects the GDIs to the rows of the digital blocks array. The RDI bits can be driven besides the GDI bits, also by the broadcast signals coming from other rows, by the analog comparator circuits, clocks, interrupts, and chaining signals from the neighboring digital blocks. RDIs are programmed by the registers *RDI_RI* (for the connections to the buses GIE and GIO), *RDI_RO* (for the connections to the buses GOE and GOO), *RDI_SYN* (for synchronization), and *RDI_IS* (for selecting the broadcast signal). Finally, register *RDI_LT* specifies logic functions based on the RDI bits.

- The interconnect to the analog blocks includes three components: the local programmable interconnect, the input part of the analog bus, and the output part of the analog bus. The local programmable interconnect links the programmable analog blocks by the *NMUX*, *PMUX*, and *RBotMux* connections. More details about these connections are also offered in Chapters 6 and 7. The input part of the analog bus connects to the input pins, the neighboring columns, and the own column of a given analog block. The output part of the analog bus connects the analog blocks to the output pins.

The clock system of the PSoC architecture consists of nine programmable clocks. The three possible clocking sources can be the internal main oscillator, the internal low speed oscillator, and an external clocking source. This allows implementing either moderate precision clocks based on the 24 MHz internal oscillator, or high precision clocks, if an external precision oscillator is used. The nine programmable clocks are *SYSCLK* (24 MHz), *SYSCLKX2* (48 MHz), *CPUCLK* (CPU clock with eight possible frequencies), *CLK32k*, *CLK24M*, *SLEEP*, and the programmable clocks *VC1*, *VC2*, and *VC3*. The clocks are programmed by the registers *OSC_CRx*.

3.7   Recommended Exercises

1. Develop an interrupt service routine, for the sequence detector application in Chapter 2, that manages the serial inputs of the detector. Discuss the highest bit rate of the detector input that can be processed by your solution without losing any data. Compare the performance of your ISR-based solution with the performance of the sampled input design discussed in Chapter 2.

2. Based on Exercise 1, assume a timing constraint that states that consecutive inputs for the sequence detector must be received in a time interval of 1 sec. An error occurs if the timing constraint is not met. Modify your design such that the timing constraint is handled with minimum CPU overhead. Propose a solution for handling the error related to timing constraint violation.

3. Summarize the structure of the PSoC bus subsystem. Identify the resources that can access each of the available buses. Identify the registers that control the characteristics of each bus.

   Assume that the speed of the buses is 12 MHz, and the processor clock speed is 24 MHz. Estimate the execution time required to read one byte from an input port, store it in the data registers of an output port, and then move it to the M8C's SRAM.

4. For 32-bit input data, propose a time-efficient solution of reading the data from the input ports, storing the data to the SRAM, moving the data to the output ports, and moving the data to the registers of the PSoC programmable digital blocks. Identify the data communication parallelism that is available for the PSoC architecture.

5. Develop a set of assembly code routines that configure, *read,* and *write* from any PSoC port. The identity of the port is passed to the routines through the X register. The data and control information are passed through the A register.

6. Capacitive sensors can serve as noncontact switches, and are very useful in harsh environments. Identify the connection mechanism of a capacitive sensor to the PSoC chip. (Additional information can be found in D. Sequine's "Capacitive Switch Scan" Application Note, AN2233a, Cypress, April 14 2004)

7. Develop a set of Assembly code routines that modify the clock frequency of the M8C microcontroller. The new clock frequency is passed to the routines through the A register.

8. For the unsigned byte multiplication algorithm in Chapter 2, estimate the power and energy consumption of the microcontroller, if the clock frequency has the following possible value 512 kHz, 1 MHz, 4 MHz, 12 MHz, and 24 MHz. The microcontroller power consumption is $P = A f$, parameter A being a constant. Interpret the results.

Bibliography

[1] PSoC Designer: Assembly Language User Guide, *Spec. #38-12004*, Cypress Microsystems, December 8 2003.

[2] PSoC Designer: C Language Compiler User Guide, *Document #38-12001 Rev.*E*, Cypress Semiconductor, 2005.

[3] PSoC EXPRESS, Version 2.0, Cypress Semiconductor, 2006.

[4] PSoC EXPRESS, *Driver Author Guide*, Version 2.0, Cypress Semiconductor, 2006, available at http://www.cypress.com.

[5] PSoC Mixed Signal Array, Technical Reference Manual, *Document No. PSoC TRM 1.21*, Cypress Semiconductor Corporation, 2005.

[6] D. Cooper, Implementing Inter-Device Communications with PSoC Express, Application Note AN2351, Cypress Semiconductor, 2006.

[7] A. Deutsch, P. Coteus, G. Kopcsay, H. Smith, C. Surovic, B. Krauter, D. Edelstein, P. Restle, On-chip wiring design challenges for gigahertz operation, *Proceedings of the IEEE*, 89, (4), pp. 529–555, April 2001.

[8] R. Ho, K. Mai, M. Horowitz, The future of wires, *Proceedings of the IEEE*, 89, (4), pp. 490–504, April 2001.

[9] D. Lewis, *Fundamentals of Embedded Software. Where C and Assembly Meet*, Upper Saddle River, NJ: Prentice Hall, 2002.

[10] J. Rabaey, A. Chandrakasan, B. Nikolic, *Digital Integrated Circuits: A Design Perspective*, Upper Saddle River, NJ: Prentice Hall, Second edition, 2003.

[11] D. Sylvester, K. Keutzer, "Impact of small process geometries on microarchitectures in systems on a chip", *Proceedings of the IEEE*, 89, 4, pp. 490–504, April 2001.

Chapter 4

Performance Improvement by Customization

The chapter focuses on design methods for optimizing system performance by customizing the architecture to the application's requirements. The basic concepts are illustrated by examples that employ PSoC's digital programmable and customized blocks.

Performance-driven customization involves developing optimized hardware circuits and software routines for the performance-critical modules of an application. A module is performance-critical if it has a significant effect on the global performance of the implementation. Customization explores performance–cost tradeoffs for the implementations of the critical modules.

This chapter presents a design methodology for reducing the execution time of algorithms executed on an architecture with one general-purpose processor and coprocessors that can be shared by several modules of an application. This architecture is popular for many embedded applications. The methodology consists of the following steps: specification, profiling, identification of the performance-critical blocks, functional partitioning, hardware–software partitioning, hardware resource allocation, mapping of the performance-critical modules to hardware, and scheduling of the modules that share the same hardware resource.

PSoC's support for customization and several illustrative design examples for performance-driven customization are also discussed. PSoC includes programmable digital blocks and blocks with dedicated functionality. Blocks can be programmable to function as timers, counters, dead-band circuits, and CRCs.

This chapter has the following structure:

- Section 1 introduces the concept of performance improvement by application-specific architecture customization.

- Section 2 presents the design methodology for customization.

- Section 3 discusses PSoC's programmable digital blocks.

- Section 4 summarizes PSoC's blocks with dedicated functionality.

- Section 5 presents concluding remarks.

A. Doboli, E.H. Currie, *Introduction to Mixed-Signal, Embedded Design*,
DOI 10.1007/978-1-4419-7446-4_4, © Springer Science+Business Media, LLC 2011

4.1   Introduction to Application-Specific Customization

The performance of an embedded system, for example, execution speed and power consumption, can be substantially improved by customizing the system's architecture to fit the application [4, 7, 9, 10, 19]. Customization involves optimizing the modules for the performance-critical parts of the application (e.g., blocks and subroutines) including the use of customized hardware circuits and software routines.

As a rule of thumb, a block or subroutine is defined as "critical" with respect to a system's overall performance attribute P, if P changes significantly with the modification of an attribute S of the block. Or, the sensitivity of P with respect to S is very large.

For example, if the attribute P is the total execution time of a system and S is the execution time of a block, the block is critical if it involves a large amount of data processing, and/or data communications with other subroutines. (More precise descriptions of performance criticality are given in Subsection 4.2.1 of the chapter.) The execution speed of an embedded system can often be increased by employing customized digital circuits for the critical parts of the application. Such circuits can be made much faster than corresponding software subroutines, because they do not have to follow the instruction cycle of a general-purpose processor (more precisely, the customized circuits do not implement the fetch-decode-execute-store instruction cycle), or equivalently, require fewer clock cycles for execution. In some applications, it is also possible to exploit the parallelism of the subroutine and to execute multiple operations simultaneously. In this case, the tradeoff is between the design cost and execution speed. The latter can often be increased substantially by relying upon dedicated hardware circuits for the performance-critical parts.

In broad terms, customizing an embedded system implementation involves the following tasks: (i) finding the performance-critical parts in the system (i.e., profiling the system), (ii) selecting the critical parts that should be implemented in hardware, (iii) determining the nature and structure of the hardware circuits that can be used to improve system performance, and (iv) implementing the overall system. The design tasks that correspond to the four generic activities depend on the specific nature of the embedded system architecture, as follows:

- *Finding the performance-critical parts of a system* requires that the designer estimate the performance characteristics of each part (e.g., blocks and subroutines) of the system, and the contribution of that part to the overall system performance. Improving the critical parts performance can often significantly enhance overall performance.

  For example, if the execution speed of the system is the primary metric of interest, the designer must determine, for each of the blocks and subroutines, the number of clock cycles needed to execute and communicate the data. In addition, it may be of interest to determine the amount of memory needed to store the data variables and to execute the code (e.g., the stack memory). In the literature [7, 9, 11] this step is referred to as *profiling*.

- *Analyzing the suitability for hardware implementation and deciding on the nature of the hardware to be employed*: The main goal is to partition the performance-critical parts into the parts that are to be implemented in hardware, and those that are to be implemented in software. The latter are executed by the general-purpose processor. The designer must estimate the resulting system performance and cost tradeoffs for each performance-critical part that is to be implemented in hardware. This step also determines the main characteristics of the hardware used, for example the structure of the circuits, the amount of resource required, and the mapping of the performance-critical parts to the physical circuits.

PSoC provides two kinds of digital blocks that allow implementation of application-specific hardware: (1) Programmable digital blocks that are programmed to implement the basic functions, such as timers, counters, deadband generators, and cyclic redundancy clocks [1]. (These blocks are discussed in Section 4.3.) (2) *Specialized digital blocks* provide dedicated functionality, such as multiply accumulate (MAC), decimator, and watchdog timers [1].

- *Implementing the overall system*: This step includes the tasks of designing the hardware circuits, developing the software routines (including the firmware routines for controlling and accessing the hardware), and integrating the entire system.

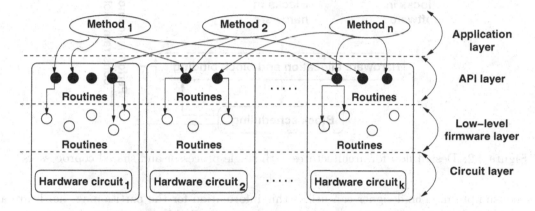

Figure 4.1: Hardware module implementation for reusability.

To improve design reusability, the customized hardware modules are based on a three-layer approach:

- The *circuit layer* includes the hardware used for the critical block, or subroutine.

- The *low-level firmware layer* incorporates the software routines, including interrupt service routines and drivers, that configure and operate the physical blocks. This layer utilizes the physical addresses of the blocks' control, status, and data registers, to implement functionality. Even if two hardware blocks have the same functionality, their firmware routines might actually differ somewhat, because the two blocks may have different physical implementations.

- The *high-level firmware (API) layer* involves the interfacing routines that are used in the control and processing routines of an embedded system to operate the hardware blocks. The hardware is described at an abstract level using symbolic identifiers and abstract data, and without involving any physical details about the hardware. These routines rely on the low-level firmware layer for actually interacting with the hardware.

Figure 4.1 illustrates the three-layer implementation of modules. This figure is based on the assumption that different implementation styles are available for a given functional block. Alternative hardware circuits, shown as blocks (i.e., *Hardware circuit$_i$*) can be used for implementing the block, with specific firmware and API routines existing for each circuit.

High-level firmware layer (API) methods provide an abstract interface to the blocks, hiding all of their implementation details. These methods are called without the designer knowing all the implementation details of the method. Only the precise functionality of the method, and the amount of resource required for its implementation must be known (e.g., hardware and memory) and the performance it delivers in terms of execution time and power

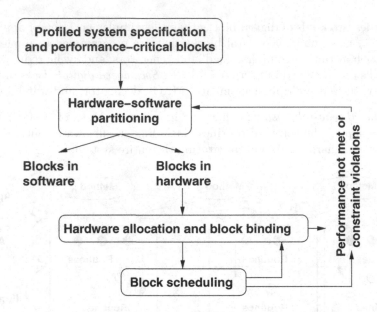

Figure 4.2: Design flow for architectures with single processor and shared coprocessors.

consumption. The designer can utilize this information for the partitioning, hardware allocation, and mapping steps to decide which specific circuit must be used in the system implementation to achieve the required performance. (These tasks are discussed in more detail in Subsection 4.2.2.)

The actual design tasks depend on the architecture employed by the embedded application:

- *Architectures with a single general-purpose processor and associated coprocessors*: The hardware used for the selected performance-critical blocks and the subroutines function as a coprocessor, and the functionality handled in the software domain is executed by the general-purpose processor, which also controls the coprocessor. Each of the selected critical parts is implemented as a dedicated circuit, and therefore, there is no sharing of hardware.

  This situation requires a design procedure that partitions the blocks and subroutines into those that are implemented in hardware circuits, and those that are to be implemented in software. This task is called *hardware–software partitioning* [7, 9, 10]. If the same block, or subroutine, can be implemented using alternative hardware circuits (e.g., different adder circuits, multiplier blocks, etc.) then a second task called *hardware resource allocation* must be performed to identify the nature of the hardware circuits that offer the best performance–cost tradeoff.

- *Architectures that have one general-purpose processor and shared coprocessors*: Although this system has only one general-purpose processor, the same hardware circuits can be shared to implement multiple, performance-critical subroutines, or blocks. In this case, the processor controls the shared coprocessors. Compared to the previously discussed architectures that did not allow sharing of the coprocessors, the cost of the system implementation can be significantly lower due to the hardware sharing, and the performance may be only marginally degraded as a result of the additional overhead introduced by sharing.

  In addition to the hardware–software partitioning, and hardware resource allocation tasks, the mapping (binding) of the critical blocks and subroutines to the hardware, and the

```
for (i = 0; i < SZ; i++)
    for (j = 0; j < SZ; j++) {
        aux = 0;
        for (m = -L; m <= L; m++)
            for (n = -L; n <= L; n++)
                aux += a[m][n] * F[i+m][j+n];
        G[i][j] = aux;
    }
```

Figure 4.3: Example of a data processing intensive application.

execution scheduling for the critical blocks that share hardware is also required. The corresponding design flow is summarized in Figure 4.2, and detailed in Subsection 4.2.2.

- *Architectures that have multiple general-purpose processors plus shared coprocessors*: This is the most general case. There may be no processor that maintains overall control. The importance of such architectures for computationally intensive applications is continuing to increase, for example in multimedia, image processing, and telecommunications.

  In this case, the design process involves not only the tasks of hardware–software partitioning, hardware resource allocation and mapping/scheduling of the customized hardware that are specific to the previous architectures, but also the tasks of allocating the general-purpose processors, mapping the software subroutines to processors, allocating the interconnect structure between the processors, mapping the data communications to the interconnect structures, and scheduling the software routines and data communications.

4.2 Design Methodology for Architecture Customization

In this section, the discussion focuses on the design methodology for embedded applications that utilize architectures with a single general-purpose processor and shared coprocessors for the performance-critical blocks and subroutines.

Figure 4.2 illustrates the design methodology for improving the execution time of embedded systems by customizing the system's architecture. This methodology is based on the following tasks: (1) specification and profiling, (2) hardware–software partitioning, (3) hardware resource allocation, (4) operation binding, and (5) operation scheduling.

4.2.1 System Specification and Profiling

The goal of the profiling step is to characterize the execution times, in terms of clock cycles and memory size requirements, of the blocks and subroutines for both data and program in the application specification, that is algorithm. Then, as shown in Figure 4.2, this information can be used to guide the design methodology. This section uses an example to illustrate the profiling step.

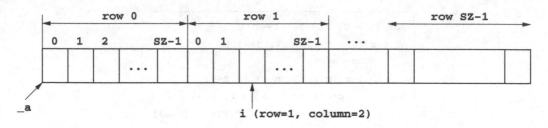

Figure 4.4: Organization of arrays in the memory.

Figure 4.3 presents a typical example of an iterative algorithm used in digital signal processing (DSP) [4, 8, 15] applications that share the following defining characteristics:

1. The processing is *data dominated*, and involves a large number of multiplication/addition computations and data transfers to/from the data arrays.

2. The number of iterations for each loop is known and constant.

3. In addition, the iterations of inner ("for") loop instructions are either uncorrelated or loosely correlated, because there are no interdependencies between iterations, and therefore the inner loop can be executed in parallel, provided that the architecture has the necessary hardware capabilities.

These three features of DSP algorithms encourage two kinds of performance optimizations of the system:

1. Optimization of the operators, individual instructions, and instruction sequences inside the loop bodies.

2. Optimization of the processing, and data accesses, across different loop iterations.

Both types of optimizations use the data collected in the profiling step.

Before discussing the parts that form the specification in Figure 4.3, the organization of the array data in the memory has to be detailed. This is important because accessing the data stored in the two-dimensional arrays involves significant overhead resulting from computation of the positions of the elements. Figure 4.4 illustrates the organization of the data arrays. For simplicity, it is assumed that the data and memory word size are the same, for example if PSoC is the target architecture, then the array elements are each one byte. The elements in the two-dimensional arrays a, F, and G in the specification are stored in consecutive memory locations, one row after another, as shown. The element $a[m][n]$ is located at the following entry i of the linear sequence:

$$i = m \times SZ + n \tag{4.1}$$

where the constant SZ is the length of a row. For example, the memory entry for the element $a[1][2]$ is at index $i = 1 \times SZ + 2 = SZ + 2$, as shown in Figure 4.4.

Profiling involves the following steps:

- *Producing the assembly code level description*: The application's assembly code is obtained by compiling (i.e., translating) the C program into assembly code. The designer can then

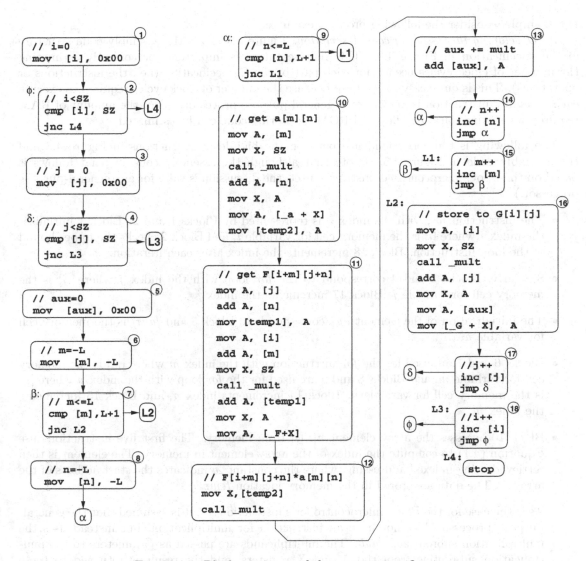

Figure 4.5: Block structure of the system specification.

determine the execution time and the program/data memory requirements by summing the number of execution clock cycles and memory bytes, respectively, required to execute each assembly code instruction.

- *Functional partitioning*: This is a hardware-related reorganization of the specification. The assembly code instructions are grouped into blocks, such that each block corresponds to a single instruction in the initial C program, and each block can be a hardware circuit with precise functionality. This is important in determining the impact of a dedicated hardware circuit with respect to performance improvement.

- *Finding the performance-criticality of the blocks*: This requires the designer to calculate the performance attributes for each block in the specification, and for the entire system. Based on this analysis, the designer can determine the performance-criticality of the blocks.

For example, consider the following profiling example.

A. Developing the assembly code description. Figure 4.5 shows the assembly code structure of the specification in Figure 4.3. This representation is important for correctly estimating the number of clock cycles needed for each part of the C specification (i.e., the instructions or operations). This is quite easy to compute because the number of clock cycles required to execute each assembly instruction is known for a general-purpose processor. Also, the amount of RAM required for storing data, and flash/EPROM for storing code can be estimated.

The following is the correspondence between the high-level instructions in Figure 4.3 and the assembly code in Figure 4.5. (Note that although the assembly code in what follows is based on the PSoC microcontroller instruction set, the discussion is valid for any general-purpose processor.)

- The instruction *for*, with the index i, is represented by Blocks 1 and 2: Block 1 initializes the index i where $[i]$ is the memory cell for variable i, and Block 2 checks the upper bound of the loop instruction. Block 18 increments the index after each iteration.

- Similarly, Blocks 3 and 4 correspond to the *for* loop with the index j where $[j]$ is the memory cell for variable j. Block 17 increments the index j.

- The initialization of the element *aux* corresponds to Block 5 and *[aux]* is the memory cell for variable *aux*.

- Blocks 6 and 7 are used for the *for* instruction with the index m where $[m]$ is the memory cell for variable m, and Blocks 8 and 9 are used for the *for* loop with the index n where $[n]$ is the memory cell for variable n. Block 14 increments index n, and Block 15 increments the index m.

- Block 10 accesses the array element $a[i][j]$ from memory. The first five instructions use Equation (4.1) to compute the index of the array element in memory. The element is then retrieved using indexed addressing, where the constant $\_a$ indicates the start address of the array a. The data are stored in the memory location *temp2*.

  As is the case for the *PSoC* microcontroller's instruction set, it is assumed that the general-purpose processor does not have an instruction for multiplication, but instead uses the multiplication subroutine, $\_mult$. The multiplicands are passed as parameters to the multiplication subroutine from the A and X registers, and the result is returned to these registers.

- Block 11 represents the assembly code for accessing the array element $F[i+m][j+n]$. Instructions 1-8 compute the index of the element in memory. Then, the value is accessed using indexed addressing with the constant $\_F$ pointing to the start address of the array F.

- Block 12 multiplies $a[i][j]$ with $F[i+m][j+n]$, and the result is added to the value of the element *aux* by Block 13.

- Block 16 assigns the value to the element $G[i][j]$. The first four instructions compute the index of the element, and the rest of the instructions assign the value of the variable *aux* to the element $G[i][j]$.

B. Functional partitioning for hardware-oriented, specification reorganization. The initial block structure of the specification is restructured, so that all the instructions in a block can be implemented using only one hardware module with a precisely defined functionality. This is

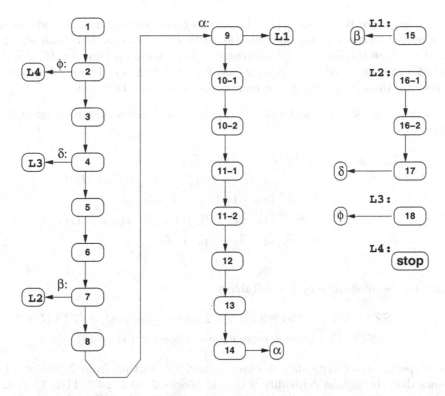

Figure 4.6: Refined block structure of the system specification.

important in determining the performance improvement that results from developing, and then utilizing, a customized hardware circuit for a set of assembly code instructions. For example, the first five instructions of Block 10, in Figure 4.5, compute the index of an array element in memory, and instructions 6 and 7 access memory to retrieve and store the array element, in a processor register. From an execution point of view, the first five instructions use the hardware circuits for computation (i.e., for multiplication and addition), and instructions 6 and 7 employ the memory circuits, and the address, data, and control signals of the memory modules. Hence, improving the execution speed of Block 10 must take into consideration the hardware required for computing the address, and the hardware for data access from memory. To determine the performance improvement offered by each customized hardware module, it is important to split Block 10 into two subblocks, such that each sub-block can be implemented by hardware with dedicated functionality.

The block structure in Figure 4.5 is reorganized as shown in Figure 4.6 to reflect the types of circuits used in processing and communication. Block 10 was split into two sub-Blocks 10-1 and 10-2. The first nine instructions of Block 11 compute the index of an array element, and were organized as the sub-Block 11-1, while the remaining instruction, defined as sub-Block 11-2 accessed memory. Similarly, Block 16 was split into the sub-Block 16-1 that computes an array element index, and Block 16-2 for reading an array element from memory.

C. Finding the performance-criticality of the blocks. Table 4.1 shows the profiling data for the refined block structure shown in Figure 4.6. The first column shows the block number, the second the number of required clock cycles, the third the number of bytes in nonvolatile memory (e.g., flash or EPROM) required to store the block's code, and the fourth the number of internal registers storing data. For Blocks 10-1, 11-1, 12, and 16-1, the profiling data include the number of clock cycles required to execute the multiplication subroutine *mult*, and the number of EPROM bytes required to store the code of the subroutine. The execution time for Blocks 10-2, 11-2, and 16-2 includes the time needed to transfer data from the memory. These values are expressed as symbols because their actual values depend on the implementation style.

The execution-time criticality of the block is computed starting from the total execution time of the application:

$$
\begin{aligned}
T_{overall} \quad = \quad & SZ^2 \left(2L+1\right)^2 (T_9 + T_{10-1} \\
& + T_{10-2} + T_{11-1} + T_{11-2} + T_{12} + T_{13} + T_{14}) \\
& + SZ^2 \left(2L+1\right)(T_7 + T_8 + T_{15}) \\
& + SZ^2 \left(T_4 + T_5 + T_6 + T_{16-1} + T_{16-2} + T_{17}\right) \\
& + SZ \left(T_2 + T_3 + T_{18}\right) + T_1
\end{aligned}
$$

$$(4.2)$$

$$(4.3)$$

or, considering the profiled clock cycles in Table 4.1:

$$
\begin{aligned}
T_{overall} \quad = \quad & SZ^2 \left(2L+1\right)^2 (136 + 3 \; mult + 2 \; memory \; access) + SZ^2 \left(2L+1\right)(33) \quad (4.4) \\
& + SZ^2 \left(76 + mult + memory \; access\right) + SZ \left(33\right) + (8)
\end{aligned}
$$

The blocks' performance-criticality is characterized by starting from Equation (4.4). This equation shows that the highest criticality is that of Blocks 9, 10-1, 10-2, 11-1, 11-2, 12, 13, and 14, for which the execution complexity is proportional to the value $SZ^2 (2L+1)^2$. For example, if $SZ = 4$ and $L = 2$, then the execution time for the seven blocks is more than 90% of the total execution time of the system. This percentage increases as the values of the two constants increase.

4.2.2 System Partitioning and Implementation

The goal of the design process is to maximize the performance of the system implementation, for example minimize the system execution time by utilizing customized hardware circuits for the performance-critical blocks. Two kinds of customization optimizations can be achieved in the process: (i) optimization of the operators, individual instructions, and instruction sequences of the critical blocks, and (ii) optimization of the data processing and communication pertaining to different critical blocks.

The targeted architecture has one processor, and one coprocessor, that can be shared among multiple performance-critical blocks and subroutines as shown in Figure 4.7. The general-purpose processor executes the blocks, and subroutines, in the software domain. In addition, the coprocessor consists of hardware circuits that implement critical blocks that are mapped to the hardware domain. The processor and coprocessor exchange data via the shared memory. These data can be mapped to multiple memory circuits, so that multiple data accesses can occur in parallel. The interconnect structure provides the needed connectivity between the processing elements and memory.

The first design optimization refers to the execution-time, optimized implementation of the operators, individual instructions, and instruction sequences pertaining to the performance-critical regions of an application. The example in Figure 4.3 is used as an illustration of this optimization. The refined block structure is shown in Figure 4.6, and the profiling data are given in

Table 4.1: Profiling data for the block structure in Figure 4.6.

Block	# Clock Cycles	# EPROM Bytes	# Register
1	8	3	-
2	13	5	-
3	8	3	-
4	13	5	-
5	8	3	-
6	8	3	-
7	13	5	-
8	8	3	-
9	13	5	-
10-1	30 + **mult**	9 + **mult**	2
10-2	5 + **memory access**	4	2
11-1	52 + **mult**	17	2
11-2	**memory access**	2	2
12	17 + **mult**	4	2
13	7	2	1
14	12	4	-
15	12	4	-
16-1	35 + **mult**	11	2
16-2	**memory access**	2	2
17	12	4	-
18	12	4	-

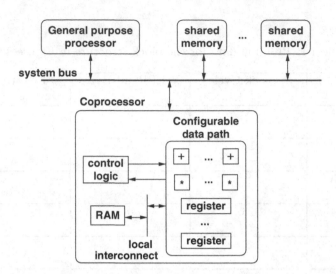

Figure 4.7: Generic architecture consisting of processor and shared coprocessor.

Table 4.1. Blocks 9, 10-1, 10-2, 11-1, 11-2, 12, 13, and 14 are the performance-critical blocks of the application. As shown in Table 4.1, of the eight blocks, the most critical from an execution time point-of-view, are Blocks 10-1, 10-2, 11-1, 11-2, and 12. These blocks are further considered as candidates for implementation as customized hardware circuits.

These five blocks provide three kinds of functionality: Blocks 10-1 and 11-1 compute the index of an array element, Blocks 10-2 and 11-2 access the data memory, and Block 12 multiplies two values. The nature of the operations involved for the three block types, and the kind of corresponding hardware circuits, are different. For example, the blocks for computing the array element indexes involve data transfers between the processor registers, addition operations, and multiplications. The blocks for accessing the data memory involve registers and interconnect to memory. Block 12 requires data registers for the operands, and a multiplication module.

Figure 4.8 shows possible hardware structures for the implementation of the performance-critical blocks:

- Figure 4.8(a) shows the structure of a possible hardware module for Block 12. The structure uses three registers to store the two operands and the result, and a multiplier circuit. In addition to the time T_{mult}, as measured in clock cycles, required to perform the multiplication operation, the circuit also introduces overhead for register loading time, and retrieving the result from the register. The total execution time is given by:

$$T_{12}^{HW} = T_{mult} + 2 \times T_{reg\ load} + T_{reg\ read} \qquad (4.5)$$

- Figure 4.8(b) shows a possible hardware structure for accessing the memory that stores the array elements. The access times for one memory word are cumulatively represented as the time $T_{mem\ read}$ for reading data from memory to a processor register, and the time $T_{mem\ write}$ for writing data from a processor register to memory. Hence,

$$T_{10-2} = T_{11-2} = T_{mem\ read} \qquad (4.6)$$

- Figure 4.8(c) illustrates a possible hardware structure for computing the index of the memory element $a[i][j]$, if the array a is a bidimensional array of size $SZ \times SZ$. This structure

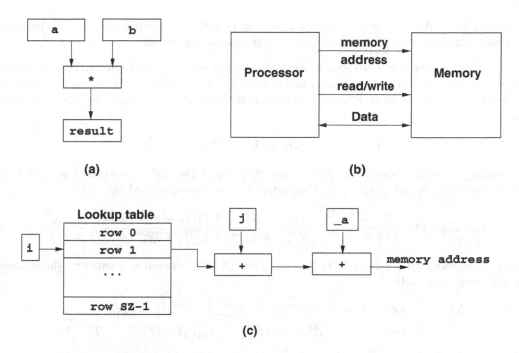

Figure 4.8: Customized hardware circuits for the performance-critical blocks.

is an alternative to the more "intuitive" implementation that uses a multiplier module to compute the starting address of row i, within the array.

Each element of the lookup table stores a pointer to the first element of a row: the first element points to the start of the first row, the second element to the start of the second row, and so on. The contents of the table are known once the array data are allocated in memory. The memory address is found by summing the starting memory address of the array a, the displacement of the row i, and the displacement, within the row i, of the column j. The resulting memory address is placed on the memory address bus. The advantage of the solution is that no multiplications are needed at the penalty of needing extra memory for storing the lookup table.

The total execution time $T_{ad\ comp}$ includes the time to load the registers holding the indexes i and j, the time to read from the lookup table, and the time to perform the two additions:

$$T_{10-1}^{HW} = 2 \times T_{reg\ load} + T_{lookup} + 2 \times T_{adder} \tag{4.7}$$

$$T_{11-1}^{HW} = 4 \times T_{reg\ load} + T_{lookup} + 4 \times T_{adder} \tag{4.8}$$

The information about the block-criticality and the execution time of the corresponding hardware circuits is used to guide the design decisions with respect to (a) hardware–software partitioning, (b) hardware resource allocation, (c) operation binding, and (d) scheduling.

A. Hardware–software partitioning. Hardware–software partitioning is the task of separating the specification blocks, and subroutines, into those that are implemented in the hardware domain (using customized circuits), and those that are executed by the general–purpose processor, so that the performance improvement is maximized, and the customization cost is less than a predefined limit. Alternatively, the scope of hardware–software partitioning can also be defined as cost minimization subject to satisfying the performance constraints.

Each of the performance-critical, customized hardware blocks and subroutines is analyzed to determine the system performance improvement. For example, if Block 11-1, is based on the hardware circuit shown in Figure 4.8(c), then the change in execution time for the system becomes:

$$\Delta T \;\; = \;\; SZ^2 \, (2L \, + \, 1)^2 \, (T_{11-1}^{SW} \, - \, T_{11-1}^{HW}) \tag{4.9}$$

The resulting execution speed-up, defined as the ratio of the software execution time to the hardware–software execution time, for the system can be approximated by:

$$Speedup \;\; \approx \;\; \frac{T_9^{SW} + T_{10-1}^{SW} + T_{10-2}^{SW} + T_{11-1}^{SW} + T_{11-2}^{SW} + T_{12}^{SW} + T_{13}^{SW} + T_{14}^{SW}}{T_9^{SW} + T_{10-1}^{SW} + T_{10-2}^{SW} + T_{11-1}^{HW} + T_{11-2}^{SW} + T_{12}^{SW} + T_{13}^{SW} + T_{14}^{SW}} \tag{4.10}$$

Similarly, if Blocks 10-1, 10-2, 11-1, 11-2, and 12 are all implemented in hardware, the change in execution time is given by:

$$\begin{aligned} \Delta T \;\; = \;\; & SZ^2 \, (2L \, + \, 1)^2 \, [(T_{10-1}^{SW} \, - \, T_{10-1}^{HW}) + (T_{10-2}^{SW} \, - \, T_{10-2}^{HW}) \\ & + (T_{11-1}^{SW} \, - \, T_{11-1}^{HW}) + (T_{11-2}^{SW} \, - \, T_{11-2}^{HW}) + (T_{12}^{SW} \, - \, T_{12}^{HW})] \end{aligned} \tag{4.11}$$

and the resulting speed-up is:

$$Speed - up \;\; \approx \;\; \frac{T_9^{SW} + T_{10-1}^{SW} + T_{10-2}^{SW} + T_{11-1}^{SW} + T_{11-2}^{SW} + T_{12}^{SW} + T_{13}^{SW} + T_{14}^{SW}}{T_9^{SW} + T_{10-1}^{HW} + T_{10-2}^{HW} + T_{11-1}^{HW} + T_{11-2}^{HW} + T_{12}^{HW} + T_{13}^{SW} + T_{14}^{SW}} \tag{4.12}$$

If the cost limit is exceeded, then the design process must consider alternative subsets of the set of performance-critical blocks, and find the subset that gives the best performance improvement, at a cost below the limit. For example, in the case of the five performance-critical blocks, the alternative subsets would be {Block 10-1, Block 11-1, Block 12}, {Block 10-2, Block 11-2, Block 12}, {Block 11-1, Block 11-2, Block 12}, and so on.

Note that in the worst case, the number of subsets that must be analyzed to find the best performance/cost tradeoff can be very large. If the set of performance-critical blocks has the cardinality N, then 2^N subsets must be analyzed. Note that this number grows exponentially with respect to N. To address this issue, blocks are analyzed in decreasing order of the ratio:

$$E_i \;\; = \;\; \frac{\Delta T_i}{\Delta Cost_i} \tag{4.13}$$

where ΔT_i is the difference in the execution time in clock cycles, between the software and hardware implementations of Block i, and $\Delta Cost_i$ is the cost increase due to the customized circuit. The factor E_i describes the performance improvement, per cost unit, that is provided by Block i. Therefore, provided that the cost limit is not exceeded, the design method maps the performance-critical blocks to the hardware domain in decreasing order of E_i. For the five performance critical blocks, this method maps hardware Block 11-1, followed by the Blocks 10-1, 12, 10-2, and 11-2 to the hardware domain.

B. Hardware resource allocation. This step identifies the nature, and number, of hardware circuits that are incorporated into the application-specific coprocessor. This is justified by the fact that there are often many alternative hardware implementations possible for the same

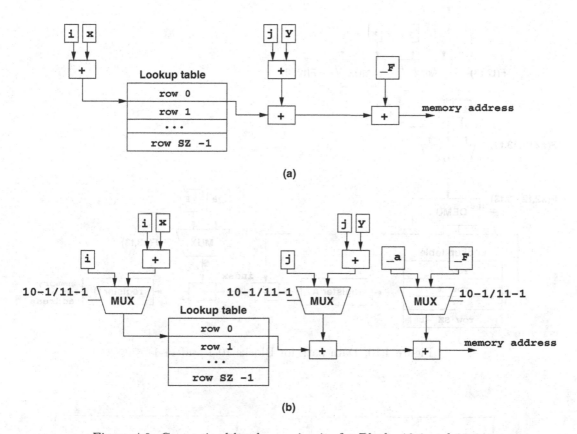

Figure 4.9: Customized hardware circuits for Blocks 10-1 and 11-1.

block functionality. These alternative implementations not only provide different execution times, and consume specific amounts of hardware, but also create distinct opportunities for sharing the hardware circuits by different functional blocks. This is an important way to reduce the total cost.

For example, consider Blocks 10-1 and 11-1 that compute the indexes of elements in two different arrays, as shown in Figures 4.5 and 4.6 In principle, these blocks perform the same computations. However, Block 11-1 also calculates the element indexes by summing the values of four variables. Figure 4.8(c) presents the customized circuit for Block 10-1, and Figure 4.9(a) introduces the circuit for Block 11-1. The only difference between the circuits is the two additional adders in the second circuit. Equations (4.7) and (4.8) express the respective execution times.

The two blocks can be modified in several ways to lower the total implementation cost. Cost can be reduced by sharing a single hardware circuit for the implementation of the two blocks, instead of having separate circuits for each of the blocks. Figure 4.9(b) shows the resulting circuit. Two multiplexer circuits are added to the initial Block 11-1 circuit, so that the correct indexes of the array elements can be selected. The third multiplexer selects the starting address of the array, represented by the constant _a for Block 10-1, and the constant _F for Block 11-1. The selection signals for the multiplexer circuits can be implemented by setting a dedicated bit, if Block 10-1 is executed, and resetting the bit, if Block 11-1 is performed. The implementation

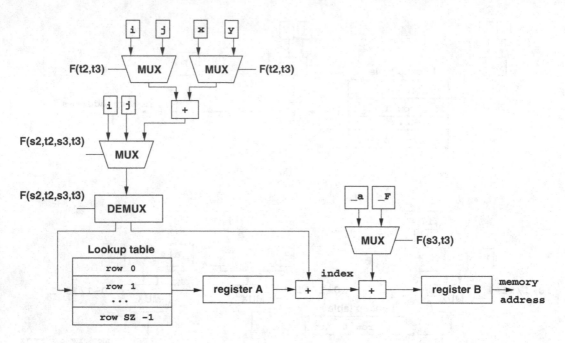

Figure 4.10: Data path for Blocks 10-1 and 11-1.

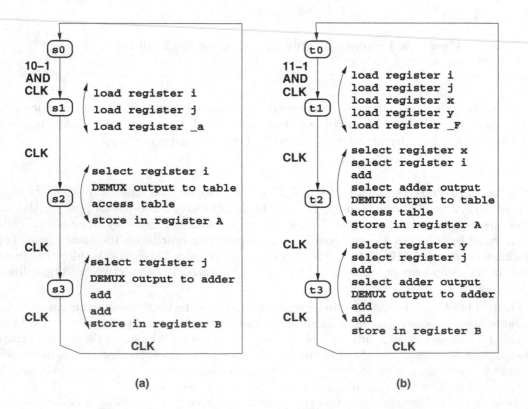

Figure 4.11: Controller circuits for Blocks 10-1 and 11-1.

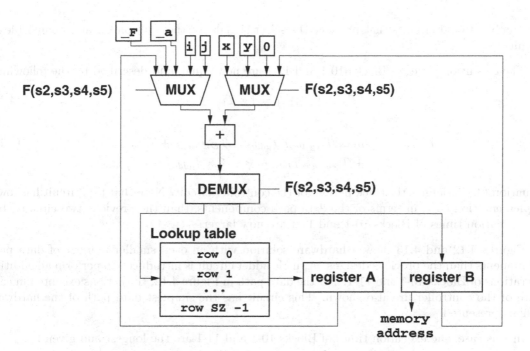

Figure 4.12: Data path for Blocks 10-1 and 11-1.

cost is reduced by the elimination of the customized circuit for Block 10-1. Although, the cost reduction is slightly diminished by the three additional multiplexers.

The tradeoff, in this case, is the increase of the execution times of the two blocks. Assuming that all of the control signals are generated in sequence, the new (worst case) execution times are now:

$$T_{10-1}^{'HW} = 2 \times T_{reg\ load} + T_{lookup} + 2 \times T_{adder} + 3 \times T_{MUX} \tag{4.14}$$

$$T_{11-1}^{'HW} = 4 \times T_{reg\ load} + T_{lookup} + 4 \times T_{adder} + 3 \times T_{MUX} \tag{4.15}$$

(assuming that all control signals are generated in sequence). In the best case, the two execution times increase only by the propagation delay of one multiplexer circuit, if the multiplexer circuits are activated in parallel.

The cost of the hardware circuits in Figure 4.9(b) can be further lowered by sharing the circuits for computing the two indexes of the array elements. The resulting circuit has a data path and controller circuits, which are shown in Figures 4.10 and 4.11. The data path circuit in Figure 4.10 uses the same adder circuit to first sum the values x and i, and then to sum values y and j.

Figure 4.11(a) is the controller for Block 10-1, and Figure 4.11(b) is the controller for Block 11-1. For Block 10-1, in the state $s1$, the circuit loads values into the data path registers. In state $s2$, the circuit accesses the lookup table. The selected value in the lookup table is stored in the A register. Finally, in state $s3$, the following two sums are computed: register $A + DEMUX$ output, and $\_a + index$. The actions corresponding to the controller states of Block 11-1 are indicated in Figure 4.11. The multiplexer and demultiplexer circuits of the data path are controlled by signals that depend on $s2$, $s3$, $t2$, and $t3$. These four signals are 1, if the controller circuit is in the corresponding state, and 0, if it is not. The functions F are the logic functions of the combinational circuits that produce the select signals for the multiplexer and demultiplexer circuits.

The execution times of Blocks 10-1 and 11-1 are now equal, and described by the following:

$$T_{10-1} = T_{11-1} = 4 \times T_{FSM} \tag{4.16}$$

$$\begin{aligned} T_{FSM} = \ & \max\{T_{reg\ load}, T_{adder} + T_{DEMUX} + T_{lookup} \\ & + T_{regload}, T_{DEMUX} + 3 \times T_{adder} + T_{reg\ load}\} \end{aligned} \tag{4.17}$$

Equation (4.17) defines the clock period for the controller circuit. Note that this circuit has lower implementation cost, in terms of the data path components, than the previous two circuits, but the execution times of Blocks 10-1 and 11-1 are now larger.

Figures 4.12 and 4.13 show a hardware solution with an even smaller number of data path components than the previous designs. A single adder circuit is introduced to perform all addition operations of Blocks 10-1 and 11-1. For the data path in Figure 4.12, the actions executed in each state of the controller are also shown. This circuit has the simplest data path of the hardware designs presented.

In this case, the execution times of Blocks 10-1 and 11-1 are the longest and given by:

$$T_{10-1} = T_{11-1} = 6 \times T_{FSM} \tag{4.18}$$

$$\begin{aligned} T_{FSM} = \ & \max\{T_{reg\ load}, T_{adder} + T_{DEMUX} + T_{lookup} \\ & + T_{regload}, T_{DEMUX} + T_{adder} + T_{reg\ load}\} \end{aligned} \tag{4.19}$$

$$\approx \ T_{adder} + T_{DEMUX} + T_{lookup} + T_{regload} \tag{4.20}$$

In addition to the alternatives for sharing hardware modules among the block implementations, another possibility for improving the system's execution time is to consider using faster adder and multiplier circuits. For example, ripple-carry adder circuits have a lower cost than look ahead adder circuits, but their execution time (time delay) is also significantly longer. Discussing efficient circuit structures for different arithmetic and logic operators is beyond the scope of this text. More details can be found in the related literature [4, 8, 15].

C–D. Operation binding/scheduling. In addition to improving the system performance by employing customized hardware for the individual blocks and subroutines, the designer can also consider exploiting the parallelism between operations in the same block, or in different blocks. This further reduces the system execution time, because multiple operations can now be executed in parallel.

Without addressing the operation parallelism of the application, its execution time, shown in Figure 4.3, cannot be reduced below the limit given by:

$$T_{limit} = SZ^2 (2L + 1)^2 (T_9 + T_{10-1} + T_{10-2} + T_{11-1} + T_{11-2} + T_{12} + T_{13} + T_{14}) \tag{4.21}$$

However, if the values of the constants SZ and L are large, this limit may still be very large.

The execution time of the application can be reduced by observing that there are no dependencies between the iterations of the innermost loop, other than all iterations adding a value to

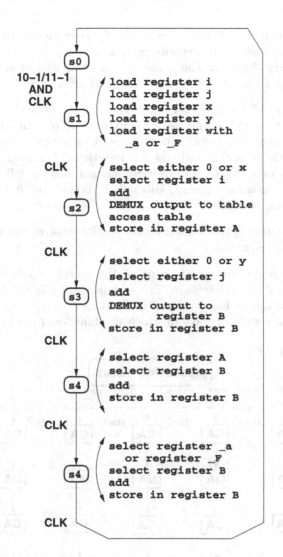

Figure 4.13: Controller for Blocks 10-1 and 11-1.

```
for (i = 0; i < SZ; i++)
   for (j = 0; j < SZ; j++) {
      auxv = 0;
      for (m = -L; m <= L; m++) {
         for (n = -L; n <= L; n++)
            aux[n] = a[m][n] * F[i+m][j+n];
         for (n = -L; n <= L; n++)
            auxv += aux[n];
      }
      G[i][j] = auxv;
   }
```

Figure 4.14: Modified data processing algorithm.

the variable *aux*. This dependency can be removed by introducing the array *aux[2L+1]*, such that the element *n* corresponds to the iteration *n*. The iteration uses this element for storing the product of the two matrix elements. The value *auxv* is computed by summing all the elements of the matrix *aux[2L+1]*. Figure 4.14 shows the modified algorithm.

If the hardware circuits offer support for parallel execution, modifying these algorithms allows the innermost iterations to be executed in parallel. Figure 4.15 presents the block structure of the application, so that the operation parallelism is exposed. The figure assumes that L = 2, hence the inner most loop of the algorithm includes five iterations. For simplicity reasons, the initial performance-critical blocks of the application were merged into the following blocks: Block *CA* computes the address of an array element, Block *MA* accesses the data memory, and Block * multiplies two data values.

In this case, implementing optimized hardware circuits involves two additional steps: operation binding and operation scheduling:

- *Operation binding* is the process of mapping the specification's functional blocks to the hardware resources of the architecture. This step is justified by the fact that the application-specific coprocessor can be based on multiple instances of the same hardware circuits, and therefore there is a variety of ways in which the functional blocks can be mapped to the circuits.

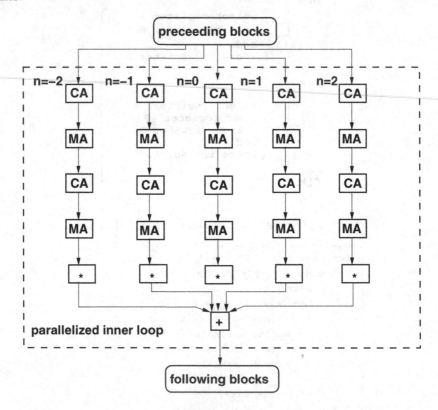

Figure 4.15: Inner loop parallelization for the modified algorithm.

- *Operation scheduling* is the activity of time sequencing the execution of blocks and subroutines mapped to the same hardware circuits.

The goal of operation binding and scheduling is to maximize the performance improvement, by minimizing the execution time of the system.

For the parallelized inner loop of the block structure in Figure 4.15, Figure 4.16 illustrates two instances of operation binding and scheduling. These instances differ in terms of the number of hardware circuits that were allocated in the coprocessor.

In Figure 4.16(a), the architecture included a distinct hardware circuit for each of the CA blocks (computing the index), MA (accessing the memory), and * (for multiplication). Because of resource limitations, only the blocks pertaining to two of the five iterations can be executed simultaneously. The time scheduling of the execution is also shown. For example, in step 0, only Block CA is executed, but in step 1, Blocks MA and CA of two different iterations occur in parallel, because they use different hardware circuits. Hence, the execution of the two iterations ends in only six steps, as compared to ten steps, if the two iterations are executed in sequence. Four iterations are performed in ten steps using parallel execution, as compared to the twenty steps required for serial execution. The total execution time of the five iterations is 15 cycles for the parallel execution, and 25 cycles if the iterations are performed sequentially. Hence, compared to sequential execution, parallel execution provides a relative speed-up of about 25%. Relative speed-up is the ratio:

$$Speedup = \frac{T_{serial} - T_{parallel}}{T_{serial}} \times 100 \ [\%] \qquad (4.22)$$

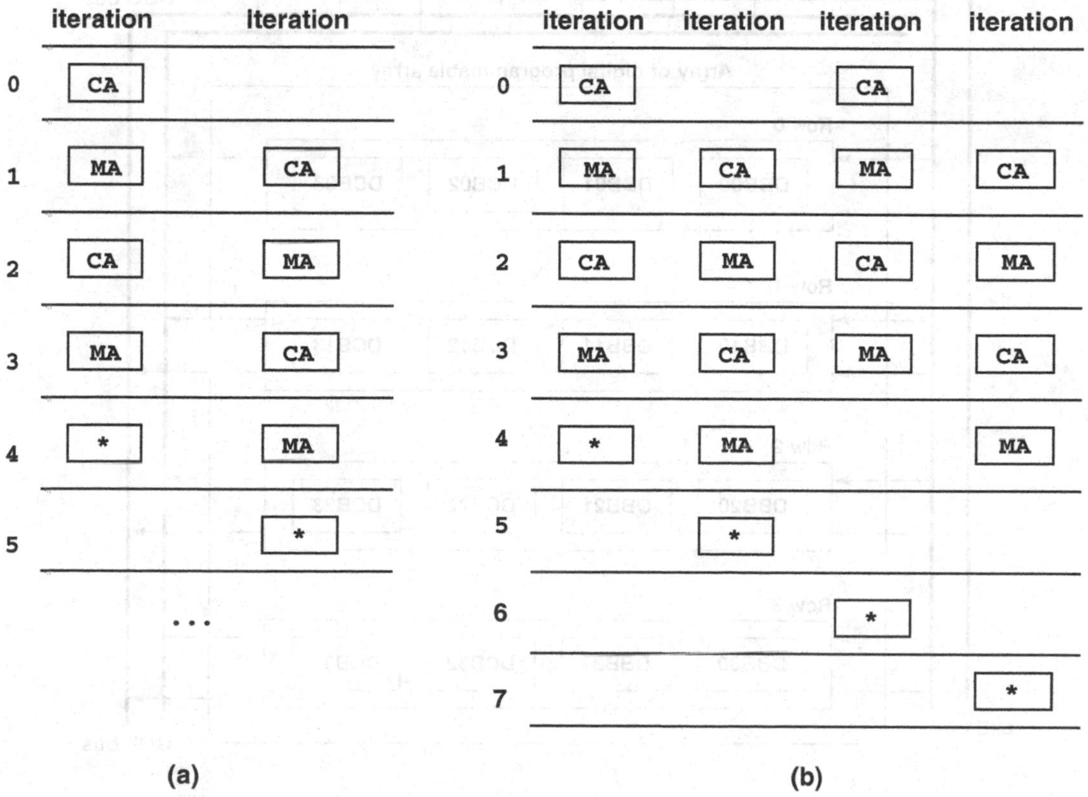

(a)

	iteration	Iteration
0	CA	
1	MA	CA
2	CA	MA
3	MA	CA
4	*	MA
5		*
	...	

(b)

	iteration	iteration	iteration	iteration
0	CA		CA	
1	MA	CA	MA	CA
2	CA	MA	CA	MA
3	MA	CA	MA	CA
4	*	MA		MA
5		*		
6			*	
7				*

Figure 4.16: Operation scheduling for different hardware resource sets.

The execution-time reduction increases slightly as a function of *L*.

The execution time of the loop iterations can be further reduced by allocating more hardware circuits to increase the number of parallel operation executions. Figure 4.16(b) illustrates this case. The coprocessor includes two additional circuits for computing the memory address, and one for multiplication. In addition, the architecture includes two memory circuits that can be accessed in parallel. The operations of the four iterations are executed in eight steps, as shown, and five iterations take only nine steps. This corresponds to a speed-up of 55%, as compared to the serial execution of the iterations, and a speed-up of 40%, as compared to the coprocessor used for the case in Figure 4.16(a).

Note that introducing one additional multiplier circuit to the coprocessor in Figure 4.16(a) does not improve the performance of the design. Also, for the coprocessor in Figure 4.16(b), the improvement would be only two execution steps for the execution of the four iterations, a speed-up of $\approx 25\%$, and none for the execution of the five iterations. Therefore, adding more hardware resources does not necessarily improve execution time.

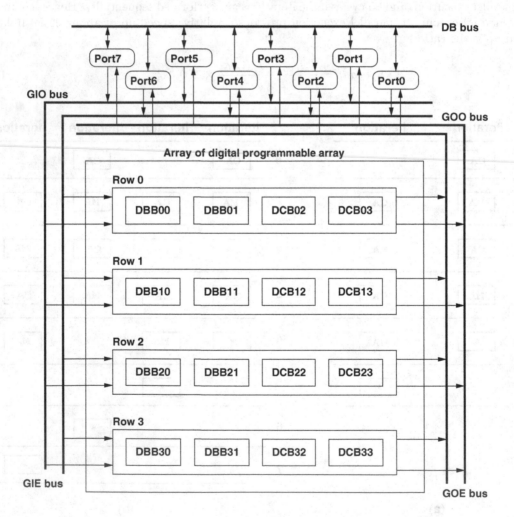

Figure 4.17: PSoC's array of programmable digital blocks PSoC [1].

4.3 Programmable Digital Blocks

In this section, PSoC hardware is customized for an application by utilizing the programmable digital blocks and PSoC's dedicated circuits.

PSoC has 16 programmable digital blocks, as shown in Figure 4.17. The blocks are organized as a two-dimensional array, as discussed in Chapter 3. Figure 4.18 describes the structure of a digital PSoC block. Each programmable block has three inputs, clock, data and auxiliary data (*Aux Data*), and four outputs, primary output (*PO*), auxiliary output (*AO*), block interrupt (*BI*), and broadcast output (*BO*).

The operation of a block is determined by programming seven registers: *CR0* and *FN* for storing the configuration and function state of a block, *IN* and *OU* for selecting the input and output signals, and *DR0*, *DR1*, and *DR2* for storing the data. In addition, the *INT* register defines the blocks interrupt mask. (cf. Chapter 3 for details on the *INT* register). Each digital PSoC block operates independently, and has a unique interrupt vector that points to a specific interrupt service routine. The physical addresses of the seven registers are shown in Tables 4.2–4.7 for the 16 blocks.

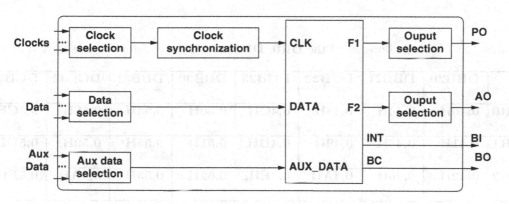

Figure 4.18: Programmable digital block [1].

Table 4.2: Physical addresses of the **DR0**, **DR1**, and **DR2** registers for rows 0 and 1 [1].

	DBB00	DBB01	DCB02	DCB03	DBB10	DBB11	DCB12	DCB13
DR0	0,20H	0,24H	0,28H	0,2CH	0,30H	0,34H	0,38H	0,3CH
DR1	0,21H	0,25H	0,29H	0,2DH	0,31H	0,35H	0,39H	0,3DH
DR2	0,22H	0,26H	0,2AH	0,2EH	0,32H	0,33H	0,3AH	0,3EH

The PSoC blocks are programmed with respect to [1]: (1) the input and clock signals, (2) the output signals, (3) the basic functionality, and (4) the data widths of the blocks:

- The data signal is selected as one of 16 different input signals. Similarly, the clock signal is chosen from one of 16 alternative clock signals, and then resynchronized with either *SYSCLK*, or *SYSCLKX2*. An auxiliary data selection circuit chooses one of four auxiliary data inputs, which is used primarily for the SPI slave function (cf. Chapter 5 for details on the SPI blocks).

- Each digital PSoC block has a primary and a secondary data output. Both outputs can be routed to one of four different destinations. PSoC blocks have two additional outputs: the block interrupt outputs for generating interrupt signals, and the broadcast output.

- PSoC blocks can be configured to be a timer, counter, deadband circuit, or a cyclic redundancy clock (CRC) circuit. PSoC communication blocks (Blocks DCB in Figure 4.17) can be either an SPI master or slave, or a UART. (These circuits are presented in Chapter 5.)

- The data width of a circuit can be expanded by chaining several digital PSoC blocks together. This chaining is achieved by using the primary output signals (also refer to Section 3.4 and Figure 3.24).

Table 4.3: Physical addresses of the **DR0**, **DR1**, and **DR2** registers for rows 2 and 3 [1].

	DBB20	DBB21	DCB22	DCB23	DBB30	DBB31	DCB32	DCB33
DR0	0,40H	0,44H	0,48H	0,4CH	0,50H	0,54H	0,58H	0,5CH
DR1	0,41H	0,45H	0,49H	0,4DH	0,51H	0,55H	0,59H	0,5DH
DR2	0,42H	0,46H	0,4AH	0,4EH	0,52H	0,53H	0,5AH	0,5EH

Table 4.4: Physical addresses of the **CR0** and **FN** registers for rows 0 and 1 [1].

	DBB00	DBB01	DCB02	DCB03	DBB10	DBB11	DCB12	DCB13
CR0	0,23H	0,27H	0,2BH	0,2FH	0,33H	0,37H	0,3BH	0,3FH
FN	1,20H	1,24H	1,28H	1,2CH	1,30H	1,34H	1,38H	1,3CH

As shown in Figures 4.18–4.20, a PSoC digital block clock signal is selected as one of 16 possible clock signals using the *Clock input* (bits 3-0) of the *IN* register. The selected clock signals are then synchronized with one of the available system clocks, either *SYSCLK* or *SYSCLK2* . The identity of the synchronization signal is determined by *AUXCLK* (bits 7-6) of the *OU* register.

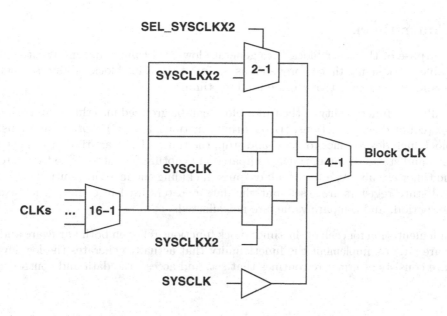

Figure 4.19: Programmable clocks of the digital blocks [1].

Table 4.5: Physical addresses of the **CR0** and **FN** registers for rows 2 and 3 [1].

	DBB20	DBB21	DCB22	DCB23	DBB30	DBB31	DCB32	DCB33
CR0	0,43H	0,47H	0,4BH	0,4FH	0,53H	0,57H	0,5BH	0,5FH
FN	1,40H	1,44H	1,48H	1,4CH	1,50H	1,54H	1,58H	1,5CH

Table 4.6: Physical addresses of the **IN** and **OU** registers for rows 0 and 1 [1].

	DBB00	DBB01	DCB02	DCB03	DBB10	DBB11	DCB12	DCB13
IN	1,21H	1,25H	1,29H	1,2DH	1,31H	1,35H	1,39H	1,3DH
OU	1,22H	1,26H	1,2AH	1,2EH	1,32H	1,36H	1,3AH	1,3EH

Similarly, the *DATA* signal to the block is selected as one of the 16 possible input signals based on the bits *Data input* (bits 7-4) of the *IN* register.

OUTEN (bit 2) of the *OU* register enables, or disables, the primary output of the PSoC digital blocks. *AUXEN* (bit 5) of the *OU* register enables, or disables the auxiliary output of the block. *Output select* (bits 1-0) selects the output data row for the primary output, and *AUX IO select* (bits 4-3) defines the output data row for the auxiliary output, as shown in Figure 4.20.

4.3.1   Timer Block

The main purpose of the timer block is to generate low-level timing and interrupt signals based on a selectable clock signal that is provided as an input [1]. This block produces programmable frequencies and pulse widths that can be used for timing.

Specifically, the functionality of the timer block can be grouped into three categories: (A) *terminal count* (count down) functions, (B) *compare* functions, and (C) *capture* functions. In addition, the block includes routines to start, and stop, the timer. The start timer routine enables the timer circuit, and starts the main timer, compare, and capture functions. The timer stop routine disables the timer circuit. As a result, all outputs, including the interrupt output, are gated low. The internal state registers are reset, but the data registers used in counting, for example the timer value, period, and compare value, are not affected.

The implementation for each of the timer block functions relies on both hardware and software. The hardware circuits implement the functionality that actually generates the low-level signals. The software consists of firmware routines that set, and access, the data and control registers of the hardware.

Table 4.7: Physical addresses of the **IN** and **OU** registers for rows 2 and 3 [1].

	DBB20	DBB21	DCB22	DCB23	DBB30	DBB31	DCB32	DCB33
IN	1,41H	1,45H	1,49H	1,4DH	1,51H	1,55H	1,59H	1,5DH
OU	1,42H	1,46H	1,4AH	1,4EH	1,52H	1,56H	1,5AH	1,5EH

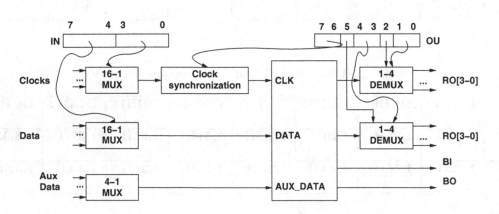

Figure 4.20: Input and output configuration [1].

A. Terminal count

Terminal count (TC) generates a programmable timing signal frequency based on the input clock signal. The related functionality includes (1) an operator that sets the period of the timer circuit, (2) the main timer function that decrements the content of the timer circuit at each clock cycle, and (3) a function that generates an interrupt signal upon reaching the value 00H. Figure 4.21 shows the inputs, outputs, and dataflow of the timer block.

The following are the three TC-related functions:

1. *Write period function*: This function sets the timer circuit's period value. This value is stored in a dedicated register, *DR1* (Data Register 1). The contents of the register do not change during the countdown process, and thus the value can be used for repeatedly initializing the timer block with the same period value. In addition to the *DR1* register, the period value is also loaded into the *DR0* register (Data Register 0), which is used for the count down function. Tables 4.2 and 4.3 provide the physical addresses of the *DR0* and *DR1* registers for the 16 programmable digital blocks of a PSoC chip.

2. *Main timer function*: This function produces a timing signal frequency equal to the clock frequency divided by the period value loaded into the *DR1* register. The pulse width is either one input clock period, or one half of the input clock period.

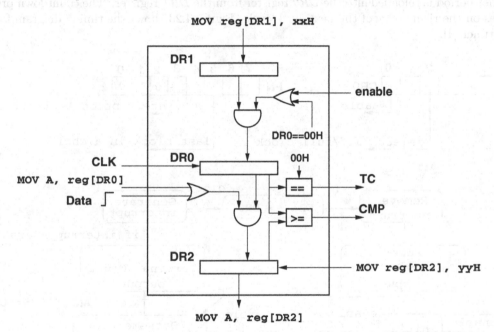

Figure 4.21: Dataflow of the timer block [1].

This functionality is implemented by the timer circuit decrementing the value of its *DR0* register on each rising edge of the clock signal. Upon reaching the value *00H* on the next positive edge of the clock, the *Terminal Count* (*TC*) output of the block is raised, the counter is reloaded with the period value stored in its *DR1* register and the count down process resumes.

3. *Terminal count interrupt function*: If the block interrupt is enabled, then the positive edge of the *TC* output generates an interrupt signal. This happens when the *DR0* register reaches the value *00H*.

Hardware circuit. The hardware circuit implements the required data transfer involving the *DR0* and *DR1* registers required by the three functions as shown in Figure 4.21, and generates the output, *TC,* and the interrupt signal. The circuit operation is programmed by the control registers, *CR0* and *FN.* Tables 4.4 and 4.5 show the physical addresses of the *CR0* and *FN* registers of the 16 programmable digital blocks.

Figure 4.22 illustrates the PSoC implementation of the main timer function [1]. The timer remains in the initial state until it is enabled by setting *enable* (bit *0*) of the control register, *CR0,* to 1.

If the timer circuit is enabled, when the rising edge of the next clock cycle occurs, the timer period stored in the *DR1* register is copied into the *DR0* register and is used for the countdown operation. Then, on the rising edge of the clock, the *DR0* register is decremented, if its value is not *00H.* Once the value in the *DR0* register becomes *00H,* the output, *TC,* of the block is set to 1. If interrupts are enabled, an interrupt signal is also produced. Bit 3 of the *FN* register enables/disables the interrupts on terminal count. Bit 2 of the *CR0* register determines the pulse width of the *TC* output, and the interrupt signal to be one full, or one half, clock cycle. Then, the timer period is reloaded into the *DR0* register from the *DR1* register. The countdown process resumes on the rising edge of the next clock cycle. Figure 4.23 shows the timing diagram for the PSoC timer [1].

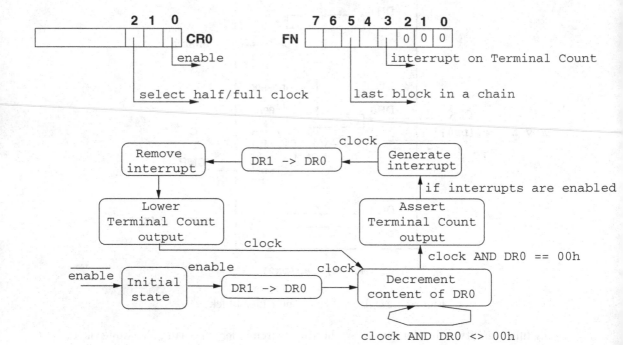

Figure 4.22: Main timer function and the corresponding control registers.

Software routines. Firmware routines set the control registers that start and stop the timer circuit, set the timer period, and enable/disable the interrupts. Figure 4.24 shows three short firmware routines related to the terminal count functions. The *Timer_Start* routine enables the timer circuit by setting bit 0 of the timer blocks control register. Figure 4.22 shows how this bit is used by the rest of the timer functionality. After setting the enabling bit, the timer circuit starts counting down on the next clock cycle. The *Timer_Stop routine* disables the timer circuit by resetting bit 0 of *CR0.* The *Timer_WritePeriod* routine writes a 16-bit value into the

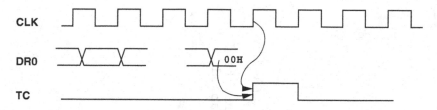

Figure 4.23: Timing diagram for the PSoC timer circuit [1].

```
Timer_Start:

_Timer_Start:

    reg[CR0]   |= 0x01
    ret
```

```
Timer_WritePeriod:

_Timer_WritePeriod:

    mov   reg[DR1_LSB], A

    mov   A, X

    mov   reg[DR1_MSB], A

    ret
```

```
Timer_Stop:

_Timer_Stop:

    reg[CR0]   &= ~0x01
    ret
```

Figure 4.24: Terminal count firmware routines [20]. Courtesy of Cypress Semiconductor Corporation.

DR1 register that stores the period value. The LSB of the new value is in the A register, and the MSB of the new value is stored in the X register.

B. Compare functionality

The compare functionality can generate an output signal of programmable frequency and pulse width, and an interrupt signal. The compare functions (1) writing the pulse width (compare) value into the *DR2* register (i.e. the *Data Register 2*) (2) reading the compare value from the *DR2* register, (3) producing the output signal, and (4) enabling/disabling the interrupts:

1. *Write compare value*: This function loads the *DR2* register with the 16-bit value used in the timers compare function. The output signals pulse width is equal to *(DR1 – DR2)*, and is programmed by selecting the appropriate register values.

2. *Read compare value*: This function returns the 16-bit value stored in the *DR2* register that is used in the compare function.

3. *Compare function*: This block compares the values of the decrementing *DR0* register, and the *DR2* register that stores the reference value. If the tested condition is true, then the auxiliary output of the timing circuit is set. The comparison operator can be programmed for either $DR0 \leq DR2$, or $DR0 < DR2$. If the value of the *DR1* register is larger than the value of the *DR2* register, the compare output is low for the first *(DR1 – DR2)* clock cycles. Then, for the next *DR2* clock cycles the output is high. Therefore, both the output period *(DR1)* and the pulse width *(DR2)* are programmable.

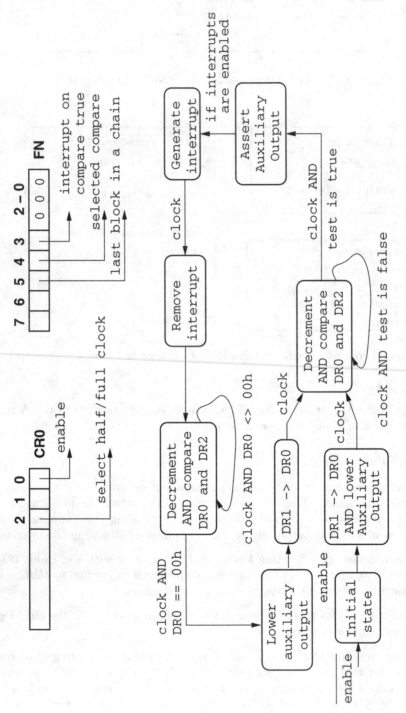

Figure 4.25: Compare functionality and the corresponding control registers.

If a capture function call occurs, then it overrides the $DR2$ register value used for comparison. (For more details, cf. the section on the capture functionality.)

4. *Enable/disable compare interrupts*: These functions enable interrupt signals, if the tested comparison condition is true. If enabled, the interrupt signal is generated on the rising edge of the compare output. Other functions disable the interrupts.

Tables 4.2 and 4.3 show the physical addresses of the $DR2$ registers of the 16 programmable digital blocks. The compare functionality is implemented using both hardware and firmware as in the case of the terminal count functionality.

Hardware circuit. The hardware circuit generates CMP, the low-level compare output, and the interrupt signal, which are the basis for the related data transfer shown in Figure 4.21.

Figure 4.25 illustrates the PSoC implementation of the compare functionality [1]. The circuit stays in the initial state provided that the timer circuit is not enabled. After enabling the timer by setting bit 0 in the $CR0$ register to 1, the $DR0$ and $DR2$ registers are compared at each clock cycle. Bit 4 of the FN register selects the comparison relationship to be either $DR0 \leq DR2$ or $DR0 < DR2$.

If the values of the $DR0$ and $DR2$ registers are equal, then the auxiliary output CMP of the circuit is set, and an interrupt signal is produced, if interrupts are enabled by setting bit 3 of the FN register. The length of the auxiliary output signal and interrupt signal are either one, or one half, clock cycles, because they are selected by bit 2 of the $CR0$ register.

Software routines. The firmware routines provide the capability to write/read the compare value stored in the $DR2$ register, to enable/disable the compare related interrupts, and to start/stop the circuit. Note that the start/stop routines are the same as those for the terminal count functionality shown in Figure 4.24.

Figure 4.24 shows the related firmware routines. The *Timer_Write CompareValue* routine loads the 16-bit value used for the comparison function into the $DR2$ register. The LSB is in the A register, and the MSB is in the X register. The *Timer_ReadCompareValue* routine returns the content of the $DR2$ register. The LSB is returned in the A register and the MSB in the X register.

C. Capture Functionality

The capture function returns the value of the $DR0$ register. There are two kinds of capture functions: hardware capture and software capture. Hardware capture occurs on the rising edge of the data input. Software capture occurs if the CPU core attempts to read the contents of the decrementing $DR0$ register. In both cases, the contents of the decrementing $DR0$ register are copied to the $DR2$ register, and the value can be accessed from there. The resulting data transfer is shown in the Figure 4.21.

Software routines. Figure 4.26 shows the firmware routine that implements the capture function. This routine, called *Timer_ReadTimerSaveCV*, returns the value of the $DR0$ register without affecting the other registers.

It first allocates three stack locations to store the counter value, that is, bytes $DR0\_MSB$ and $DR0\_LSB$, and the flags register. The X register points to the bottommost of the three stack locations allocated. Then, the values of the configuration register, $CR0$, and compare register, bytes $DR2\_MSB$ and $DR2\_LSB$, are stored on the stack, above the allocated locations. Next, two

```
Timer_ReadTimer:
_Timer_ReadTimer:

mov   A, reg[DR0_LSB]
mov   A, reg[DR2_MSB]
mov   X, A
mov   A, reg[DR2_LSB]
ret
```

```
Timer_ReadTimerSaveCV:
_Timer_ReadTimerSaveCV:

mov    X, SP                              .SetupStatusFlag:
add    SP, 3                              mov    [X], A
mov    A, reg[CR0_LSB]                    M8C_DisableGInt
push   A                                  pop    A
mov    A, reg[DR2_LSB]                    mov    reg[DR2_MSB], A
push   A                                  pop    A
mov    A, reg[DR2_MSB]                    mov    reg[DR2_LSB], A
push   A                                  pop    A
mov    A, reg[DR0_MSB]                    mov    reg[CR0_LSB], A
mov    [X+1], A                           pop    A
mov    A, reg[DR0_LSB]                    pop    X
mov    [X+2], A                           reti
mov    A, 0
tst    reg[CPU_SCR0], CPU_SCR0_GIE_MASK
jz     .SetupStatusFlag
mov    A, FLAG_GLOBAL_IE
```

Figure 4.26: Capture function related firmware routines [20]. Courtesy of Cypress Semiconductor Corporation.

bytes of the counter register (bytes $DR0\_MSB$ and $DR0\_LSB$) are saved in the stack entries $X + 1$ and $X + 2$, respectively. Flags are stored in the stack entry pointed to by the X register. After disabling the interrupts, the $DR2$ and $CR0$ registers are restored to their saved values. Finally, the A and X registers return the value of the counter register: byte $DR0\_MSB$ is returned in the A register and byte $DR0\_LSB$ is returned in the X register.

4.3.2   Counter Block

The functionality of the counter block is very similar to that of the timer block. As in the case of the timer block, a counter has one clock input, one data input, one primary output, and one auxiliary output. The main functions are also similar. However, the counter circuits do not implement the capture functionality of the timer block.

The counter block uses three data registers: the $DR0$ register for counting, the $DR1$ register for storing the period value, and the $DR2$ register for storing the compare value. The period value is loaded into the $DR1$ register only if the counter circuit is disabled. The value is also automatically copied into the $DR0$ register.

Figure 4.27 shows the PSoC implementation for the countdown function of the block [1]. The main difference, with respect to the terminal count functionality of the timer block in Figure 4.22, is that the content of the $DR0$ register is decremented only if the $Data$ input is high. If the $Data$ input is low, then decrementing stops, the two output signals are set to low, and the values of the three registers of the counter circuit $DR0$, $DR1$, and $DR2$, are saved.

Upon decrementing the value of the $DR0$ register to $00H$, the counter circuit sets the auxiliary output to high, and issues an interrupt signal if interrupts are enabled. Similar to the timer block, interrupts are enabled by setting bit 3 of the FN register to 1. The auxiliary output and interrupt signals are always applied for one clock cycle.

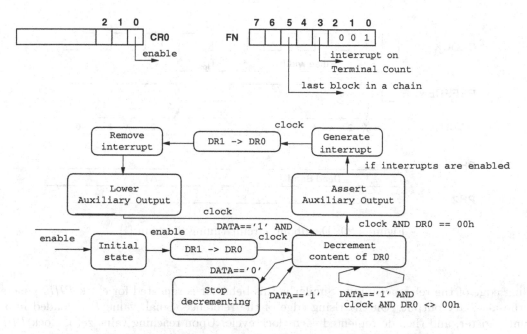

Figure 4.27: The counter functionality and the related control registers.

4.3.3 Deadband Block

Deadband circuits generate two nonoverlapping clocks, called *PHI1* and *PHI2*, that are separated by a value called the *deadband*. Deadband circuits are implemented using PSoC digital blocks, for which the configuration and function registers are as shown in Figure 4.28. The deadband generator has three inputs: viz., the clock signal, the reference signal (the input *DATA* of the digital block), and the *KILL* signal (the input *Aux DATA* of the block). The primary circuit output generates clock *PHI1*, and the secondary output clock *PHI2*. The input reference can be produced by a PWM or by toggling a bit using the Bit–Bang interface [1]. The bit is the primary output of the previous digital block, or the bit-bang clock register.

Figure 4.29 shows the operation of the enabled deadband circuit. After the reference signal is asserted, the next rising edge of the clock cycle causes the count register of the PSoC block to be loaded with the period value, P. Then the count value is decremented on each clock cycle, and after reaching value zero, clock *PHI1* becomes high. Clock *PHI1* is lowered to zero on the

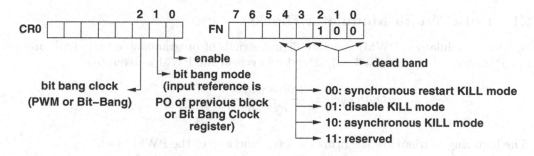

Figure 4.28: Configuration and function registers for the deadband circuit.

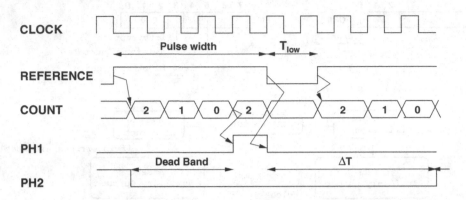

Figure 4.29: Deadband circuit timing diagram [1].

falling edge of the reference input. Similarly, this behavior is repeated for clock *PHI2* generated at the auxiliary output. On the rising edge of the reference signal, value P is loaded into the count register, and then decremented each clock cycle. Upon reaching value zero, clock *PHI2* is set high until the time of the falling edge of the reference signal.

Deadband and ΔT are defined as:

$$Deadband = P + 1 \tag{4.23}$$
$$\Delta T = T_{low} + P + 1 \tag{4.24}$$

These values are determined by the characteristics of the reference signal and period value P.

If the circuit is disabled then all outputs are low, the *DR0*, *DR1*, and *DR2* registers save their state, and the internal state of the circuit is set to a reset state. The asynchronous signal, *KILL*, applied to the data input of the block disables both outputs immediately.

4.4   Customized PSoC Digital Blocks

PSoC has three types of hardware blocks with customized functionality: (1) pulse width modulation, (2) multiply-accumulate, and (3) decimator circuits.

4.4.1   Pulse Width Modulator Blocks

Pulse width modulators (PWM) generate digital signals of programmable period, W and pulse width, P as defined in Figure 4.30(a). The duty cycle of the PWM is given by:

$$Duty\ cycle = \frac{Pulse\ width}{Period} = \frac{W}{P+1} \tag{4.25}$$

The following routines are the firmware-level functions of the PWM block:

- *PWM_Start*: This function enables the PWM.

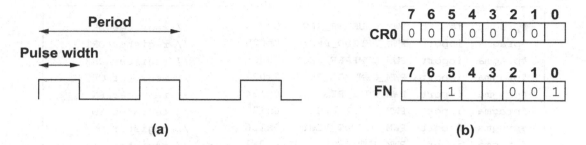

Figure 4.30: Pulse width modulator.

- *PWM_Stop*: This function disables the PWM. As a result, the circuit outputs are lowered to 0.

- *PWM_WritePulseWidth*: This function sets the pulse width of the output signal. If the loaded value is W then the pulse width is $W \times T_{clock}$. T_{clock} is the period of the clock signal to the PWM.

- *PWM_WritePeriod*: This function sets the period of the output signal. If the loaded value is P, then the period of the signal is actually $(P + 1) \times T_{clock}$, where T_{clock} is the period of the PWM clock signal.

- *PWM_bReadCounter*: This function returns the current value of the counter register.

- *PWM_bReadPulseWidth*: This function returns the value of the pulse width.

- *PWM_EnableInterrupts*: This function enables interrupts when the terminal count reaches zero.

- *PWM_DisableInterrupts*: This function disables interrupts when the terminal count reaches zero.

The PWM is based on digital PSoC blocks programmed to operate as counters [1]. One digital PSoC block is used for an 8-bit PWM, and two chained, digital, PSoC blocks are required for a 16-bit PWM. Figure 4.30(b) shows the configuration of the *FN* and *CR0* control registers:

- Bits 2-0 in the *FN* register configure the digital PSoC block as a counter.

- Bit 3 determines the type of interrupt, that is, interrupt on a terminal count, if the bit is set to 0, and interrupt on a true compare value, if the bit is set to 1.

- Bit 4 sets the compare function, which is either $\leq$, if the bit is set to 0, or $<$, if the bit is set to 1.

- Bit 5 set to 1 configures the digital PSoC block as standalone, because it is an 8-bit PWM.

- Bit 6 enables the primary output of the PSoC block to drive the broadcast net, if the bit value is 1, or disables the driving if the bit value is.

- Bit 7 is set to 0, if the data input is noninverted, and if set to 1, the data input is inverted.

- Bit *0* in the *CR0* register is set to 1, if the PWM is enabled, and to 0, if the PWM is disabled.

```
#pragma  ioport  PWM_COUNTER_REG:  0x024        // register DR0
#pragma  ioport  PWM_PERIOD_REG:   0x025        // register DR1
#pragma  ioport  PWM_COMPARE_REG:  0x026        // register DR2
#pragma  ioport  PWM_CONTROL_REG:  0x027        // register CR0
#pragma  ioport  PWM_FUNC_REG:     0x124        // register FN
#pragma  ioport  PWM_INPUT_REG:    0x125        // register IN
#pragma  ioport  PWM_OUTPUT_REG:   0x126        // register OU
#pragma  ioport  PWM_INT_REG:      0x0E1        // register INT


#define  PWM_Start_M    PWM_CONTROL_REG |= 0x01
#define  PWM_Stop_M     PWM_CONTROL_REG &= ~0x01

 PWM_Start:                         PWM_Stop:
_PWM_Start:                        _PWM_Stop:
 RAM_PROLOGUE RAM_USE_CLASS_1       RAM_PROLOGUE RAM_USE_CLASS_1
 PWM_Start_M                        PWM_Stop_M
 RAM_EPILOGUE RAM_USE_CLASS_1       RAM_EPILOGUE RAM_USE_CLASS_1
 ret                                ret


 PWM_WritePulseWidth:               PWM_WritePeriod:
_PWM_WritePulseWidth:              _PWM_WritePeriod:
 RAM_PROLOGUE RAM_USE_CLASS_1       RAM_PROLOGUE RAM_USE_CLASS_1
 mov  reg[PWM_COMPARE_REG], A       mov  reg[PWM_PERIOD_REG], A
 RAM_EPILOGUE RAM_USE_CLASS_1       RAM_EPILOGUE RAM_USE_CLASS_1
 ret                                ret
```

Figure 4.31: PWM firmware level routines [20]. Courtesy of Cypress Semiconductor Corporation.

In addition to *the FN* and *CR0* registers, the *DR0, DR1, IN, OU,* and *DR2* registers are also used in the implementation of the PWM. The *DR0* register is the down counter. The *DR1* register stores the period value, and the *DR2* register stores the pulse width of the output signal.

Similar to the counter circuit, bits 7-4 in the *IN* register select one of the 16 possible data inputs, and bits 3-0 select one of the 16 possible clock signals. The *OU* register follows the general bit structure of the other programmable blocks.

Figure 4.31 shows the implementation of the firmware functions for a digital PSoC block. This figure shows the physical addresses of the eight PSoC block registers for configuration, function, data, input and output selection, and interrupt mask. Routine *PWM_Start* starts the operation of the PWM by enabling bit 0 of the configuration *CR0* register. Similarly, routine *PWM_Stop* stops the PWM by disabling bit 0 of *CR0* register. The *PWM_WritePulseWidth* routine loads the pulse width, expressed as multiples of the selected clock cycles, into the *DR2* register, which is used as a compare register for the PWM. The routines parameter is passed through the A register.

```
PWM_bReadCounter:                        push   A
_PWM_bReadCounter:                       mov    reg[PWM_INPUT_REG], 0x00
bPWM_ReadCounter:                        M8C_SetBank0
_BPWM_ReadCounter:                       mov    A, reg[PWM_COUNTER_REG]
                                         mov    A, reg[PWM_COMPARE_REG]
bOrigCompareValue:    EQU    0           push   A
bOrigClockSetting:    EQU    1           mov    A, [X+bOrigCompareValue]
wCounter:             EQU    2           mov    reg[PWM_COMPARE_REG], A
STACK_FRAME_SIZE:     EQU    3           M8C_SetBank1
                                         mov    A, [X+bOrigClockSetting]
RAM_PROLOGUE RAM USE_CLASS_2             mov    [PWM_INPUT_REG], A
mov    X, SP                             M8C_SetBank0
mov    A, reg[PWM_COMPARE_REG]           PWM_Start_M
push A                                   pop    A
PWM_Stop_M                               ADD    SP, -(STACK_FRAME_SIZE-1)
M8C_SetBank1                             RAM_EPILOGUE RAM_USE_CLASS_2
mov    A, reg[PWM_INPUT_REG]             ret



PWM_bReadPulseWidth:                     RAM_PROLOGUE RAM USE_CLASS_1
_PWM_bReadPulseWidth:                    mov    A, reg[PWM_COMPARE_REG]
bPWM_ReadPulseWidth:                     RAM_EPILOGUE RAM USE_CLASS_1
_bPWM_ReadPulseWidth:                    ret
```

Figure 4.32: PWM firmware routines (continued) [20]. Courtesy of Cypress Semiconductor Corporation.

The *PWM_WritePeriod* routine sets the period of the PWM by loading the *DR1* register with the period value. The parameter of the routine is passed through the A register.

Figure 4.32 shows the firmware routines for reading the pulse width and the current value of the counter register stored in the *DR0* register. Figure 4.33 shows the firmware routines for enabling and disabling interrupts. Figure 4.34 presents the API method for setting the duty cycle of the PWM,

---

Example (Software implementation of a PWM). Figure 4.35 shows the main elements of the software implementation of a PWM. The circuit is used to turn an LED on/off based on the values in the *PRT2DR* register. Initially, the LED is turned on by loading a 1 in the *PRT2DR* register.

Then, the value of the PWM pulse width, i.e., *0xFF* in this example, is loaded into memory at address 4.

The LED remains on for a time interval equal to the execution time of the *delay_loop1* loop. The pulse width of the PWM emulation is given by:

$$Pulse\ width = [(7+5) \times (R_5 + 1) + (7+5+8)] \times (R_4 + 1) \times clock\ cycles \qquad (4.26)$$

```
#define PWM_EnableInt_M  M8C_EnableIntMask(PWM_INT_REG, 0x02)
#define PWM_DisableInt_M M8C_DisableIntMask(PWM_INT_REG, 0x02)


PWM_EnableInt:                           PWM_DisableInt:
_PWM_EnableInt:                          _PWM_DisableInt:
RAM_PROLOGUE RAM_USE_CLASS_1             RAM_PROLOGUE RAM_USE_CLASS_1
PWM_EnableInt_M                          PWM_DisableInt_M
RAM_EPILOGUE RAM_USE_CLASS_1             RAM_EPILOGUE RAM_USE_CLASS_1
ret                                      ret
```

Figure 4.33: PWM firmware routines (continued) [20]. Courtesy of Cypress Semiconductor Corporation.

```
void PWMSetDutyCycle (BYTE bID, BYTE bDutyCycle) {

    BYTE Period,
    BYTE CompareValue;
    if (bDutyCycle > 100) {
        bDutyCycle = 100;
    }
    Period = PWMGetPeriod (bID);
    CompareValue = ((int) (Period + 1) * (int)bDutyCycle)/100;
    PWMSetPulseWidth (bID, CompareValue);
}
```

Figure 4.34: PWM firmware routines (continued) [20]. Courtesy of Cypress Semiconductor Corporation.

The same equation should be used to evaluate the length of time that the LED is off. R_4 and R_5 represent the values loaded at memory addresses 4 and 5, respectively. The execution time is 7 clock cycles for a decrement instruction, 5 cycles for a jump instruction, and 8 clock cycles for a move instruction.

If memory entries at addresses 4 and 5 are initialized to zero, the minimum pulse width is 32 clock cycles. Similarly, the minimum PWM period is 64 clock cycles. This means that the code cannot be used, if shorter pulse widths, or faster PWM signals, are needed. Also, the values of pulse width and period are controlled by two variables. The variable stored at location 5 provides a finer tuning, whereas the variable at location 4 gives a coarser adjustment. Fine tuning is in

```
include "m8c.inc"
include "memory.inc"
include "PSoCAPI.inc"


export _main


_main:                                        mov     reg[PRT2DR], 0x00
                                              mov     [4], 0xFF
   mov     reg[PRT2DR], 0x01           delay_loop2:
   mov     [4], 0xFF                          mov     [5], 0xFF
delay_loop1:                                  innerloop2:
   mov     [5], 0xFF                          dec     [5]
innerloop1:                                   jnz     innerloop2
   dec     [5]                                dec     [4]
   jnz     innerloop1                         jnz     delay_loop2
   dec     [4]                                jmp     _main
   jnz     delay_loop1                        .terminate
```

Figure 4.35: PWM implementation as assembly code routines.

steps of 12 clock cycles, and coarse tuning is in steps of 32 cycles. The code has to be modified, if finer adjustment steps are needed for PWM pulse width, or period.

4.4.2 Multiply ACcumulate

Multiply ACcumulate (MAC) circuits can provide two distinct functions: fast multiplication of two operands, and multiplication followed by summing, that is accumulating, the resulting products.

Figure 4.36 shows the structure of a MAC that is connected to the system bus. Its operation does not require an additional clock or enable signal. For fast multiplication, the 8-bit multiplicands are loaded into the MUL_X and MUL_Y registers, and the 16-bit product is available in the register pair, MUL_DH and the MUL_DL. The MUL_DH register stores the more significant byte, and MUL_DL register the less significant byte. The two multiplicands are signed integers, and the product is also a signed integer.

For the multiply accumulate function, the two multiplicands are loaded into registers MAC_X and MAC_Y. The product is summed with the previous results, and stored as a 32-bit, signed number in the ACC_DR3, ACC_DR2, ACC_DR1, and ACC_DR0 registers. These registers are reset by writing into the MAC_CL1 or MAC_CL0 register. The type of MAC function, that is

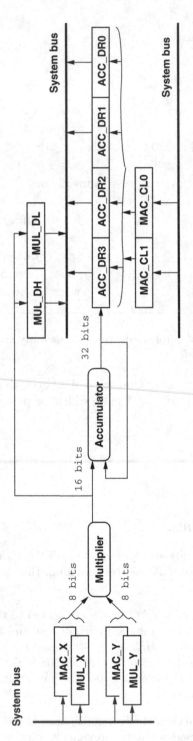

Figure 4.36: Multiply accumulate (MAC) circuit [1].

```
#include <m8c.h>
#include "PSoCAPI.h"

void main {
   char a[] = {5, 4, 3, 2, 1, 2, 1, 2, 1, 2, 3, 4, 5, 6, 7, 8};
   char b[] = {1, 2, 3, 4, 5, 6, 7, 8, 9, 8, 7, 6, 5, 4, 3, 2};
   char sp;
   int i;

   sp = 0;
   for (i = 0; i < 16; i++)
      sp += a[i] * b[i];
}
```

Figure 4.37: C code for scalar product algorithm.

either multiplication, or multiply and accumulate, is selected by loading the multiplicands into the respective register pair.

Example (Scalar product of two vectors). This example illustrates the computational speed benefits offered by the use of MAC, as compared to software routines, by calculating the scalar product of two vectors, a and b. The two vectors can have dimensions of 16, 64, or 256. This example compares three implementations of the scalar product algorithm using C language, assembly code, and assembly code using the MAC for multiplication and summing the partial results.

A. C program for the scalar product of two vectors

Figure 4.37 shows the C code for the scalar product algorithm. The code produced by the C compiler resides in flash memory. This includes the program instructions, and the constant values, required for initializing vectors a and b.

Figure 4.38 highlights the main steps of the code. The data stack space is first allocated for all local variables. Each stack entry is one byte. Variable i is stored at the bottom of the stack. Two stack entries, i.e., two bytes, are allocated for the variables, as variable i of type *int*. Vector a is stored next, and occupies sixteen stack locations entries (i.e., one entry for each vector element). Vector b is stored on top of vector a, and also has sixteen allocated entries. Finally, variable sp is allocated on top of the stack. Figure 4.38(a) shows the stack allocation of the local data variables.

Next, the code for the compiled program initializes the stack entries for vectors a and b with the constant values stored in flash. Figure 4.38(b) shows the code fragment for initializing vector a. Similar code initializes vector b. The first instruction, at address 0318, sets the current page pointer *CUR_PP*, stored in register 208, to point to SRAMs page zero. Then, the virtual register, *_r1*, also located in SRAMs page 0, but at address 03, is loaded with value 80, virtual register *_r0* situated in SRAMs page 0, at address 04, is loaded with value 1, and virtual register *_r3*

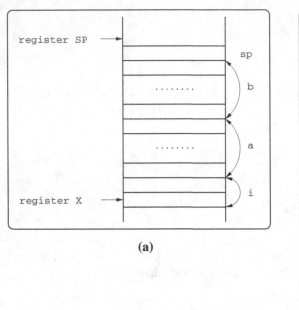

(a)

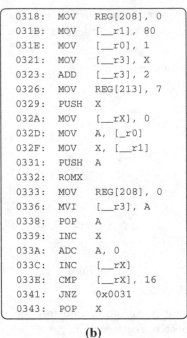

```
0318:   MOV    REG[208], 0
031B:   MOV    [__r1], 80
031E:   MOV    [__r0], 1
0321:   MOV    [__r3], X
0323:   ADD    [__r3], 2
0326:   MOV    REG[213], 7
0329:   PUSH   X
032A:   MOV    [__rX], 0
032D:   MOV    A, [_r0]
032F:   MOV    X, [__r1]
0331:   PUSH   A
0332:   ROMX
0333:   MOV    REG[208], 0
0336:   MVI    [__r3], A
0338:   POP    A
0339:   INC    X
033A:   ADC    A, 0
033C:   INC    [__rX]
033E:   CMP    [__rX], 16
0341:   JNZ    0x0031
0343:   POP    X
```

(b)

Figure 4.38: Assembly code generated by a C compiler.

located in SRAM page 0, at address 01, points to the first, that is, the bottommost stack entry assigned to vector a. The instruction at address 0326 sets the *MVW_PP* register, stored in a register at address *213*, to point to SRAM page 7. The *MVW_PP* register is used to efficiently move data from the accumulator into SRAM by using indirect addressing. Then, the contents of the X register, pointing to the bottommost stack entry storing local variables, is saved on the stack. The virtual register *__rX* is initialized with the value 0.

Data transfer from flash memory. At address *032D*, the A register is loaded with the contents of the virtual register *__r0*, and the X register with the contents of the virtual register *__r1*. The A register is stored on the stack, because the following instruction *ROMX* alters its contents. Registers A and X are used to determine the address in flash, from which instruction *ROMX*, at address *0332*, reads data. The A register contains the most significant byte, and the X register, the least significant byte. Address *0x0150* is for first read operation. The read data are returned in the A and X registers. The move immediate instruction, *MVI*, at address *0336*, stores the data in the A register into a page pointed to by the *MVW_PP* register (SRAM page 7) and location register $X + 2$. Then, the contents of the A register are popped from the stack and the index of the next vector elements is incremented. Virtual register *__rX* is also incremented, and execution jumps back to address *0x033*, if there are elements in vector A that still have to be read. Finally, the POP X instruction, at address *0343*, restores the contents of the X register to point to the bottommost stack element allocated for the local variables.

Scalar product of two vectors. Figure 4.39 shows the compiler-produced assembly code required to compute the scalar product of two vectors, in this example. The code first initializes the stack entries holding the loop index i, to zero. These are the entries pointed to by the X register, and the entry on top of it. Next, the virtual registers are set so that they point to the SRAM page holding vectors a and b. Virtual register *__r0* points to page 7, where vector b is

```
036D:    MOV     [X+1], 0            039C:    MOV     A, [X+0]
0370:    MOV     [X+0], 0            039E:    ADC     A, [__r2]
0373:    MOV     REG[208], 0         03A0:    MOV     REG[212], A
0376:    MOV     [__r0], 7          03A2:    MVI     A, [__r3]
0379:    MOV     [__r1], X         03A4:    MOV     [__r2], A
037B:    ADD     [__r1], 18        03A6:    PUSH    X
037E:    MOV     A, [X+1]           03A7:    MOV     A, [__r0]
0380:    ADD     A, [__r1]         03A9:    MOV     X, [__r2]
0382:    MOV     [__r1], A         03AB:    LCALL   __mul8
0384:    MOV     A, [X+0]           03AE:    POP     X
0386:    ADC     A, [__r0]         03AF:    ADD     [X+34], A
0388:    MOV     REG[212], A         03B1:    INC     [X+1]
038A:    MVI     A, [_r1]          03B3:    ADC     [X+0], 0
038C:    MOV     [__r0], A         03B6:    MOV     A, [X+1]
038E:    MOV     [__r2], 7         03B8:    SUB     A, 16
0391:    MOV     [__r3], X         03BA:    MOV     A, [X+0]
0393:    ADD     [__r3], 2         03BC:    XOR     A, 128
0396:    MOV     A, [X + 1]          03BE:    SBB     A, 128
0398:    ADD     A, [__r3]         03C0:    JC      0x0373
039A:    MOV     [__r3], A
```

Figure 4.39: Assembly code generated by a C compiler (continued).

stored. Then the virtual register $\_r1$ is loaded and points to the start of the vector b (register X + 18). Then, virtual register $\_r1$ is updated to point to the current element $b[i]$. This element occupies the $i-1$ entry on top of the stack entry holding $b[0]$. The current SRAM page number for MVI instruction is computed, and stored in the MVR_PP register located in register 212. For this example, the MVR_PP register points to SRAM page 7. The value of element $b[i]$ is loaded into the A register, and then stored in the virtual register $\_r0$.

A similar set of steps is executed to access vector element $a[i]$. Virtual register $\_r2$ points to the SRAM page number, and virtual register $\_r3$ to the element $a[i]$. After storing the current SRAM page in the MVR_PP register (register 212), element $a[i]$ is moved to the A register, and the virtual register $\_r2$. At this point, both operands are available. Element $a[i]$ is stored in register $\_r2$ and element $b[i]$ in register $\_r0$. Next, the value of the X register is stored on the stack, because the multiplication routine $\_mul8$ affects the register value. Then, value $b[i]$ is stored in the A register, and the $a[i]$ value in the X register. The two multiplicands for the multiplication routine $\_mul8$ are passed by using the values in the A and X registers.

This partial product is stored in the A register and this partial product is added to the location at address X register + 34, storing the scalar product. The X register is restored to the saved value. Index i is incremented and the code iterates until all of the pairs of vector elements $a[i]$ and $b[i]$ are multiplied together. The final scalar product is stored at entry $X + 34$ in the data stack.

B. Assembly code for scalar product without using MAC.

Figure 4.40 shows the Assembly code for the scalar product of two vectors without using the hardware MAC. The code is much shorter and faster than the code produced by the C compiler.

```
MOV     [3], 0          MOV     [23], 3         _mul8:
MOV     [5], 5          MOV     [24], 4             CMP     [X], 0
MOV     [6], 4          MOV     [25], 5             JZ      accu
MOV     [7], 3          MOV     [26], 6             AND     F, 251
MOV     [8], 2          MOV     [27], 7             RRC     [X]
MOV     [9], 1          MOV     [28], 8             JNC     shift
MOV     [10], 2         MOV     [29], 9             ADD     A, [X+16]
MOV     [11], 1         MOV     [30], 8         shift:
MOV     [12], 2         MOV     [31], 7             ASL     [X+16]
MOV     [13], 1         MOV     [32], 6             JMP     _mul8
MOV     [14], 2         MOV     [33], 5         accu:
MOV     [15], 3         MOV     [34], 4             ADD     [37], A
MOV     [16], 4         MOV     [35], 3             MOV     A, 0
MOV     [17], 5         MOV     [36], 2             INC     X
MOV     [18], 6         MOV     [37], 0             INC     [3]
MOV     [19], 7         MOV     X, 5                CMP     [3], 16
MOV     [20], 8         MOV     A, 0                JNZ     _mul8
MOV     [21], 1         MOV     reg[0xD1], 0    .terminate
MOV     [22], 2                                     JMP     .terminate
```

Figure 4.40: Assembly code for scalar product without using MAC.

The values for vectors a and b are stored directly into SRAM starting from address 5 to and ending at 20, for vector a, and address 21 to 36 for vector b. Index i is stored at location 3, and the scalar product is computed using the location at address 37. The X register is set to point to the first element in vector a, and the A register is used to compute the product $a[i] \times b[i]$. If the value of element $a[i]$ is zero, the multiplication algorithm ends, and execution jumps to the section labeled *accu*. The product value stored in the A register is added to the memory entry storing the scalar product (address 37). Then, the X register is incremented to point to the next element in vector a and index i is incremented. If there are still elements to be multiplied, then the algorithm jumps to address *_mul8*, and continues the computation.

The code for multiplying two unsigned numbers starts at address *_mul8*, and the algorithm is shown in Figure 2.11 of Chapter 2. Note that in this algorithm, the values of vectors a and b are destroyed during multiplication. The end result of the scalar product is stored at address 37.

C. Assembly code for scalar product using MAC

Figure 4.41 shows the assembly code, using MAC for the scalar product algorithm. First, the MAC is reset, including the *ACC_DR3*, *ACC_DR2*, *ACC_DR1*, and *AC_DR0* register. This is accomplished by writing to the *MAC_CL0* register located at address *EEH*. The partial products $a[i] \times b[i]$ are computed by writing operand $a[i]$ into the *MAC_X* register at address *ECH*, and operand $b[i]$ into the *MAC_Y* register located at address *EDH*. Instruction *MOV reg[EDH],0* is introduced after each partial product to allow the MAC to finish its computation. After all partial products have been computed, the value of the scalar product is stored in the four registers: *ACC_DR3* located at address *EEH*, *ACC_DR2* at located address *EFH*, *ACC_DR1* located at address *ECH*, and *AC_DR0* located at address *EDH*.

```
MOV     reg[EEH], 0          MOV     reg[EDH], 6          MOV     reg[EDH], 0
MOV     reg[ECH], 5          MOV     reg[EDH], 0          MOV     reg[ECH], 4
MOV     reg[EDH], 1          MOV     reg[ECH], 1          MOV     reg[EDH], 6
MOV     reg[EDH], 0          MOV     reg[EDH], 7          MOV     reg[EDH], 0
MOV     reg[ECH], 4          MOV     reg[EDH], 0          MOV     reg[ECH], 5
MOV     reg[EDH], 2          MOV     reg[ECH], 2          MOV     reg[EDH], 5
MOV     reg[EDH], 0          MOV     reg[EDH], 8          MOV     reg[EDH], 0
MOV     reg[ECH], 3          MOV     reg[EDH], 0          MOV     reg[ECH], 6
MOV     reg[EDH], 3          MOV     reg[ECH], 1          MOV     reg[EDH], 4
MOV     reg[EDH], 0          MOV     reg[EDH], 9          MOV     reg[EDH], 0
MOV     reg[ECH], 2          MOV     reg[EDH], 0          MOV     reg[ECH], 7
MOV     reg[EDH], 4          MOV     reg[ECH], 2          MOV     reg[EDH], 3
MOV     reg[EDH], 0          MOV     reg[EDH], 8          MOV     reg[EDH], 0
MOV     reg[ECH], 1          MOV     reg[EDH], 0          MOV     reg[ECH], 8
MOV     reg[EDH], 5          MOV     reg[ECH], 3          MOV     reg[EDH], 2
MOV     reg[EDH], 0          MOV     reg[EDH], 7          MOV     A, reg[EEH]
MOV     reg[ECH], 2
```

Figure 4.41: Assembly code for scalar product using MAC.

Table 4.8: Execution time in clock cycles for different implementation styles

Vector size	C Code without MAC	C Code with MAC	Assembly Code without MAC	Assembly Code with MAC
16	8958	6043	2861	390
64	45177	23659	11932	1580
256	–	–	52268	6188

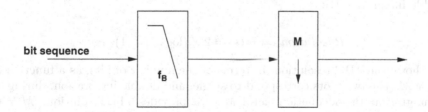

Figure 4.42: Dataflow of the decimation operation.

D. Execution times

Table 4.8 summarizes the execution times for each of the implementations discussed. In addition to the listed execution times, 1494 clock cycles are needed for initializing the embedded system. Vector sizes of 256 could not be handled by the version of the C compiler used for this example. Note that using MAC for vectors of length 16, the code produced by the C compiler required

2915 fewer clock cycles than when MAC was used. This represents a 32% saving in the execution time of the algorithm. For vectors of length 64, the execution time, when using MAC, is 21,618 clock cycles less than without MAC, which is a 47% improvement in execution time. For vectors of length 16, the execution time for the algorithm written in assembly code is 6097 clock cycles shorter, which is a 68% saving in time, as compared to the C code without MAC. The time saving increases to 33,245 clock cycles for vectors of length 64, which represents a 73% improvement. Finally, for vectors of length 16, the algorithm written in assembly code using MAC requires 8560 fewer clock cycles than the C algorithm without MAC. Similarly, for vectors of length 64, it needs 43,597 fewer clock cycles. These savings in time represent 95% and 96%, respectively, of the execution time of the C program.

4.4.3 Decimator Blocks

Decimator blocks are used to remove the frequency components above a given frequency value f_B, and then to reduce the data rate of the signal by the factor M, called the down sampling factor. Figure 4.42 presents the dataflow of the decimation operation. Chapter 7 discusses incremental, analog-to-digital converters, and Chapter 9 discusses oversampled ADCs, which are two popular applications of decimator blocks.

The transfer function for decimator circuits is defined as

$$H(z) = \left(\frac{1}{M}\right)^2 (1 - z^{-M})^2 \left(\frac{1}{1 - z^{-1}}\right)^2 \tag{4.27}$$

PSoC provides two types of decimator blocks: type 1, and type 2. Type 1 circuits implement only the integration part of the transfer function $H(z)$, whereas the differentiation part is computed in software. Type 2 provides a full hardware implementation of the transfer function.

The resolution of the analog-to-digital converters that use PSoCs decimator circuits can be estimated using the following two relationships: for single integration [1]:

$$Resolution\ (\#\ bits) = 1.5 \times (\log_2 M - 1) \tag{4.28}$$

and for double integration [1]:

$$Resolution\ (\#\ bits) = 2 \times (\log_2 M - 1) \tag{4.29}$$

Figure 4.43 shows the ADC resolution, in terms of the number of bits, as a function of the decimation factor M. The two plots correspond to single, and double integration, during decimation. For single integration, the decimation factor $M = 32$ provides 6 bit resolution, $M = 64$ provides 6.5 bit resolution, and $M = 256$ gives 10.5 bit resolution. For double integration, the decimation factor $M = 32$ provides 8 bit resolution, the factor $M = 64$ provides 10 bit resolution, and the factor $M = 256$ gives 14 bit resolution.

Type 1 Decimators

This decimator circuit can be programmed to perform a single or a double integration operation [1]. The structure of the circuit is shown in Figure 4.44. The input data are applied to the one-bit input, *DATA*. If the input bit is 0, then the value -1 is added to the accumulated value, otherwise, the value 1 is added. The accumulated value is 16-bits long and is stored in the A0 register for

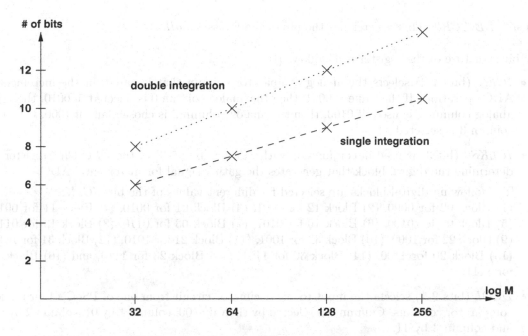

Figure 4.43: ADC performance as a function of the decimation factor M.

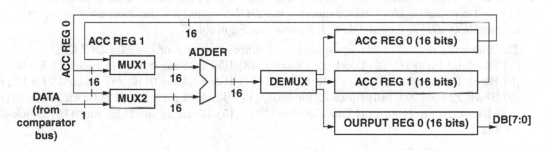

Figure 4.44: Structure of the PSoC type 1 decimator circuit [1].

the first integration operation, i.e., the sum of $DATA$ + A0 register. The 16-bit A1 register stores the accumulated value for the second integration operation (the sum A0 register + A1 register). The output register contains the sum of the two integration operations.

Type 2 Decimators

Type 2 decimator circuits offer a full hardware implementation of the integration and differentiation steps in a decimator. The circuit structure is shown in Figure 4.45.

The type 1, and type 2, decimators are programmed by using the DEC_CR0, DEC_CR1, and DEC_CR2 control registers. Note that PSoCs decimator circuits are designed for analog-to-digital conversion (ADC) only, and therefore the control registers bit structures are based on this fact.

Register DEC_CR0: This register has the physical address *0,E6H*.

The bit structure of the register is as follows [1]

- *IGEN* (bits 7-4) selects the analog comparator column that is gated for the incremental ADC operation. If the value is 0001, then the analog column 0 is selected, if 0010, then the analog column 1 is used, if 0100, then the analog column 2 is chosen, and if 1000, then the column 3 is selected.

- *ICLKS0* (bit 3) is used in conjunction with *ICLKS* (bits 5-3) of the *DEC_CR1* register to determine the digital block that generates the gating signal for incremental ADC.

  The following digital blocks are selected for different values of the bits *ICLKS*:
  (1) Block 02 for 0000, (2) Block 12 for 0001, (3) Block 01 for 0010, (4) Block 11 for 0011, (5) Block 00 for 0100, (6) Block 10 for 0101, (7) Block 03 for 0110, (8) Block 13 for 0111, (9) Block 22 for 1000, (10) Block 32 for 1001, (11) Block 21 for 1010, (12) Block 31 for 1011, (13) Block 20 for 1100, (14) Block 30 for 1101, (15) Block 23 for 1110, and (16) Block 33 for 1111.

- *DCOL* (bits 2-1) selects the input to the decimator circuit from one of PSoC's four analog comparator columns. Column 0 is selected by the value 00, column 1 by 01, column 2 by 10, and column 3 by 11.

- *DCLKS0* (bit 0) is used in conjunction with the *DCLKS* (bits 2-0) of the *DEC_CR1* register to select the digital block that generates the *DCLK* signal for the decimator registers, that is, the *DEC*, *DIFF 0*, and *DIFF 1* registers in Figure 4.45.

  The following digital blocks are selected for different values of the bits *DCLKS*:
  (1) Block 02 for 0000, (2) Block 12 for 0001, (3) Block 01 for 0010, (4) Block 11 for 0011, (5) Block 00 for 0100, (6) Block 10 for 0101, (7) Block 03 for 0110, (8) Block 13 for 0111, (9) Block 22 for 1000, (10) Block 32 for 1001, (11) Block 21 for 1010, (12) Block 31 for 1011, (13) Block 20 for 1100, (14) Block 30 for 1101, (15) Block 23 for 1110, and (16) Block 33 for 1111.

Register DEC_CR1: The physical address of this register is 0, E7H.

The bit structure of the register is as follows [1]:

- *ECNT* (bit 7) set to the value 0 disables the decimator as a counter for incremental ADC, and instead configures the circuit to operate together with a $\Delta\Sigma$ ADC. If the bit is 1 then the decimator is enabled for an incremental ADC. The bit is available only for decimators of type 1.

- *IDEC* (bit 6) selects the digital block latch control between noninverted, if the bit is set to 0, and inverted, if the bit is set to 1.

- *ICLKS* (bits 5-3) are used in conjunction with the *ICLKS0* (bit 3) of the *DEC_CR0* register to determine the digital block that generates the gating signal for the incremental ADC operation.

- *DCLKS* (bits 2-0) are used in conjunction with *DCLKS0* (bit 0) of the *DEC_CR0* register to select the digital block that generates the signal *DCLK* for the decimator registers.

Register DEC_CR2: The physical address of this register is *1, E7H*.

The bit structure of this register is as follows [1]:

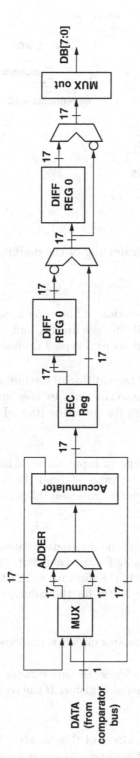

Figure 4.45: Structure of the PSoC type 2 decimator circuit [1]

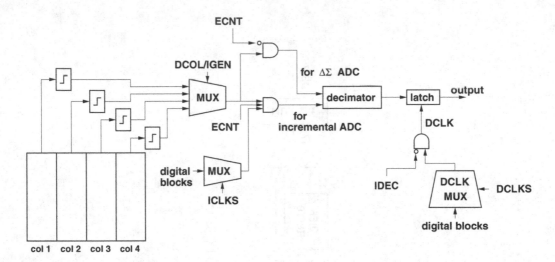

Figure 4.46: Programming of the decimator circuit.

- *Mode* (bits 7-6) programs the operation of the type 2 decimator circuits: The decimator functions as a type 1 decimator, if the bits are 00, and the circuit is used for incremental ADC, if the bits are 01, and functions as a type 2 decimator, if the bits are 10.

- *Data out shift* (bits 5-4) programs the shifting of the output bits: No shifting of bits occurs for the value 00, shifting of all bits to the right by one position for the value 01, shifting of all bits to the right by two positions for the value 10, and shifting of all bits to the right by four positions for the value 11.

- *Data format* (bit 3) defines the type of input data: if the bit is 0, then an input bit 0 is interpreted by the decimator as the value -1, and the input bit 1 as the value 1. If the control bit is programmed to 1 then the input bit 0 has the value 0, and for the input bit 1, the value is 1.

- *Decimation rate* (bits 2-0) programs the decimation rate of the circuit as: (1) the circuit is off for the value 000, (2) the rate is 32 for the value 001, (3) the rate is 50 for the value 010, (4) the rate is 64 for the value 011, (5) the rate is 125 for the value 100, (6) the rate is 128 for the value 101, (7) the rate is 250 for the value 110, and (8) the rate is 256 for the value 111.

The 16-bit output register of the decimator circuit is implemented using two 8-bit registers [1]:

- *Register DEC_DH*: The physical address of this register is *0, E4H*. The register contains the more significant byte of the output register. If the register is written to, contents of the accumulator registers is cleared.

- *Register DEC_DL*: The physical address of this register is *0, E5H*. This register stores the least significant byte of the output register. As in the previous case, writing to this register clears the accumulator registers.

4.5 Conclusions

This chapter has focused on design methods for optimizing system performance by customization of the architecture for the application. The concepts are illustrated by referring to PSoC's programmable and customized digital blocks.

Performance-driven customization involves developing optimized modules for the performance-critical modules of an application, including customized hardware circuits and software routines. A module is performance-critical, if it has a significant effect on the global performance of the implementation. For example, with respect to the total system execution time, a module is performance-critical if it involves a large amount of processing and/or data communications with other subroutines. For developing optimized modules, architecture customization explores different performance–cost tradeoffs for the critical modules.

Any design methodology for architecture customization includes four general steps: (i) finding the performance-critical parts of an application, (ii) selecting the critical parts that are subjected to implementation in hardware, (iii) determining the nature and structure of the optimized hardware circuits, and (iv) implementing the overall system design. The design methodologies differ depending on the type of the targeted architecture: architectures with one general-purpose processor and dedicated coprocessors, with one general-purpose processor and shared coprocessors, and with multiple general-purpose processors plus shared coprocessor.

A design methodology for reducing the execution time of algorithms executed on an architecture with one general-purpose processor and coprocessors that are shared by multiple modules of the implementation has also been presented. This architecture is popular for many embedded applications. The analyzed application is inspired by digital signal processing, and was selected because of the following characteristics: it performs a large number of computations, the number of loop iterations is constant and known, and loop iterations can be executed in parallel, in as much as there are few data dependencies between the iterations.

The discussed design methodology consists of the following: specification, profiling, identification of the performance-critical blocks, functional partitioning, hardware–software partitioning, hardware resource allocation, mapping of the performance-critical modules to hardware, and scheduling of the modules that share the same hardware resource.

The second part of the chapter treated the support offered by the PSoC architecture for customization, and explained several design examples that illustrate the design methodology for performance-driven customization. The support provided by the PSoC architecture includes programmable digital blocks and blocks with dedicated functionality.

Programmable digital blocks can implement the following functions: timer, counter, deadband, and CRC. The operation of a block is defined by programming seven registers: registers IN and OU define the programmable input, output, and clock, registers FN and $CR0$ set the functionality, and registers $DR2$, $DR1$, and $DR0$ store the data involved in the circuit operation. In addition, register INT controls the interrupt generation by a programmable block. This chapter defined the functionality of the blocks, and described the firmware and API routines for operating the programmable blocks. (Refer to [1] for details on CRC blocks.)

The dedicated PSoC functionality and associated programming for the pulse width modulator (PWM), multiply accumulate (MAC), and decimator were also treated in detail in this chapter. (Refer to [1] for details on sleep and watchdog timers.)

A comprehensive discussion was presented of two examples of performance-driven architecture customizations: the software implementation of PWM, and the speed-optimized implementation of an algorithm for computing the scalar product of two vectors. Two additional design examples are presented in Chapter 10.

4.6 Recommended Exercises

1. Using the PSoC programmable digital blocks, develop a technique for measuring the signal frequency. The goal of this exercise is to maximize the precision of the measurement.[1]

2. Develop an execution-time efficient implementation of the transfer function:

$$H(z) = 2 \times \frac{0.0488 \; + \; 0.0976z^{-1} \; + \; 0.0488z^{-2}}{1 \; - \; 0.9428z^{-1} \; + \; 0.3333z^{-2}}$$

Estimate the maximum frequency of the signals that can be filtered with your implementation. Provide solutions for increasing the signal frequency that can be processed.

3. Develop a cost efficient implementation of the transfer function:

$$H(z) = 2 \times \frac{0.1464 \; + \; 0.1464z^{-2}}{1 \; - \; 0.5858z^{-1} \; + \; 0.4142z^{-2}}$$

Estimate the maximum frequency of the signals that can be filtered with your implementation. Provide solutions for increasing the signal frequency that can be processed. What is the corresponding cost increase?

4. Provide an execution time efficient implementation of the transfer function

$$H(z) = 2 \times \frac{0.1464 \; - \; 0.2929z^{-1} \; + \; 0.1464z^{-2}}{1 \; + \; 0.1716z^{-2}}$$

Estimate the maximum frequency of the signals that can be filtered with your implementation. Provide solutions for increasing the signal frequency that can be processed.

5. Design a cost-efficient implementation for a filter bank that includes the three transfer functions in exercises 2, 3 and 4. Minimize the execution time of the implementation. The solution should exploit all opportunities for hardware sharing among the implementations for the three transfer circuits.

6. Develop a time-efficient design for a bitstream sequence recognizer. The input data are received serially at an input port. The application must recognize if a predefined sequence of length 8 appeared in the input. Estimate the execution time of your design, and predict the maximum input bit rate that can be handled by your implementation without bit loss.

   Extend the design for sequences of lengths 16 bits and then 32 bits.

7. Propose a design for a FIFO (First-In-First-Out) buffer. The buffer should handle data of length B bytes, where the length is a parameter for the design. Estimate the execution time of your implementation. Suggest solutions for speedingup the design.

8. Develop (a) a cost-efficient and (b) a time-efficient algorithm for computing the square root of an unsigned integer number. Compare the two solutions.

9. Program the CRC algorithm in C language, and then in assembly code. Estimate the execution time, and code size, for the two algorithms. Identify the C language constructs that lead

---

[1](See Application Note AN2283, "Measuring Frequency," Cypress Semiconductor Corporation.

to significant execution overhead. Propose solutions that minimize the execution time of the C program. Compare the execution times of the two software procedures with the execution time of the CRC hardware module that is based on PSoC, programmable, digital blocks.

10. Propose a general approach for scheduling the operations pertaining to parallel loop iterations. Assume that there are no data dependencies between operations in different iterations. Consider an example of 5 parallel iterations and explain the scheduling result of your algorithm.

11. Propose a general method for binding operations to hardware circuits and variables to registers. The goal is to find the minimum amount of hardware that guarantees satisfying a predefined timing constraint on a ADFG (Acyclic Dataflow Graph). Illustrate your algorithm on an ADFG with more than 10 nodes that represent operations on at least 5 variables.

12. Is it possible to develop a hardware-only, PSoC-based implementation for the differentiation step of the decimator module, assuming that only Type 1 decimator blocks are available? If so, describe your design and if not, explain why not.

13. The design example in Section 4.2 minimized the execution time for a given cost of the hardware circuits. How does the design change, if the goal is to find the minimum cost implementation that meets a specified execution time requirement?

14. Analyze the advantages and limitations of the cost function defined in expression (4.12). Propose more complex cost functions that address the limitations.

15. Compute the expressions of the logic functions F in Figures 4.10 and 4.12.

16. Develop firmware routines for managing and operating the deadband block. Use the firmware routines in a C program to generate nonoverlapping signals with different frequencies, and deadband values.

17. Build a 32-bit PWM module and develop the corresponding firmware routines for starting and stopping the module, setting the pulse width and period, enabling and disabling interrupts, and reading the pulse width and counter.

18. Identify possibilities of optimizing the execution time of the code in Figure 4.39(b) and Figure 4.40. Find the reasons for the difference in execution time between the C code and the assembly code, without MAC.

19. Develop a set of firmware routines for operating the type 1 decimator blocks.   •

20. For the international data encryption algorithm (IDEA), propose an implementation that is optimized for execution time. The algorithm is described in Chapter 10. Compare your solution with the solution discussed in Chapter 10.

Bibliography

[1] PSoC Mixed Signal Array, Technical Reference Manual, *Document No. PSoC TRM 1.21*, Cypress Semiconductor Corporation, 2005.

[2] Pseudo Random Sequence Generator, *CY8C29/27/24/22/21xxx Data Sheet*, Cypress Semiconductors, September 21 2005.

[3] 8-bit and 16-bit Pulse Width Modulators, *CY8C29/27/24/22/21xxx, CY7C6xxxx, and CYWUSB Data Sheet*, Cypress Semiconductors, October 3 2005.

[4] J. Ackenhusen, *Real-Time Signal Processing: Design and Implementation of Signal Processing Systems*, Upper Saddle River, NJ: Prentice Hall, 1999.

[5] M. Chiodo et al., A case study in computer-added codesign of embedded controllers, *Journal of Design Automation of Embedded Systems*, 1, (1), pp. 51-67, 1996.

[6] D. Comer, *Computer Networks and Internets with Internet Applications*, third edition, Upper Saddle River, NJ: Prentice Hall, 2001.

[7] G. De Micheli, R. Ernst, W. Wolf, *Readings in Hardware/Software Co-Design*, San Francisco: Morgan Kaufmann, 2002.

[8] P. Diniz, E. da Silva, S. Netto, *Digital Signal Processing*, Cambridge: Cambridge University Press, 2002.

[9] R. Ernst, J. Henkel, and T. Benner, Hardware-software cosynthesis of microcontrollers, *IEEE Design & Test*, 10, (4), December 1993, pp. 64-75.

[10] R. Gupta, G. De Micheli, A Cosynthesis Approach to Embedded System Design Automation, *Journal of Design Automation of Embedded Systems*, 1, (1), pp. 69-120, January 1996.

[11] A. A. Jerraya, J. Mermet, System-Level Synthesis, A. A. Jerraya, J. Mermet (editors), Boston: Kluwer Academic Publishers, 1999.

[12] O. Ozbek, Estimating PSoC Power Consumption, *Application Note AN2216*, September 21 2004.

[13] M. Raaja, Binary to BCD Conversion, *Application Note AN2112*, Cypress Semiconductors, February 12 2003.

[14] V. Sokil, Hardware Bitstream Sequence Recognizer, *Application Note AN2326*, Cypress Semiconductors, December 8 2005.

[15] D. Stranneby, *Digital Signal Processing: DSP and Applications*, Boston: Newnes, 2001.

[16] S. Sukittanon, S. Dame, 3-Channel Filterbank in PSoC, *Application Note AN2315*, Cypress Semiconductors, September 26 2005.

[17] J. Valvano, *Embedded Microcomputer Systems. Real Time Interfacing*, third edition, London: Thomson 2007.

[18] D. Van Ess, Measuring Frequency, *Application Note AN2283*, Cypress Semiconductors, May 26 2005.

[19] W. Wolf et al. Hardware-Software Codesign: Principles and Practices, Boston: Kluwer Academic Publishers, 1999.

[20] PSoC Express, Version 2.0, Cypress Semiconductor, 2006.

Chapter 5

Programmable Data Communication Blocks

This chapter presents a design methodology for implementing performance-optimized communication subsystems for embedded applications. Serial communication modules based on the SPI and UART standards are detailed, and their implementation using PSoC's programmable digital block is explained.

Communication channels are important for connecting multiple embedded systems, or a system to its peripherals. A channel has high-level primitives for configuring the channel, sending/receiving data on the channel, and managing the channel operation and properties.

This chapter defines the main characteristics of the channel implementation units (CIUs) that can serve as abstract channels. To help devise a systematic methodology for mapping abstract channels to predefined CIUs, CIUs are described as tuples with four components: the primitives, the required hardware resources, their performance, and specific constraints. The tuple information is used to identify the CIUs (from a predefined library) that are implementation alternatives for an abstract channel. The design methodology presented results in optimized implementations for the communication channels in an embedded application by: (i) selecting for each abstract channel the set of feasible CIUs, (ii) finding the abstract channels that share the same CIU, and (iii) building hardware/software implementations for the channels that cannot be mapped to the existing CIUs. The last part of the chapter focuses on the SPI and UART communication modules, and their hardware–software implementation based on the programmable communication blocks in PSoC. Several examples illustrate the use of the SPI and UART firmware routines for communication.

This chapter has the following structure:

- Section 1 introduces the concept of abstract data channels.

- Section 2 defines CIUs and their main properties.

- Section 3 discusses the design methodology for implementing abstract channels by reusing CIUs from a predefined library.

- Section 4 details PSoC's SPI block.

- Section 5 presents PSoC's UART modules.

A. Doboli, E.H. Currie, *Introduction to Mixed-Signal, Embedded Design*,
DOI 10.1007/978-1-4419-7446-4_5, © Springer Science+Business Media, LLC 2011

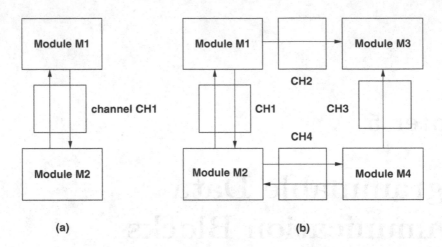

Figure 5.1: Data communication between modules.

- Section 6 presents the chapter conclusions.

5.1   Abstract Communication Channels

Communication blocks (e.g., SPI [2, 3] and UART [4]) provide interfacing capability between modules through serial, or parallel, bit transmission. Communication blocks are important not only for interfacing the embedded system to other systems and to the external peripherals, but also for connecting (integrating) the system's modules. Before discussing the hardware–software implementation of communication blocks in more detail, it is appropriate to consider some high-level data communications concepts, for example abstract channels and their related high-level primitives.

In a simple case, two systems, or modules, exchange data, such as characters, integers, floating point numbers, structures, and so on, by calling a set of high-level primitives pertaining to the abstract communication channels defined for the two communicating systems:

- *Abstract data channel*: An abstract data channel is established between each pair of communicating modules to transmit data of a certain type, such as characters, integers, long integers, etc. The two modules can send and receive data on the channel. A module can have multiple abstract channels for communication with several modules and for sending/receiving different data types.

  Even though this definition may seem restrictive, it simplifies the presentation of abstract channels, and captures the main aspects of their design. A more general definition would be a single channel for communicating data of different types between two modules, and also single channels that support the broadcasting of data from one module to multiple modules. The concepts presented for the restricted channel definitions can be extended to the more general channel types.

  Figure 5.1 illustrates several abstract channel examples. Note that each channel represents a given module pair that only communicates a specific type of data on the channel. In Figure 5.1(a), the modules *M1* and *M2* use the channel *CH1* for communicating characters. Figure 5.1(b) shows the modules *M1* and *M2* communicating through the channel *CH1*, the

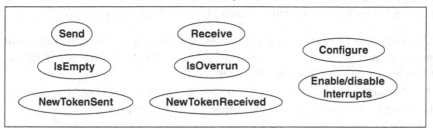

Figure 5.2: High-level data communication primitives.

modules *M1* and *M3* using the channel *CH2*, and modules *M2* and *M3* sending/receiving data on the channel *CH3*.

- *High-level primitives*: The high-level primitives of a channel allow the two communicating modules (i) to send and (ii) receive data on the channel, (iii) to verify whether the channel is ready for sending or receiving data, and (iv) to manage the operation of the channel. This set of primitives is summarized in Figure 5.2.

  1. *Send*: This primitive sends a data value from one module to the other. In the *non-blocking send* function, there is no confirmation signal returned from the receiving module to the sending module to confirm the successful receipt of the data value. In the *blocking send* function, the sender does not transmit another data value unless the receiver confirms the receipt of the previous one.

     Hence, for the nonblocking send function, data might be lost, if the sending module transmits several data values before the receiver gets a chance to access (receive) the data. On the positive side, there is less interaction (synchronization) between the two modules as compared to that of the blocking send. This simplifies the design because no acknowledgment signal must be generated by the transmitter for the sender, and also increases the degree of concurrency between the two modules. Higher concurrency results in shorter execution times for the communication modules due to the shorter busy–waiting times of the modules.

  2. *NewValueSent*: This function returns the logic value *true*, if the current data value was transmitted, and the value *false*, otherwise.

  3. *IsEmpty*: This primitive returns the logic value *true*, if the buffer used for sending data are empty. If the physical data transmission is ongoing, or has not started, the function returns the value *false*.

     Note that the function is not the complement of the function *NewValueSent* if different buffers are used to store/transmit data, as is the case for PSoC's SPI and UART modules.

  4. *Receive*: This function receives a new data value transmitted on the specified channel. If new data are available, then the function returns the new data value. Otherwise, it returns a predefined value that signals that new data have not been received. This definition corresponds to the *nonblocking receive* function because the module execution

can proceed if new data are not available. In contrast, the *blocking receive* function suspends the module execution until the new data are received.

Compared to the blocking receive, the nonblocking receive function reduces the idle waiting time of the receiving module, but can result in data loss if the module is not sufficiently fast in retrieving the received data.

5. *NewValueReceived*: This routine returns the logic value *true*, if a new data value was received by the module, and the value *false*, otherwise.

6. *IsOverrun*: This routine returns the logic value *true*, if the received data value was overwritten by a new data value, before the receiving module accessed the data. Otherwise, it returns the value *false*.

7. *ConfigureCommunication*: These routines configure the parameters of the transmission process, including the starting and stopping conditions for communication on the channel.

8. *Enable/Disable Interrupts*: These functions enable/disable the interrupts that signal specific transmission conditions. For example, interrupts can be produced, if the transmission of a data value was completed, or a new data value was received at the input. Interrupts offer the necessary support to implement nonblocking send/receive primitives, for example they indicate if a new data communication can be initiated for a nonblocking send operation, or a new data value is available at the input for nonblocking transmit primitives.

Example (Two communicating modules). This example presents the pseudocode for two communicating modules. Each module receives a data value from the other module, performs computations, and then communicates a data value to the other module. The intermodule communication uses an abstract channels, and the related high-level primitives.

The functionality of the communicating modules *M1* and *M2* in Figure 5.1(a) can be expressed as shown in Figure 5.3. First, the modules are configured for communication by using the *ConfigureCommunication* routines. Then, the module continuously polls the function *NewValueReceived* until a new data value is received on the channel. After accessing the data value by calling the function *Receive*, the module performs its computations. Finally, it checks whether the transmission buffer is empty and therefore available to send a new data value. The routines *IsEmpty* and *NewValueSent* are utilized for this purpose. After the buffer becomes available and a new data value transmission can be started, the data value is sent by calling the primitive *Send*. This step completes one iteration. Then, the module proceeds with a new iteration, as shown in the figure.

Although the communication procedure between the two modules is simple to describe and implement, it has two main disadvantages. First, the modules might stay idle for long time intervals during their polling steps before receiving and sending data. This results in a low utilization rate of the modules, and longer execution times for the modules. The utilization rate can be increased by replacing polling by nonblocking communication primitives and interrupt-based implementation of the primitives.

Secondly, data loss can occur as a result of the difference in the execution speed of the two modules. In this example, there is no data loss, even if one module is much slower than the other

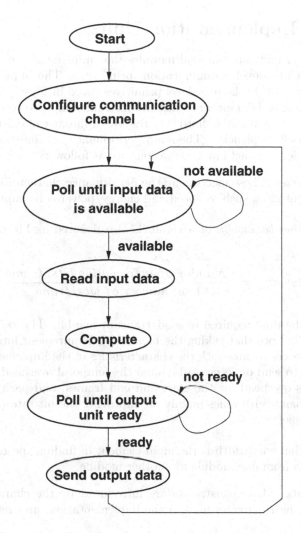

Figure 5.3: Pseudocode of a communicating module.

module. This is because the module executions *synchronize* at the beginning, and end, of each iteration. Synchronization is achieved at the polling steps. For example, if module *M1* is faster than module *M2*, at the beginning of each iteration, it must wait until module *M2* completes its current iteration, so that it can send the data value for module *M1*. However, for the case in which module *M1* only produces data for module *M2*, without receiving any data values from module *M2*, data loss can result unless blocking send functions are used. Thus an acknowledgement signal that is produced by module *M2*, and received by module *M1* is required.

5.2   Channel Implementation Units

Channel implementation units are physical modules that implement the theoretical concepts of abstract channels, and high-level communication, primitives. The implementation of abstract channels must not only provide the high-level primitives listed in Figure 5.2, but also (i) deliver the required performance, and (ii) meet any design constraints imposed by the physical implementation of the system, such as available hardware resources, number of existing I/O pins, existing supply voltages and clock frequencies. The main performance attributes and design constraints describing communication channel implementations are as follows:

- *Performance metrics*: These metrics are the performance characteristics of a data communication implementation, such as speed and energy/power consumption.

The *communication bandwidth* of a channel, also denoted as bit rate, or throughput, is defined as [17]:

$$Bandwidth \quad = \quad \frac{Number\ of\ information\ bits/frame}{total\ number\ of\ bits/frame} \times \frac{1}{bit\ time} \qquad (5.1)$$

The bit time is the time required to send/transmit one bit. The value 1/bit time is called the *baud rate* [17]. Note that taking the bandwidth requirement into consideration during the design process constrains both the characteristics of the implementation (e.g., the time (speed) required to send or receive a bit) and the temporal overhead of the communication protocol, such as overhead for the start and end frames, and parity bits. Increasing the communication bandwidth relies mainly on improving the bit rate (and decreasing the bit time) of the channel.

The communication bandwidth is the main element in finding the *communication delay* of transmitting data from one module to another module.

- *Design constraints*: These constraints are introduced by the characteristics of the environment and hardware circuits used in the implementation, and refer to three important issues:

  1. *Global clock*: The presence, or absence, of a global clock for the communicating modules determines whether these channels are *synchronous*, or *asynchronous*, respectively. In synchronous communication, a global clock is used by both modules for data transmission and reception. The global clock is either a dedicated resource, or an output signal generated by one of the modules for use by other modules, such as in the SPI protocol which is discussed in Subsection 5.4. In asynchronous communication, no common clock is involved; instead the modules use specific control bits to signal the starting/ending of a data communication, and transmit the data bits at precise intervals of time. Examples of asynchronous protocols are the UART protocol presented in Subsection 5.5, and the 1-wire protocol [9].

     Asynchronous protocols have higher overhead due to the extra information that needs to be sent between the modules and therefore lower bandwidths if more relaxed timing intervals are used for transmitting one bit. However, they provide more flexibility in as much as they do not require a global clock.

  2. *Number of physical links*: The number of physical links available for data communication determines how many bits can be sent simultaneously between the modules.

Serial protocols transmit one–bit at a time, whereas parallel protocols send multiple bits in parallel. Hence, the bandwidth of parallel protocols is higher, but at the expense of more physical interconnects and I/O pins. For SoCs with a small number of pins this can be an important limitation.

Serial protocols are further distinguished as *full-duplex*, *half-duplex*, and *simplex* connections [17]. In full-duplex, bits are transferred in both directions simultaneously, that is from module *M1* to module *M2*, and vice versa. Half-duplex communication allows communication in both directions with the constraint that bits are transmitted only in one direction at a time. Finally, simplex connections allow transmission in one direction.

Although the transmitting bandwidths in a single direction might be similar for all three cases, full-duplex supports execution overlapping of the primitives send and receive. This improves the concurrency of the two communicating modules at the expense of using one more physical link.

3. *Noise level*: For communication in noisy environments where data may be altered, redundant bits are computed by the sender and sent along with the data so that the receiver can determine if the received data are correct. In the simplest case, the redundant information is a *parity* bit: for even parity, the parity bit is computed such that the number of the transmitted bits 1 is even, and odd, if odd parity is used. Other error-checking mechanisms include the cyclic redundancy check (CRC) and checksum [17]. The penalty for maintaining the integrity of communication is obviously the increase in communication overhead in terms of reducing the bandwidth and the hardware required for error checking.

The goal of an abstract channel implementation is to find an optimized design that (i) provides the communication primitives required by the application, (ii) can be built with the hardware available and (iii) meets all performance requirements, and (iv) satisfies any other design (e.g., cost, size, number of used I/O pins, etc.). In many cases, the implementations utilize standard specifications (e.g., SPI, UART, PCI, etc.) which allow easier interfacing to peripherals. The available implementation modules are stored in a library, and reused as required. The intuitive design approach is to identify, for each abstract channel, all library modules that can correctly implement the channel, and then select the best solution from the chosen modules. The identification procedure verifies the matching of the properties of an abstract channel and of the implementations in the library. The matching procedure requires a precise definition of the properties of implementation modules.

By definition, the channel implementation unit (CIU) CI_i provides (i) a set of high-level communication primitives, (ii) requires a certain amount of hardware resources, (iii) is capable of delivering a given performance, and (iv) must meet a set of design constraints, such as size, number of I/O pins, external clocks, and supply voltages. Therefore, the channel implementation unit can be expressed formally as the tuple:

$$CI_i \;=\; < Primitives_i, HWresources_i, Performance_i, Constraints_i > \qquad (5.2)$$

For example, the PSoC's SPI module provides a set of high-level primitives for configuring the module, verifying the status of a communication, data sending and receiving, and interrupt enabling and disabling. These primitives, detailed in Subsection 5.4, form the set *Primitives* of the module. The set *HWresources* includes all the hardware resources needed for the implementation of the SPI module, for example one of the four PSoC programmable communication blocks and interconnection wires. The constraint set includes three I/O pins and a 3.3V supply voltage. The attribute *Performance* includes the communication speed of about 1 Mbits/sec. The above definition is general and extensible to other attributes, such as the power consumption of the module [11].

The channel implementation units available for embedded system design are part of a predefined library, *ChLibrary*:

$$CHLibrary \quad = \quad \{CI_1, CI_2, \cdots, CI_M\} \tag{5.3}$$

For example, for the PSoC architecture, Subsections 5.4 and 5.5 present two channel implementation units, for the SPI block and the UART block. Also, several other PSoC-based implementations are available in the technical literature, including the implementation of the **1-wire** protocol [14], the design of a specialized 9-bit UART serial protocol [9], and various communication circuits with coding and error correction [10, 13, 15, 16]. These CIUs form the library *CHLibrary* available for developing PSoC-based, embedded system implementations. The space spanned by the four elements of the CI_i tuples is defined by the library *CHLibrary*, and is the union of the corresponding parameters of the modules in the library.

5.3   Hardware–Software Implementation of Channels

Implementing the abstract channels in a system specification requires addressing the following problems [6]:

1. Selecting the channel implementation units that are feasible for each abstract channel.

2. Finding the abstract channels that can share an implementation unit.

3. Developing the hardware–software implementation of each channel implementation unit.

In addition, the implementations must meet the performance requirements of each abstract channel (e.g., the communication delay, average and peak loads) and satisfy all the design constraints, such as the number of required I/O pins, clock frequencies, and buffer memories, while minimizing the total hardware cost of the solution.

Several design methodologies can be employed to address the design problem [6, 8]. Given its effectiveness, a simplified version of the channel design methodology proposed by Daveau et al. [6] is appropriate. The methodology has the following refinement steps:

1. *Channel implementation unit allocation*: The step selects the channel implementation units, from the channel library *CHLibrary* that can be used for each of the abstract channels in the specification. Hence, this step identifies the set of viable design solutions, because each of the identified CIUs can potentially be used as the abstract channel.

   Because abstract channels can be developed in different ways, this step identifies the CIU candidates that (i) provide the necessary primitives for the abstract channels, (ii) deliver the performance required by the application, (iii) use the available hardware resources, and (iv) do not violate the design constraints.

2. *Abstract channel mapping*: This task identifies the abstract channels that share the same CIU. This reduces the amount of hardware used in the design and the cost of the design.

It may also help meet the design constraints, for example if there are not enough physical interconnects, or I/O pins, available to build multiple CIUs.

Handling channel implementation mapping is a complex task. It involves finding (i) the scheduling (arbitration) scheme by which the single CIU services the multiple abstract channels, (ii) inserting buffers for storing the data of an abstract channel that cannot be immediately serviced by the CIU, and (iii) adding glue and control logic for multiplexing the shared CIU. While addressing the three issues, the design process must estimate the performance of the channel implementation, the amount of used hardware, and the satisfaction of all design requirements.

3. *Developing channel implementation units*: In some design flows, this step might actually precede channel allocation and mapping. The CIUs for different communication schemes are designed and stored in the library, *CHLibrary*, for future use. However, it is possible that a new CIU must be developed at the end of the channel allocation and mapping steps, because none of the implementations in the library are actually feasible solutions. For example, none of the CIUs can offer the needed communication bandwidth. In this case, the software and the hardware for the needed CIU must be codesigned.

This activity also involves the developing of the data structures and routines that manage the multiple abstract channels, and their channel implementation units.

Figure 5.4 illustrates the methodology for the first two steps. Sections 5.4 and 5.5 detail the implementation of two channel implementation units for the PSoC architecture, the SPI and the UART modules.

1. *Channel implementation unit allocation.* There are several possible channel implementations that can serve as the specific abstract channel AC_j. The channel implementation CI_i is called a *valid* implementation of the channel AC_j as long as (i) the primitives offered by the implementation cover all the primitives of the abstract channel, (ii) the performance of the implementation is equal, or superior, to the performance required by the abstract channel, (iii) there are sufficient hardware resources available for the implementation, and (iv) the design requirements of the implementation do not conflict with the design requirements of the other modules, or the characteristics of the environment.

The four conditions are summarized formally as follows [6]:

$$APrimitives_j \subseteq Primitives_i \tag{5.4}$$

$$APerformance_j \leq Performance_i \tag{5.5}$$

$$HWresources_i \subseteq available\ hardware \tag{5.6}$$

$$Constraints_i \cap \{\cup_k Constraints_k \cup Environment\} \neq \emptyset \tag{5.7}$$

The channel implementation allocation procedure considers all of the abstract channels, ACj, in a system specification, and identifies all the channel implementations that are valid for the channel ACj by verifying conditions (5.4)-(5.6). The resulting set is denoted as the set $ValidCI(AC_j)$. Note that conditions (5.4)-(5.6) depend only on the individual channel AC_j and the currently analyzed implementation CI_i. Thus, the results for CIU allocation do not depend on the order of analyzing the abstract channels in the specification.

2. *Abstract channel mapping.* The design task involves: (i) the initial channel mapping is produced for all the abstract channels in the specification by mapping the channels to a dedicated

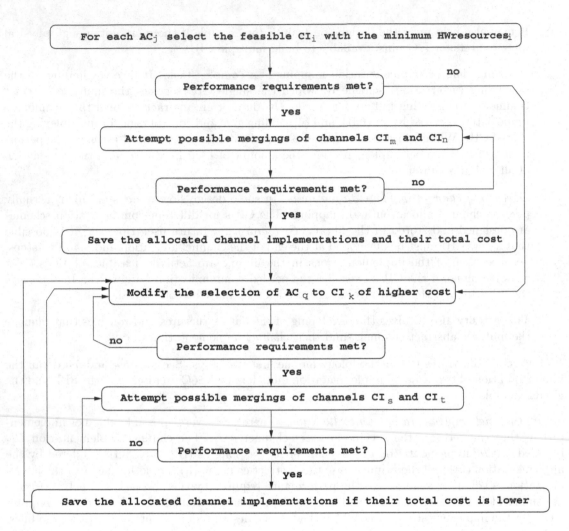

Figure 5.4: Successive refinement steps for communication design.

CIUs and (ii) an improvement step in which alternative mappings are created in the attempt to lower the overall implementation cost by having multiple abstract channels sharing the same CIU.

The initial implementation solution for the communication subsystem selects the feasible channel implementation CI_i that uses the minimum amount of hardware resources $HWresources_i$, and thus the channel implementation with minimum hardware cost, for each abstract channel AC_j. In addition, it must not conflict with the constraints already imposed by the previously selected CIUs (condition (5.7)). The bandwidth of the channel implementation must be sufficiently high to accommodate the average and peak load requirements and the communication delay constraint of the abstract channel. The communication delay estimation must also consider any overhead introduced by the protocols and the hardware-software implementation of the channel, CI_i.

In addition, all design constraints must be satisfied by the selected CIUs. This condition (5.7) involves the constraints due to the other selected CIUs, and therefore introduces correlations among the selected channel implementations. For example, selecting a very fast, but resourcewise expensive, CIU constrains the nature of the CIUs that can be used for mapping the remaining abstract channels within the resource constraints of the design. Therefore, the order of mapping the abstract channels to CIUs is important.

Several strategies can be considered for building the initial mapping. The first strategy, called "most constrained channels first, is to consider the abstract channels in the decreasing order of their performance constraints, for example the abstract channels with high bandwidth requirements, first. Each of the abstract channels is mapped to the lowest cost CIU that delivers the needed performance.

The second strategy, called best bandwidth and cost ratio first, computes the following ratio:

$$R_{CI_j} = \max\left\{\frac{Performance_i}{HW resources_i}\right\} \tag{5.8}$$

for each abstract channel CI_j, where the maximum value is computed considering all the CIU$_i$ that meet the performance requirement of the abstract channel. Then, the abstract channels are mapped in the decreasing order of their R_{CI_j} values.

The one-to-one mapping of the abstract channels to channel implementation can result in violating the design constraints, e.g., the number of required I/O pins may be higher than the number of available pins. Also, the solution may not represent the minimum hardware cost, because the cost can be lowered by sharing one channel implementation for several abstract channels.

To address the design constraint and hardware cost issues, the improvement step of the methodology in Figure 5.4 attempts to map multiple abstract channels to the same channel implementation. This is achieved by "merging" two distinct channel implementations CI_m and CI_n, so that all the abstract implementations initially mapped to the two channels are now mapped onto a single channel implementation. The merging step obviously reduces the hardware cost of the channel implementation. However, the single channel may not offer enough bandwidth to satisfy the communication load and delay requirements of all the mapped abstract channels. This can be addressed by "switching", the shared CIU, to a CIU of higher hardware cost but with superior bandwidth. Repeated channel merging and switching steps are conducted and analyzed until the implementation solution is satisfactory for the entire application.

Two channel implementation units CI_m and CI_n can be merged and replaced by one of the two CIUs, for example CIU CI_m, if the following conditions are met:

$$Primitives_n \subseteq Primitives_m \tag{5.9}$$

$$Compose(APerformance^m, APerformance^n) \leq Performance_m \ req'd. \tag{5.10}$$

Relationship (5.9) states that the channel CI_m should have all the primitives of channel CI_n. The relationship (5.10) denotes that the channel implementation must be capable of meeting the communication speed constraints of all the abstract channels $APerformance^m$ mapped to the channel CI_m and all the abstract channels $APerformance^n$ on the channel CI$_n$. The operator $Compose$ finds the resulting performance after merging the two channels. The performance depends on the scheduling policies used for executing the channel primitives on the CIU. For example, the round-robin policy treats all abstract channels identically, and in sequence, one channel after another. A different approach is to assign a priority to the abstract channels, and "execute" the high priority channel first.

Note that the conditions expressed by (5.6), about having sufficient hardware available, and (5.7), on meeting the design constraints, were not mentioned in the above equation set because they are implicitly satisfied in as much as channel CI_m met the conditions when it was selected. However, for applications where this is an invalid assumption, the two conditions would have to be analyzed.

3. Developing channel implementation units. This step designs the hardware circuits and develops the firmware routines for the CIU.

Because a CIU was previously defined as the tuple:

$$CI_i \;=\; <Primitives_i, HWresources_i, Performance_i, Constraints_i> \tag{5.11}$$

this step implements the primitives $Primitives_i$ as a mixture of hardware circuits and software routines to provide the performance $Performance_i$. The hardware used is expressed as $HWresources_i$. Two case studies are shown for CIU implementations: Subsection 5.4 presents the PSoC-based, hardware/software implementation of the block SPI, and Subsection 5.5 discusses the implementation of the UART block.

Figure 5.5 illustrates the data structures and routines that manage the data communication on multiple abstract channels and CIUs. The data structures create the necessary associations between the abstract channels and CIUs, high-level primitives of an abstract channel and the CIU firmware.

This figure shows that the same set of firmware routines can handle identical CIUs, such as the X channel type. Multiple identical CIUs are used if the channel allocation and mapping confirm that similar abstract channels ought to be mapped to distinct CIUs to meet the bandwidth requirement of the application. A different set of firmware routines handle CIUs of different types, such as the Y channel type. The management subsystem uses two tables:

1. *Symbolic channel names table*: This table links the symbolic names of the abstract channels, for example channel $CH1$, to the physical addresses of the corresponding CIUs, such as the addresses of the hardware control registers, associated RAM/ROM blocks, associated pins, clocks, and so on. The information in this table is chosen after executing the channel allocation and mapping steps.

2. *Firmware routine pointers table*: The entries in this table point to the firmware routines that are needed to implement the high-level primitives of an abstract channel. The firmware routine is selected based on the nature of the high-level primitive and the identity (type) of the actual CIU, which is indicated by the corresponding entry in the symbolic channel names table.

This figure exemplifies the calling of the primitive *send* using the abstract channel $CH1$ to transmit the data value A. The figure assumes that the abstract channel uses the CIU of type X that relies on the hardware block 1 for its physical implementation. This mapping of the abstract channel $CH1$ was determined in the channel allocation and mapping steps shown in Figure 5.4. The high-level primitive *send* maps the channel identifier $CH1$ to the physical addresses corresponding to the CIU, such as the data and control registers of the hardware block 1. Also, the addresses of the firmware routines of the CIU needed for the primitive *send*

send (channel CH1, token A)

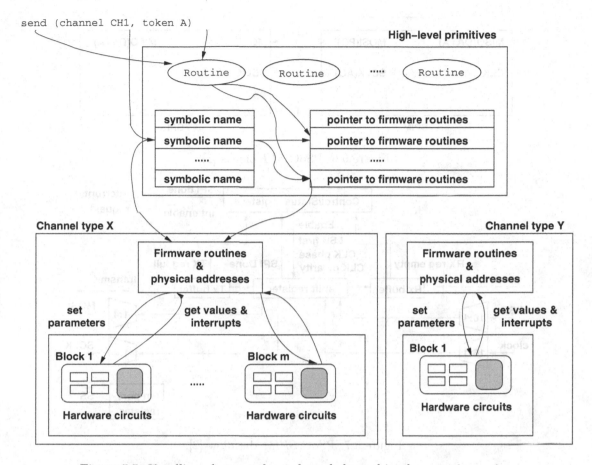

Figure 5.5: Handling abstract channels and channel implementation units.

are identified using the firmware routine pointers table. Then, the firmware routines are called using the physical addresses of the CIU to configure, start, and stop the communication on the channel, and also transmit the data value, A, to the hardware block 1 that actually sends the bits of the data value.

5.4 Channel Implementation Unit: SPI Block

The hardware/software implementation of the serial peripheral interface (SPI) block [1, 2, 3] is based on the SPI protocol that implements a full-duplex, synchronous, serial transmission between two blocks. One of the blocks is configured as the SPI master (SPIM) and the other block as the SPI slave (SPIS).

Full-duplex, serial communication occurs as follows: the SPIM transmits serial bits to the SPIS, while, simultaneously, the SPIS sends serial bits to the SPIM. In addition to bi-directional data transfer, the SPI protocol uses two additional signals generated by the SPIM for the slave block: the clock $SCLK$ and the select signal SS. Figure 5.6 illustrates the SPI protocol signals. (The hardware circuits for the SPI blocks are discussed in Subsection 5.4.1, and the software routines are presented in Subsection 5.4.2.)

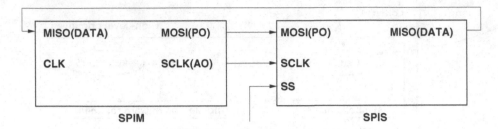

Figure 5.6: PSoC *SPI* blocks [1].

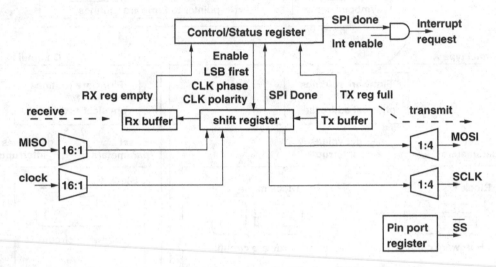

Figure 5.7: PSoC *SPIM* dataflow [2].

5.4.1 Hardware Circuit

Figure 5.7 illustrates the dataflow of the master module *(SPIM)*, Figure 5.8 shows the dataflow of the slave *(SPIS)*, and Figure 5.9 presents the functioning of the SPI hardware using PSoC's programmable digital communication blocks [1, 2, 3]. The related control/status register and data registers are also shown in these figures.

A PSoC programmable digital communication block is configured as an SPI module *(SPIM or SPIS)* by setting bits 2-0 (called the "bits Function") of the FN register to the value 110. The physical addresses of the FN registers are listed in Tables 4.4 and 4.5. If bit 3 of the same register is set to 0, then the module is a SPIM, and if the bit is set to 1, then the module is a SPIS.

The operation of the blocks SPI is shown in Figure 5.9, and is as follows:

- *Clocks*: The circuit uses an internal clock *INT_CLK* with a period twice as long as the period of the input clock *CLK* (the input clock is presented in Figures 5.6, 5.7, and 5.8):

$$T_{INT\_CLK} = 2\,T_{CLK} \tag{5.12}$$

The internal clock is also used by the SPIM to generate the clock, *SCLK*, that is used for operation by the SPIS, as shown in the Figures 5.6–5.8. The clock *SCLK* corresponds to the auxiliary output *(AO)* of the PSoC digital communication block. The clock *SCLK* is

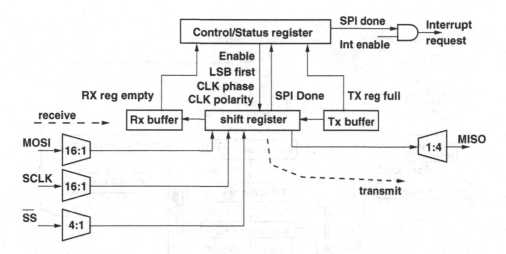

Figure 5.8: PSoC *SPIS* dataflow [3].

produced after the circuit SPIM is enabled, and stops when the circuit is disabled. The clock can be programmed to be idle on 0 (noninverted clock), or on 1 (inverted clock). Non-inverting is programmed by setting the bit 1 (bit *Clock Polarity*) of the control/status register CR0 to 0, and for inverting to 1.

- *Serial transmission and reception*: Circuit functioning for data transmission and reception is summarized in Figure 5.9. The following registers and flags are involved in the communication process.

The SPIM and SPIS modules use the DR0, DR1, and DR2 data registers of the PSoC's programmable digital communication blocks for serial transmission: the DR1 register is the transmit register (TX), the DR2 register is the receive register (RX), and the DR0 register is the shift register (SR) as shown in Figures 5.7 and 5.8. The physical addresses of the registers are given in Tables 4.2 and 4.3.

The following fields of the CR0 register define the status (flags) of the SPIM or SPIS blocks [1].

1. *RX Reg Full*: The bit 3 (*RX Reg Full* bit) is 0, if the register RX is empty, and 1, if a new byte was loaded into the register RX. The bit is reset after reading from the register RX of the corresponding block SPI.

2. *TX REG EMPTY*: Bit 4 (*TX REG EMPTY* bit) of the CR0 register is 0, if there is a byte buffered in the TX register. If data are loaded into the TX register, the bit is cleared.

3. *SPI Complete*: Bit 5 (bit *SPI Complete*) is set to 1, if the byte was shifted out. If the serial transmission is ongoing, or if no transmission is active, the bit is 0. The bit is reset after reading the CR0 register.

4. *Overrun*: Bit 6 (bit *Overrun*) indicates overrun status of the RX register has occurred. If the bit is 0 then there was no overrun, and if the bit is 1, an overrun and hence, data loss has occurred. The bit is reset after reading the CR0 register.

The functioning of the blocks SPI is discussed next. The operation is illustrated in Figure 5.9. The corresponding signal waveforms are shown in Figure 5.10.

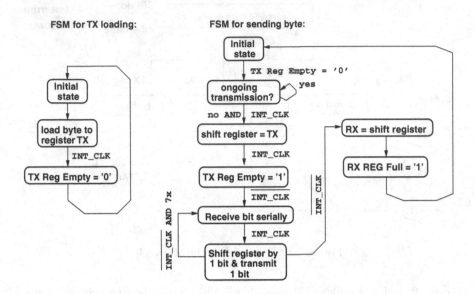

Figure 5.9: *SPI* circuit operation.

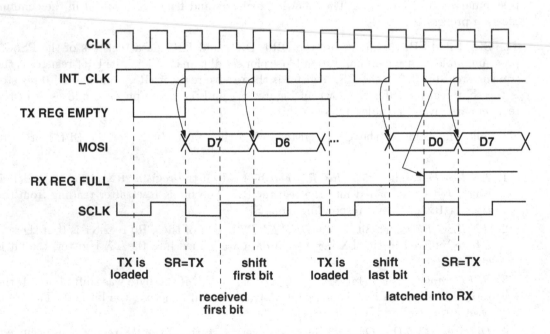

Figure 5.10: Timing diagrams of the *SPIM* circuit [1].

The SPIM and SPIS blocks are enabled by setting bit 0 (*Enable*) of the circuit's CR0 register to 1. The circuit is not enabled, if the bit is 0. The circuit moves to the initial state.

To initiate a serial transmission, a byte is loaded into the SPIM block's TX register, if a transmission is not already in progress. Once the data are loaded into the register, the *TX REG EMPTY* bit is cleared, as shown in the Figure 5.10.

The rising/falling edges of the clock, *INT_CLK*, are used to simultaneously store one input bit, and shift out one output bit. For example, storing (latching) of input bits can be programmed to occur on the rising/falling edge of the clock, and the shifting out of bits to occur on the falling/rising edge. This mode is called 0, if the clock polarity is noninverting, and Mode 1 for an inverting clock polarity [2, 3]. If bit 2 (*Clock Phase* bit) of the CR0 register is set to 0, then latching occurs on the rising edge, otherwise, it occurs the falling edge. If the clock polarity is noninverting this mode of operation is called Mode 3, and if the polarity is inverting then it is denoted as Mode 4 [2, 3]. The figures shown are all for Mode 0.

If no transmission is in progress then the value of the TX register is copied into the shift register (SR). Then, if the *Clock Phase* bit is set to 0, on the following falling edge of the clock, the first bit is shifted in, and on the next rising edge, the first bit is shifted out. After receiving and shifting out eight data bits, the *RX REG FULL* bits and *TX REG EMPTY* are asserted on the falling and, respectively on the rising edge of the clock, as shown in the timing diagram in Figure 5.10. Once the eighth bit has been received, the content of the shift register is copied into the register RX register. If the content of the RX register has not already moved then it is lost, which asserts the *OVERRUN* bit half a cycle before the last bit is received.

The clock signal *SCLK* is generated by SPIM for SPIS. In Mode 0, the rising edge of the clock *SCLK* is used by SPIS to store the input bits, and the falling edge for shifting out the output bits, which are then latched by SPIM. This defines the setup time of the input shift register to be one half of the period of the *INT_CLK* clock:

$$t_{setup} = 0.5 \, T_{INT\_CLK} \qquad (5.13)$$

The functioning is similar for other modes selected by the corresponding settings of the *Clock Phase* and *Clock Polarity* bits.

The SPIM and SPIS blocks can be configured so that they start the serial transmission with either the most significant bit, or the least significant bit. This is achieved by programming the bit 7 (LSB First bit) of the CR0 register. If the bit is 0, then the most significant bit is shifted out first, otherwise the serial transmission starts with the least significant bit.

- *Interrupts*: If enabled, interrupts are produced for two conditions: (1) interrupts are generated when the register TX becomes empty, by default, and (2) when the *SPI Complete* signal is produced. The first interrupt type can be used to signal that the buffer TX is available for transmitting a new byte. The second interrupt indicates that a new byte was received, and is available in the buffer RX. The two interrupt signals are depicted in Figures 5.7 and 5.8.

The two interrupts are selected by setting bit 4 (*Interrupt Type* bit) of the corresponding FN register. If the bit is set to 0, then an interrupt occurs on *TX REG EMPTY*, and if the bit is set to 1 then an interrupt occurs on *SPI Complete*.

- *Slave Select*: The input signal *Slave Select* (SS) must be enabled, in addition to setting the bit *Enable*, for starting the operation of an SPIS module.

The input SS can be controlled by a signal generated by a dedicated block, an I/O pin, or by firmware, but not by the related SPIM block, because SPIM does not have an output for this purpose. The control of the SS signal is programmed by bit 5 ($AUXEN$) of the DCBxxOU register for PSoC's digital communication block. If the bit is 1, then the signal SS is active, and the software-controlled slave selection occurs. In addition to this bit, Bits 4-3 (AUX IO $Select$) of the register DCBxxOU must be set to the value 00. If Bit $AUXEN$ is set to 0, then the signal SS is defined by hardware, as selected by the bits AUX IO $Select$: row input 0 for the bits 00, row input 1 for the bits 01, row input 2 for the bits 10, and row input 3 for the bits 11.

Depending on the mode set for the SPIS, the signal SS must be handled in different ways, so that the setup timing requirement expressed by Equation (5.13) is met. This guarantees the correct functioning of the SPIS. For modes 0 and 1, the signal SS must be disabled after the communication of a byte, and then re-enabled at least T_{setup} before the next byte is communicated. This guarantees the correct reception of the first bit, which is latched on the rising clock edge for the two modes. For modes 2 and 3, the latching is on the falling edge, and therefore the setup timing constraint is automatically satisfied without having to disable and re-enable the signal SS.

The operation of a SPIS is similar to that of a SPIM with the difference that the register TX must be loaded before the falling edge of the input SS.

Example (Performance tradeoffs for the SPI communication). The example analyzes the timing overhead involved in the communication through the block SPI, if the control of the slave SPIS is accomplished by software or by a dedicated hardware circuit.

Figure 5.11 illustrates the first case, in which the signal SS is activated and deactivated by setting the corresponding bits of the register $DCBxxOU$ of the related PSoC communication block. Figure 5.11(a) shows the transmission of the first byte by the SPIM. Five clock cycles at 24 MHz are needed to load the byte in the register TX of the block SPIM, and then nine clock cycles at 12 MHz are needed to transmit serially the byte. Thus, about 1 (0.95) μsec are required to transmit the first byte. This provides a bit rate of about 1 (1.05) Mbits/sec.

Figure 5.11(b) presents the timing overhead, if several bytes are communicated using blocks SPI programmed in modes 0 or 1, and software for activating and deactivating the signal SS of the SPIS. Ten clock cycles at 24 MHz are needed to activate the signal SS for the slave. The loading of the register TX with a new bytes is in parallel with the serial transmission of the previous byte. Finally, the signal SS is deactivated in ten clock cycles at 24 MHz. Thus, the total timing requirement is about 1.57 μsec, and the bit rate is about 0.63 Mbits/sec.

Finally, Figure 5.12 shows the timing overhead if several bytes are communicated, and a dedicated hardware circuit is used to produce the SS signal for the SPIS block. In this case, the total time to transmit one byte is approximately 1.00 μsec to send one byte, or 0.92 Mbits/sec. This is an improvement of about 46% in communication time (speed) at the expense of using an additional PSoC digital block to generate the SS signal. This overhead can be further reduced, if the instruction, MOV, for loading the received data into the RX register is executed in parallel

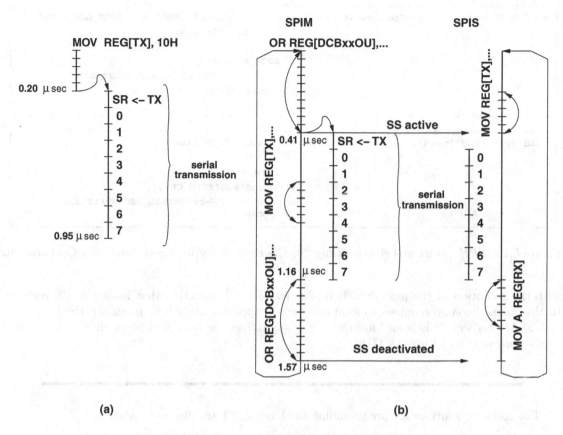

Figure 5.11: *SPI* communication.

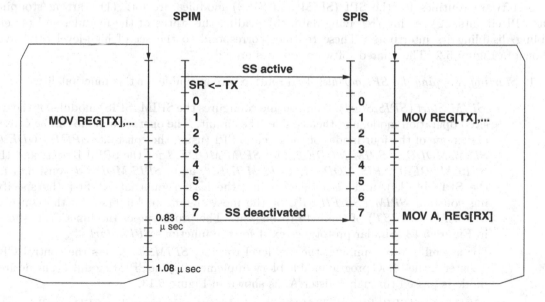

Figure 5.12: Improved *SPI* communication.

```
void SPIM_Start (BYTE bConfiguration);        mov  A, SPIM_MODE_2 | SPIM_LSB_FIRST
                                              call SPIM_Start
                                              ...
                                           SPIM_Start:
                                              or   A, bfCONTROL_REG_START_BIT
                                              mov  REG[SPIM_CR_0], A
                                              ret
```

```
void SPIM_Stop (void);                           call SPIM_Stop
                                                 ...
                                              SPIM_Stop:
                                                 and REG[SPIM_CR_0],
                                                         ~bfCONTROL_REG_START_BIT
                                                 ret
```

Figure 5.13: *SPIM* start and stop routines [2]. Courtesy of Cypress Semiconductor Corporation.

with the reception of the next data byte, similar to the instruction that loads the TX register. In this case, the time required to send one byte is approximately 0.83 μsec, and the bit rate is ≈ 1.20 *Mbits/sec*. This is an improvement of $\approx$ 91% of the communication time, as compared to the case shown in Figure 5.11(b).

The software routines for programming the blocks SPI are discussed next.

5.4.2   Software Routines

The software routines for the SPI (SPIM and SPIS) modules are for: (1) starting/stopping the SPI circuits, (2) sending/receiving data, (3) reading the status of the module, and (4) enabling/disabling its interrupts. These routines correspond to the set of high-level primitives shown in Figure 5.2. The related software routines are [2]:

1. *Starting/stopping the SPI module*: Two routines are available for this functionality:

   - *SPIM_Start (SPIS_Start)*: This routine configures the SPIM (SPIS) module for the desired operation mode, and then enables the circuit. The operation mode can be chosen (1) as one of the four modes of the circuits SPI (using the constants *SPIM_MODE_0*, *SPIM_MODE_1*, *SPIM_MODE_2*, and *SPIM_MODE_3* for the SPIM blocks, and the *SPIS_MODE_0*, *SPIS_MODE_1*, *SPIS_MODE_2*, and *SPIS_MODE_3* constants for the SPIS blocks), and (2) either sending the least significant bit first (by specifying constant *SPIM_LSB_FIRST*), or the most significant bit first (by the constant *SPIM_MSB_FIRST*) of the byte to be sent. The prototype of the function is shown in Figure 5.13. Similar prototypes exist for the functions *SPIS_Start* [3].

     For assembly programming, the low level routine, *SPIM_Start*, sets the control CR0 register of the PSoC programmable block implementing the SPIM module. The desired mode is passed through register A, as shown in Figure 5.13.

   - *SPIM_Stop (SPIS_Stop)*: This function disables the corresponding SPIM (SPIS) module by resetting the *Enable* bit of the corresponding programmable block's CR0 reg-

```
          void SPIM_SendTxData (BYTE bSPIMData);              mov  A, bSIMData
                                                              call SPIM_SendTxData
                                                              ...

                                                          SPIM_SendTxData:
                                                              mov  REG[SPIM_RX], A
                                                              ret
```

```
          BYTE SPIM_bReadRxData (void);                       call SPIM_bReadRxData
                                                              mov  bRxData, A
                                                              ...

                                                          SPIM_bReadRxData:
                                                              mov  A, REG[SPIM_RX]
                                                              ret
```

Figure 5.14: *SPIM* send and receive routines [2]. Courtesy of Cypress Semiconductor Corporation.

```
          BYTE SPIM_bReadStatus (void);                       call SPIM_bReadStatus
                                                              and  A, SPIM_DONE
                                                              ...

                                                          SPIM_bReadStatus:
                                                              mov  A, REG[SPIM_CR0]
                                                              ret
```

Figure 5.15: *SPIM* read status function [2]. Courtesy of Cypress Semiconductor Corporation.

ister. The prototype for this function and its Assembly language implementation are shown in Figure 5.13.

2. *Sending/Receiving data*: The related software routines are as follows.

- *SPIM_SendTxData (SPIS_SendTxData)* : This routine sends one byte from the SPIM (SPIS) to the slave (master) device. The function has one parameter, which is the byte to be transmitted, and does not return any value [2]. The prototype of the function is shown in Figure 5.14.

  This low-level routine, also shown in Figure 5.14, actually loads the byte into the TX register of the SPIM (SPIS) module. The data for this routine are passed through the register A, as shown in Figure 5.14.

- *SPIM_bReadRxData (SPIS_bReadRxData)*: This low-level routine reads a byte sent by the slave (master) module for the master (slave), and returns the value to the caller. The prototype of the function is given in Figure 5.14.

  This routine places the byte received from the slave module into the A register. Figure 5.14 shows this process. The low-level routine reads the value in the RX register of the circuit SPIM (SPIS).

```
#include "PSoCAPI.h"
char Message[] = "Hello world!";
char *pStr = Message;

void main() {
    SPIM_Start (SPIM_MODE_0 | SPIM_MSB_FIRST);
    while (*pStr != 0) {
        while (!(SPIM_bReadStatus() & SPIM_TX_BUFFER_EMPTY));
        SPIM_SendTxData (*pStr);
        pStr++;
    }
}
```

Figure 5.16: Program using the *SPIM*-related routines [2]. Courtesy of Cypress Semiconductor Corporation.

3. *Reading the status of the module*: The status of the blocks SPI can be accessed by using the following function:

- *SPIM_bReadStatus (SPIS_bReadStatus)* : As shown by the function prototype in Figure 5.15, this routine returns one byte, which is the content of the control/status register, CR0, of the corresponding PSoC block. The assembly-level routine for accessing the status information is also illustrated in the figure.

  To access the individual status flags, a bitwise-*AND* instruction must use the returned value and the following masks as operands: (i) mask 0x20 to access the flag *SPI Complete*, (ii) mask 0x40 for the flag *Overrun*, (iii) mask 0x10 for the flag *TX REG EMPTY*, and (iv) mask 0x08 for the flag *RX REG FULL*.

4. *Enabling/disabling interrupts*: There are two routines for this functionality.

- *SPIM_EnableInt (SPIS_EnableInt)*: This routine enables the SPIM (SPIS) interrupt on the *SPI Complete* condition. The function does not have any parameters, and does not return a value.

- *SPIM_DisableInt (SPIS_DisableInt)*: This function disables the SPIM (SPIS) module's interrupt.

5. *Enabling/disabling the Select signal of SPIS*: The two functions are as follows.

- *SPIS_EnableSS*: The related block SPIS is enabled provided that the *SS* input of the block is controlled by software. This routine does not have any parameters, and does not return a value.

- *SPIS_DisableSS*: This function disables the input *SS* of the PSoC communication block used for the SPIS, if the input is controlled by software. The routine does not have any parameters, and does not return a value.

Example (SPIM communication). This example presents a C program that transmits the characters of a character string using the high-level primitives of the block SPIM [2]. The code is shown

```
#include "PSoCAPI.h"

void main() {
    BYTE bData;
    SPIS_Start (SPIS_MODE_0 | SPIS_MSB_FIRST);
    while (1) {
        while (!(SPIS_bReadStatus() & SPIS_SPI_COMPLETE) ) ;
        bData = SPIS_bReadRxData ();
        ...
        SPIS_SetupTxData (bData);
    }
}
```

Figure 5.17: Program using the *SPIS* related routines [2]. Courtesy of Cypress Semiconductor Corporation.

in Figure 5.16. The character string is stored by the variable *Message*, and the pointer *pStr* is used to traverse the characters in the string. The hardware circuit of the block SPIM is configured and enabled using the *SPIM_Start* routine (Mode 0 is used for the serial transmission, and the most significant bit of a byte is sent out first). The variable *pStr* traverses each of the string characters until reaching the character '\0' that signals the end of the string. For each character, the program first waits until the TX register of the SPIM is empty. The *TX REG EMPTY* flag signals, if the buffer is empty: the value returned by the function *SPIM_bReadStatus* is bitwise *AND*-ed with the mask 0x10 (constant *SPIM_TX_REG_EMPTY*). After the TX buffer becomes empty, the character pointed to by the variable *pStr* is transmitted by calling the *SPIM_SendTxData* routine. Then, the variable *pStr* is incremented to point to the next character, and the loop repeats until the character '\0' is reached and the program ends.

Example (SPIS communication): This second example discusses a C program for receiving bytes by using the SPIS module. The program illustrates the use of the high-level primitives related to the SPIS block. The initial instruction configures the block by calling the *SPIS_Start* routine. Then, inside the infinite loop, the code waits until the next byte has been "completely" received by polling the *SPI Complete* flag. The byte is retrieved by calling the *SPIS_bReadRxData* function, and stored in the variable, *bData*. The *SPIS_SetupTxData* function is called to transmit a byte to the SPI master module.

5.5 Channel Implementation Unit: UART Block

The universal asynchronous receiver transmitter block (UART) is an 8-bit, duplex, RS-232 compliant, serial communication block [1, 4]. The transmitted frames are 11-bits long, if the optional parity bit is included, or 10-bits long, if no parity bit is transmitted. The structure of the frame

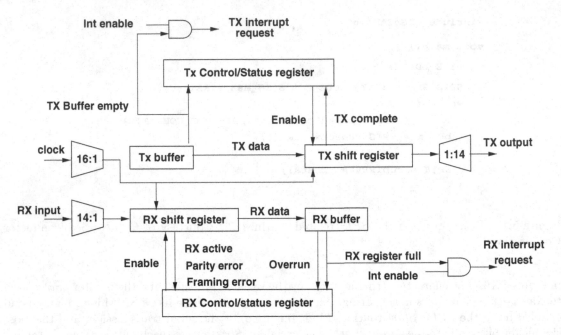

Figure 5.18: Block structure of the block UART [4].

is the following: a start bit is followed by eight data bits, an optional parity bit, and finally the stop bit.

The implementation of the UART block includes both hardware circuits and software routines. The hardware circuits implement the UART functionality by using two PSoC programmable communication blocks [1]. Bits 2-0 of the block's FN register must be set to 101. In addition, bit 3 must be set to 0, for the PSoC programmable block to operate as a receiver, and to 1 as a transmitter. The physical addresses of the FN registers are shown in Table 4.4. In addition, the circuits can be programmed to select clock frequency, and the UART's interrupt conditions.

The software routines offer a high-level interface to initialize the circuit, enable its operation, set its communication parameters, and send/receive a data byte.

5.5.1   UART Hardware Circuit

The hardware includes circuits for the transmit block and the receive block each of which is capable of operating independently.

A. Transmit block

This circuit uses two data registers and one control register: the transmit (TX) register uses the DR1 register of the programmable communication blocks, the DR0 register as the shift register, and the CR0 register as the control register. The physical addresses of the data registers are in Tables 4.2 and 4.3, and the addresses of the control registers are given in Tables 4.4 and 4.5. The structure of the block is shown in Figure 5.18.

The operation of the transmit block is illustrated in Figure 5.19. The operation is controlled by the internal clock, INT_CLK, which is obtained by dividing the UART's external clock, CLK, by eight, that is the frequency of the clock, CLK, has to be eight times larger than the desired

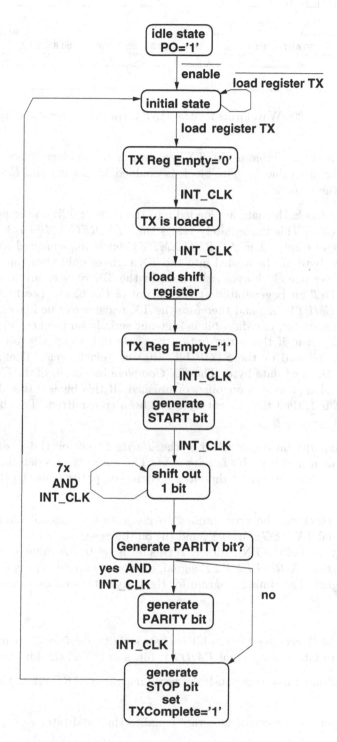

Figure 5.19: UART receiver operation [4].

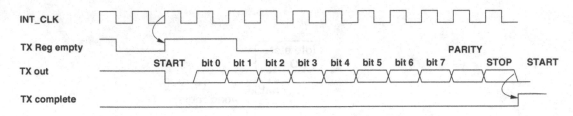

Figure 5.20: Waveforms for the UART transmit operation [4].

baud rate. If no data are available in the TX register, or if the block is not enabled, then the block output is set to the value 1. The block is enabled by setting the *Enable* bit (bit 0) of the CR0 register to the value 1.

After enabling the block, the data are loaded into the register TX, on the positive edge of the internal clock, *INT_CLK*. This immediately resets the *TX REG EMPTY* bit, indicating that the register is no longer empty. The *TX REG EMPTY* bit is implemented as bit 4 of the CR0 register. Thus, a new byte can be loaded into the TX register only after the *TX REG BIT* bit is set to 1 again. After one clock cycle, the data in the TX register are loaded into the shift register and the *START* bit is generated at the output of the block. Loading the shift register resets the *TX REG EMPTY* bit, and therefore the TX register can be loaded with a new byte. After generating the start bit, one data bit is sent out serially on the next eight rising edges of the clock, *INT_CLK*. Then, if the parity option was selected, the parity bit is transmitted on the next rising edge, followed by the *STOP* bit, after one clock cycle. Then, the transmission process continues for the next data byte. The *TX Complete* bit (bit 5) of the CR0 register signals whether the transmission process is ongoing, or finished. If this bit is 0 then bits are still being shifted out, and if it is 1, then the current byte has been transmitted. This bit is cleared if the CR0 register is read.

Generation of the parity bit is controlled by the *Parity Enable* bit (bit 1) of the CR0 register. If this bit is set to 1 then the parity bit is generated. The parity type is determined by the *Parity Type* bit (bit 2) of the CR0 register. If the bit is 0 then even parity is used, otherwise odd parity bit is generated.

The transmitter block can be programmed to generate two kind of interrupts: (i) on the generation of the signal *TX REG EMPTY*, and (ii) on the generation of the *TX Complete* signal which are selected by bit 4 of the FN register. If this bit is set to the value 0, then an interrupt is generated on setting the *TX REG EMPTY* signal, otherwise an interrupt is produced on setting the *TX Complete* signal. The timing diagram for the transmit operation is shown in Figure 5.20.

B. Receive Block

The UART receiver block receives 11 (10)-bit packets serially, the first bit being the *START* bit, followed by eight data bits, one optional *PARITY* bit, and the *STOP* bit.

- Reception is initiated asynchronously by receiving the *START* bit at the input (the bit is encoded as 0).

- Then, each input bit is received at a rate equal to the baud rate.

- The *PARITY* bit can be programmed for checking for even, or odd, parity. For even parity there must be an even number of 1 bits in each of the 11 (10)-bit packets received, and for odd parity, there must be an odd number of 1 bits in each packet.

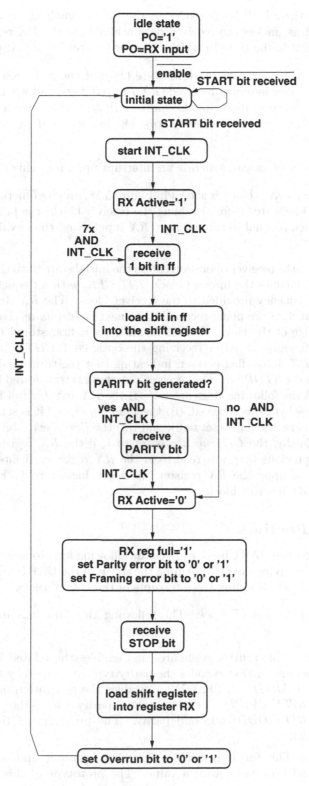

Figure 5.21: UART receiver operation [4].

- The receiving of the *STOP* bit (encoded as the bit 1) signals the end of the packet. This sets the status flags, makes the received data available in the RX register, and resets the receiver block back to the state in which it is ready to receive a new packet.

- The status flags of the receiver block indicate (1) that the reception is active (*RX Active* flag), or (2) that a new byte was received (*RX Reg full* flag), and whether the reception was correct. In addition, these flags can also indicate that (3) there was a framing error (*Framing error* flag), (4) a parity error (*Parity error* bit) has occurred, or (5) an overrun situation occurred (bit *Overrun*).

- The UART receiver block can generate an interrupt upon receiving a new data byte.

The dataflow of the receiver block is shown in Figure 5.18, and its functionality is summarized in Figure 5.21. In the idle state (before enabling the receiver block) the primary output (*PO*) of the UART receiver block is equal to the receiver (*RX* input), and the auxiliary output (*AO*) has the value 1.

After being enabled, the receiver operation enters the initial state until the *start* bit is received at the *RX* input. This enables the internal clock, *INT_CLK,* with a frequency eight times lower than the input clock frequency provided to the receiver block. The *RX Active* flag is set to the value 1, indicating that data are being received. The next eight bits received represent the data byte: on the rising edge of the clock, *INT_CLK*, each bit is first stored in the input flip flop, and then into the shift register. After receiving the optional *PARITY* bit, on the next clock cycle, *INT_CLK*, the *RX Active* flag is reset, indicating that the transmission has ended. Then, the next bit received is the *STOP* bit. The received data byte is transferred from the shift register to the *RX* register and the following flags are set: (i) the *RX* register full bit is set to the value 1 to indicate that a new byte was received, (ii) the *Framing error* bit is set to the value 1, if the *STOP* bit was 0 (otherwise the flag is set to 0), and (iii) the *Parity* error bit is set to 1, if a parity error was detected. Finally, the *Overrun* flag is set to 1, if the *RX* register was loaded with a new byte, before the previous byte was read from the *RX* register. If interrupts were enabled, an interrupt is produced upon the *RX* register full flag being set to 1. Figure 5.22 shows the waveforms for the UART receiver block.

5.5.2   Software Routines

The software routines for the UART block provide the following functions [4]: (1) starting/stopping the block, (2) sending/receiving data, (3) reading the status of the UART receiver and transmitter subblocks, and (4) enabling/disabling the interrupts of the UART Block:

1. *Starting/stopping the UART block*: The following two functions implement the related functionality:

   - *UART_Start*: This routine configures and enables the related UART block and has one input parameter that encodes the parity type to be used by the module. The constant 0x00 (*UART_PARITY_NONE*) disables the generation of the parity bit, 0x02 (*UART_PARITY_EVEN*) defines even parity for the block, and 0x06 (*UART_PARITY_ODD*) sets odd parity. The prototype of this function is shown in Figure 5.23.

   - *UART_Stop*: This function disables the related module and does not have any parameters, and does not return a value. The prototype of this function is shown in Figure 5.23.

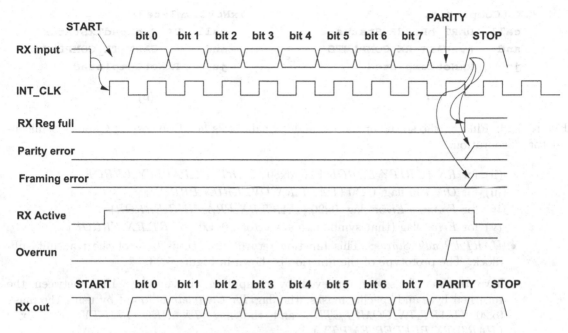

Figure 5.22: Waveforms for the UART receiver operation [4].

```
void UART_Start (BYTE bParitySetting);      void UART_Stop (void);

void UART_SendData (BYTE bTxData);          void UART_EnableInt (void);

BYTE UART_bReadRxData (void);               void UART_DisableInt (void);

BYTE UART_bReadTxStatus (void);            BYTE UART_bReadRxStatus (void);

void UART_SetTxIntMode (BYTE bTxIntMode);   void UART_IntCntl (BYTE bMask);
```

Figure 5.23: UART related functions [4]. Courtesy of Cypress Semiconductor Corporation.

2. *Sending/receiving data*: The corresponding software routines are as follows.

- *UART_SendData*: This routine transmits a byte that is passed as an argument to the routine. The function loads the byte to the TX register, and starts the serial communication process. The ending of the serial communication is indicated by the *TX Complete* flag being set. The prototype of this function is shown in Figure 5.23.

- *UART_bReadRxData*: This function returns the data byte that was received, and stored in the RX register. The prototype of the function is given in Figure 5.23.

3. *Reading the status of the UART receiver/transmitter*: The status of the receiver and transmitter subblocks is retrieved by calling the following two functions.

- *UART_bReadRxStatus*: This function returns the status byte of the receiver subblock. The prototype of the function is shown in Figure 5.23.

  The individual flags are retrieved by computing a bitwise *AND* of the byte returned by the function and the following constants:

  (i) for *RX REG FULL* flag, the mask 0x08 (*UART_RX_REG_FULL*)

```
RxNotCompleted:                         TxNotCompleted:
  call   UART_bReadRxStatus               call   UART_bReadTxStatus
  and    A, UART_RX_COMPLETE              and    A, UART_TX_COMPLETE
  jz     RxNotCompleted                   jz     TxNotCompleted
```

                    (a)   (b)

Figure 5.24: Finding the status of the Rx and Tx subblocks [4]. Courtesy of Cypress Semiconductor Corporation.

   (ii) for *RX PARITY ERROR* Flag,0x80 (*UART_RX_PARITY_ERROR*)
   (iii) for *Overrun* flag, 0x40 (*UART_RX_OVERRUN_ERROR*)
   (iv) for *Framing Error* flag, 0x20 (*UART_RX_FRAMING_ERROR*)
   (v)   for *Error* flag (that symbolizes any error), 0xE0 (*UART_RX_ERROR*)

- *UART_bReadTxStatus*: This function returns the status byte of the transmit sub-block. The prototype of the function is shown in Figure 5.23.

  Two distinct flags are retrieved by computing the bitwise *AND* between the returned byte and specific masks: the flag *TX Complete* is found by using the mask 0x20 (*UART_TX_COMPLETE*), and the flag *TX REG EMPTY* by 0x10 (*UART_TX_BUFFER_EMPTY*).

4. *Enabling/disabling the interrupts of the block UART*: The following four routines are defined for this purpose.

   - *UART_EnableInt*: This routine enables the two interrupts of a UART block. The prototype of the function is shown in Figure 5.23.
   - *UART_DisableInt*: This function disables all the interrupts of a UART block. The prototype of the function is shown in Figure 5.23.
   - *UART_SetTxIntMode*: This routine programs the interrupt mode of the *TX* sub-block.The prototype of the function is shown in Figure 5.23. If the input parameter is set to 0x00 (*UART_INT_MODE_TX_REG_EMPTY*) then an interrupt is generated after the *TX REG EMPTY* flag is asserted, and if set to 0x01 (*UART_INT_MODE_TX_COMPLETE*) then after the *TX Complete* flag has been set an interrupt is produced.
   - *UART_IntCntl*: This function separately enables/disables the interrupts of the receiver and transmitter subblocks. The prototype of the function is shown in Figure 5.23. The receiver and transmitter interrupts are enabled by passing the following constants *UART_ENABLE_RX_INT* and *UART_ENABLE_TX_INT*, respectively, as parameters to the function. The interrupts are disabled by calling the routine with the following parameters: the constant *UART_DISABLE_RX_INT* for the receiver, and the constant *UART_DISABLE_TX_INT* for the transmitter.

Example [4]: This example demonstrates how to find the status flags for the Rx and Tx subblocks. The related assembly code instructions are shown in Figure 5.24.

Figure 5.24(a) illustrates the code for polling the status *RX Complete* flag until the next byte is received. Calling the *UART_bReadRxStatus* subroutine returns the status of the *RX*

```
        mov   A, UART_PARITY_NONE
        call  UART_Start
        call  Counter8_Start
.WaitForData:
        call  UART_bReadRxStatus
        and   A, UART_RX_Complete
        jz    .WaitforData
        and   A, UART_RX_ERROR
        jz    .GetData
        mov   A, 0x00
        jmp   .TxData
.GetData:
        call  UART_bReadRxData
.TxData:
        call  UART_SendData
        jmp   .WaitForData
```

Figure 5.25: Receiving and sending data using the *UART* block [4]. Courtesy of Cypress Semiconductor Corporation.

subblock. The specific bit is found by computing the bitwise *AND* of the status byte and the *UART_RX_COMPLETE* mask. Figure 5.24(b) presents the code for polling the status flag *TX Complete* until the ongoing transmission ends.

Example [4]: This second example discusses the assembly code for an application that receives bytes using the UART communication protocol, and then returns the received byte, if the byte was correctly received, or the value 0x00, if an error has occurred during transmission.

The code first configures and enables the block by calling the *UART_Start* subroutine. The module then waits until the byte has been received in the RX buffer. If no error has occurred during the reception, then the data are retrieved from the RX buffer and stored into the A register by calling the *UART_bReadRxData* routine. If an error has occurred then the constant 0x00 is loaded into the A register. The data are returned by calling the *UART_SendData* subroutine.

5.6   Conclusions

This chapter has presented a design methodology for implementing performance optimized communication subsystems for embedded applications. Serial communication modules based on the SPI and UART standards have been detailed, and their implementation, using PSoC's programmable digital block, explained.

Communication channels offer a set of high-level primitives for data communication between two systems. Channels are important for connecting multiple embedded systems, or a system to its peripherals. High-level primitives are used to configure the channel, sending/receiving data on the channel, and managing the channel operation and properties. An example was presented that illustrates the use of primitives to provide the functionality of the communicating modules.

This chapter has also defined the main characteristics of channel implementation units (CIUs) for abstract channels, including performance attributes, or example the bandwidth of CIUs, and constraints, such as the number of required links, availability of global clocks, and noise level of the environment. To help in devising a systematic methodology for mapping abstract channels to predefined CIUs, CIUs are described as tuples with four components: (i) offered primitives, (ii) needed hardware resources, (iii) their performance values, and (iv) specific constraints. Then, the tuple information can be used to identify the CIUs, in a predefined library, that are implementation alternatives for a specific abstract channel.

The design methodology presented finds optimized implementations for the communication channels in an embedded application by conducting the following activities: (i) selecting the set of feasible CIUs for each abstract channel, (ii) finding the abstract channels that share the same CIUs, and (iii) building hardware/software implementations for the channels that cannot be mapped to the existing CIUs.

The remainder of this chapter focused on the *SPI* and *UART* communication modules, and their hardware/software implementation based on PSoC's programmable communication blocks. The SPI protocol is an example of a serial, full-duplex, synchronous data transmission between a master and a slave. Data transmission and reception are simultaneous, and clocked by a clock generated by the master. UART is a serial, eight-bit, full-duplex, asynchronous protocol. The transmitted frames are 11 bits long, and include a parity bit for detecting erroneous data transfers. Several examples illustrated the use of the SPI and UART firmware routines, and propose solutions that minimize the timing overhead of SPI data transmission.

5.7   Recommended Exercises

1. Extend the concept of abstract data channels to define prioritized abstract data channels, in which data channels have specific priorities assigned to them. If two abstract channels are mapped on the same CIU then the channel with higher priority is served first, if the two channels need to communicate simultaneously. Discuss the high-level primitives that must be introduced for handling channel priorities. Detail the implementation details required for the CIUs.

2. Using the flow shown in Figure 5.3, write C programs for two communicating modules. One module reads data from its input port, and sends it to the other module, which stores the data in the flash memory. Use the UART protocol for the application. Identify the maximum rate of the data at the input port, so that no data are lost in your design.

3. Study the PSoC based implementation of the 1-wire protocol as described in the Application Note by W. Randall 1-Wire User Modules, AN2166, Cypress Microsystems, May 24 2004.

4. For the communication protocols presented in the chapter bibliography (SPI, UART, 1-wire, 9-bit protocol), identify the set of primitives, performance attributes, hardware requirements, and constraints for each of the protocols.

5. Propose an algorithm that implements the Compose operator required for the channel merging step. The goal should be to satisfy the timing requirements of the application while meeting the constraints set for the hardware resources available for CIUs.

6. Following the generic data structure shown in Figure 5.5, indicate the data structure needed to implement the communicating modules shown in Figure 5.1(b). Assume that the abstract data channels *CH1*, *CH2*, and *CH3* share the same CIU, and channel *CH4* is mapped to a distinct CIU.

7. Study the I2C protocol for PSoC as described in [1].

8. Write C code that implements the data communication shown in Figure 5.1(b). Assume that modules *M1* and *M2* communicate using the SPI protocol, while UART protocol is used for the rest of the communications. Estimate the timing overhead that is due to data communication. Identify the number of I/O pins that are used for communication considering that each module is executed on a separate PSoC chip.

9. For the two modules shown in the Figure 5.1, write C programs for the modules, such that one byte is sent from module *M1* to module *M2* using UART then the next byte is sent using SPI, and so on. Your solution should reuse the same hardware blocks to implement the SPI and UART protocols (dynamic reconfiguration). Estimate the timing overhead that is due to dynamic reconfiguration, and compare this overhead with the execution time of the data communication.

10. For the structure shown in Figure 5.1(b), write a C function that would broadcast a byte from module *M1* to the other three modules. After sending out a byte, module *M1* waits until an acknowledgement signal is received from all modules, indicating that the byte was received.

11. For Exercise 10, extend the solution such that modules *M1* waits for a predefined time period T to receive the acknowledgment signal from the other modules. In the case, where the

acknowledgment signal is not received, _M1_ assumes that the communication was not successful, and resends the data to the module that did not acknowledge.

12. For Exercise 10, propose an implementation of the broadcast function that is flexible in handling any number of targeted modules, provided that there is a path from module _M1_ to the other modules.

Bibliography

[1] PSoC Mixed Signal Array, Technical Reference Manual, *Document No. PSoC TRM 1.21*, Cypress Semiconductor Corporation, 2005.

[2] SPI Master, *CYC29/27/24/22/21xxx, CY7C6xxxx, and CYWUSB Data Sheet*, Cypress Semiconductor Corporation, October 3 2005.

[3] SPI Slave, *CYC29/27/24/22/21xx, CY7C6xxxx, and CYWUSB Data Sheet*, Cypress Semiconductor Corporation, October 3 2005.

[4] UART, *CYC29/27/24/22/21xxx, CY7C64215, and CYWUSB Data Sheet*, Cypress Semiconductor Corporation, October 3 2005.

[5] D. Comer, *Computer Networks and Internets with Internet Applications*, Upper Saddle River, NJ: Prentice Hall, Third edition, 2001.

[6] J.-M. Daveau, G. F. Marchioro, T. Ben-Ismail, A. A. Jerraya, Protocol selection and interface generation for HW-SW codesign, *IEEE Transactions on Very Large Scale Integration Systems*, 5, (1), pp. 136-144, March 1997.

[7] G. De Micheli, R. Ernst, W. Wolf, *Readings in Hardware/Software Co-Design*, San Francisco: Morgan Kaufmann, 2002.

[8] A. A. Jerraya, J. Mermet, *System-Level Synthesis*, A. A. Jerraya, J. Mermet (editors), Boston: Kluwer Academic Publishers, 1999.

[9] A. Kagan, Implement 9-Bit Protocol on the PSoC UART, *Application Note AN2269*, Cypress Semiconductors, September 20 2005.

[10] V. Kremin, RC5 Codec, *Application Note AN2091*, Cypress Microsystems, February 5 2003.

[11] O. Ozbek, Estimating PSoC Power Consumption, *Application Note AN2216*, September 21 2004.

[12] M. Raaja, Binary to BCD Conversion, *Application Note AN2112*, Cypress Semiconductors, February 12 2003.

[13] M. Raaja, Manchester Encoder using PSoC, *Application Note AN2281*, Cypress Semiconductors, June 2 2005.

[14] W. Randall, 1-Wire User Modules (Introduction), *Application Note AN2166*, Cypress Microsystems, May 24 2004.

[15] V. Sokil, Hardware Bitstream Sequence Recognizer, *Application Note AN2326*, Cypress Semiconductors, December 8 2005.

[16] V. Sokil, Serial Bit Receiver with Hardware Manchester Decoder, *Application Note AN2325*, Cypress Semiconductors, December 8 2005.

[17] J. Valvano, Embedded Microcomputer Systems. Real Time Interfacing, London, Third edition: Thomson, 2007.

Chapter 6

Continuous-Time, Analog Building Blocks

This chapter[1] presents a discussion of basic concepts of continuous-time analog circuits, including circuit operation and performance characterization in the presence of nonidealities. Also discussed are the uses of PSoC's programmable, continuous-time, reconfigurable analog blocks as application-specific analog circuits.

The first section introduces the defining characteristics of operational amplifiers (OpAmps). OpAmps are described by a set of nonlinearities, including finite, frequency-dependent gain, poles and zeros, distortion, offset voltages, output saturation, slew rate, common mode gain, and noise. Understanding the impact of nonidealities on OpAmp performance is very important, not only for circuit design (which is not the focus of this text), but also for predicting the performance of analog and mixed-signal subsystems.

The discussion uses OpAmp structural macromodels as a vehicle for describing the functionality/performance of OpAmps and OpAmp-based blocks. Macromodels are an efficient, yet accurate enough, method of capturing the nonidealities of real OpAmps for system design. System simulation, for design and verification, using macromodels is also fast and applicable to complex subsystems, in contrast to transistor-level simulation, which is precise, but very slow.

The second section introduces OpAmp-based analog blocks, that is inverting and noninverting amplifiers, instrumentation amplifiers, summing and difference amplifiers, integrators, and comparators, both with and without hysteresis. Estimates of the impact of the OpAmp nonidealities on the performance of the basic blocks is also discussed. The third section introduces PSoC's programmable, continuous-time, analog blocks and their structure. The control programmable registers that are used to configuring the analog block are also discussed in detail.

This chapter has the following structure:

- Section 1 defines a broad set of OpAmp nonidealities and presents a systematic method for building OpAmp macromodels.

- Section 2 focuses on the OpAmp-based basic block and estimates performance.

- Section 3 introduces PSoC's programmable continuous-time analog blocks.

---

[1]Sections of this chapter are reproduced from references [12] and [17]. Courtesy of Cypress Semiconductor Corporation.

A. Doboli, E.H. Currie, *Introduction to Mixed-Signal, Embedded Design*,
DOI 10.1007/978-1-4419-7446-4_6, © Springer Science+Business Media, LLC 2011

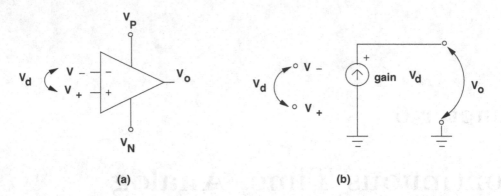

(a)   **(b)**

Figure 6.1: Ideal OpAmp symbol and macromodel.

- Section 4 provides the chapter conclusions.

6.1 Introduction to Operational Amplifiers

Operational amplifiers, widely known as OpAmps, are the basic building blocks in analog design. The primary use of OpAmps is to amplify (multiply) an input signal by a constant value. The input signal is usually a voltage, but can also be a current.

The output signal of an OpAmp, usually a voltage, is given by the following relationship:

$$V_o = gain \times V_d \tag{6.1}$$

where V_o is the output signal of the operational amplifier, V_d is the differential input signal ($V_+ - V_-$), and *gain* is the constant gain of the OpAmp. This amplifier is also referred to as a *linear* operational amplifier, due to the linear relationship between its input and output signals.

6.1.1 Ideal OpAmps

By definition, an *ideal* OpAmp is described by the following properties:

- *Gain:* The gain of the OpAmp is assumed to be infinite and not a function of frequency, or input amplitude. The OpAmp's gain at frequency zero is called its DC gain, and denoted as $gain_0$.

- *Input impedance:* The ideal OpAmp has infinite input impedance that does not depend on frequency. Therefore, the input current of an ideal OpAmp is zero.

- *Output impedance:* The ideal OpAmp has zero output impedance that is also independent of frequency.

Figure 6.1 shows the symbol and model for an ideal OpAmp.

6.1.2 Real OpAmps

Real OpAmps deviate significantly from that of ideal linear operational amplifiers with respect to gain, phase, frequency, linearity and noise. The following are the more important nonidealities of a real OpAmp:

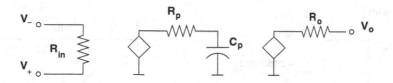

Figure 6.2: OpAmp macromodel.

- *Finite gain*: The gain of real OpAmps is a function of frequency which can be expressed in terms of poles and zeros. The DC gain is finite.

- *Poles and zeros*: The behavior of real, i.e., nonideal, operational amplifiers depends on the frequency. The transfer function of an OpAmp as a function of frequency (defined as the ratio V_o/V_d), can be expressed as:

$$\frac{V_o}{V_d} = \frac{a_n s^n + a_{n-1}s^{n-1} + a_{n-2}s^{n-2} + ... + a_0}{b_m s^m + b_{m-1}s^{m-1} + b_{m-2}s^{m-2} + ... + b_0} \qquad (6.2)$$

$$= \frac{a_n(s - z_n)(s - z_{n-1})...(s - z_0)}{b_m(s - p_m)(s - p_{m-1})...(s - p_0)} \qquad (6.3)$$

where $s = j\omega$, $\omega = 2\pi f$, f is the frequency, p_m, p_{m-1}, ... , p_0 are the poles of the amplifier, and z_n, z_{n-1}, ..., z_0 are its zeros. Thus the magnitude and phase the gain for real OpAmps are both functions of frequency.

The DC gain is given by:

$$gain_0 = \frac{a_0}{b_0}. \qquad (6.4)$$

The transfer function of an OpAmp is often approximated using only first-order polynomials of the variable s:

$$gain(f) = \frac{V_o}{V_d} = \frac{a_1}{b_1}\left[\frac{s - z_d}{s - p_d}\right] \qquad (6.5)$$

where p_d is the dominant pole, and z_d is the dominant zero. If the circuit does not have zeros, or if the zeros have little influence on the transfer function for the frequency domain of interest, then the gain is:

$$gain(f) = \frac{V_o}{V_d} = \frac{a_0}{b_1}\left[\frac{1}{s - p_d}\right] \qquad (6.6)$$

or

$$gain(f) = \left[\frac{gain_0}{1 - j\frac{f}{f_{3db}}}\right] \qquad (6.7)$$

where $gain_0$, the gain at frequency zero, is:

$$gain_0 = -\frac{a_0}{b_1\,p_d} \qquad (6.8)$$

f_{3dB} is called the 3 dB frequency of the circuit [2], and is defined as

$$f_{3dB} = \frac{p_d}{2\pi} \qquad (6.9)$$

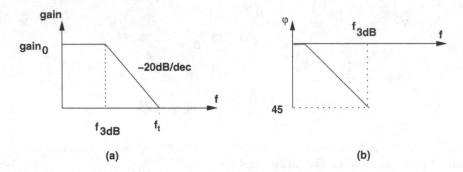

(a)   (b)

Figure 6.3: Magnitude and phase response for the single-pole OpAmp model.

Equation (6.7) represents the single-pole model of an OpAmp, and Figure 6.2 shows the circuit description of this model. The pole is expressed as:

$$p_d = -\frac{1}{R_p \, C_p} \tag{6.10}$$

Figure 6.3(a) is a graph of the variation of the gain magnitude with the frequency, f, for the single-pole circuit model. The OpAmp gain remains close to the DC-gain $gain_0$ until the frequency gets close to the frequency f_{3dB}. When $f = f_{3dB}$, the OpAmp gain (expressed in dB) is 3 dB lower than the DC-gain $gain_0$. The slope of the plot is -20 dB/dec for frequencies greater than f_{3dB}. Finally, the frequency f_t at which the circuit gain is one is called the *unity gain frequency* [2].

Example (Transfer function of the single-pole OpAmp model): The transfer function of the single-pole OpAmp macromodel is given by:

$$\frac{V_o}{V_d} = \frac{gain_0}{1 - j\frac{f}{f_{3dB}}} \tag{6.11}$$

The magnitude response can be expressed as:

$$gain(f) \quad = \quad |\frac{V_o}{V_d}| \tag{6.12}$$

When $f = f_{3db}$, the magnitude of the gain is equal to:

$$gain(f_{3db}) \quad = \quad \frac{gain_0}{\sqrt{2}} \tag{6.13}$$

Or expressed in dB,

$$gain(f_{3db}) \quad = \quad gain_0 - 3\,dB \tag{6.14}$$

Imposing the condition that the *magnitude*$(f_t) = 1$, and solving for the unknown frequency f_t, yields:

$$f_t \quad = \quad f_{3dB} \, \sqrt{gain_0^2 - 1} \tag{6.15}$$

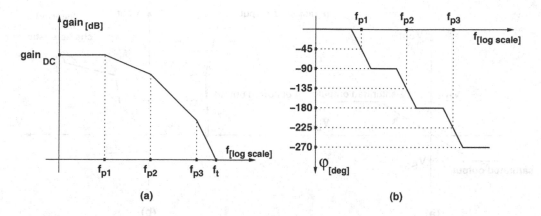

Figure 6.4: Magnitude and phase response for the three OpAmp model.

for the unity gain frequency. The phase response is given by:

$$\varphi(f) = \arctan \frac{f}{f_{3\mathrm{dB}}} \tag{6.16}$$

Thus, the phase, φ, is 45 degrees when $f = f_{3\mathrm{dB}}$. Figure 6.3 (b) illustrates the phase response.

More complex circuit models involve two, or more, poles, assuming that the poles are well separated. As shown in Figure 6.4, each additional pole changes the slope of the magnitude plot by -20 $\frac{dB}{dec}$, and the phase response by 90 degrees.

- *Input impedance*: The input impedance of real OpAmps is a function of frequency and may be very large, but not infinite. This can be modeled as an input resistance R_{in} in series with a capacitance, C_{in}, as shown in Figure 6.7. A more accurate modeling results, if the *common-mode input impedance* is also considered. The common-mode input impedance is defined as the input impedance, with respect to ground, when the two differential inputs of an OpAmp are tied together. The common-mode input impedance can be modeled by adding two capacitors, C_{cm}, each connected to one of the differential inputs and ground, as shown in Figure 6.7.

  Less precise OpAmp models may include only a very large input resistance, R_{in}, to represent the input impedance, as shown in Figure 6.2.

- *Output impedance*: The output impedance of real OpAmps is small, but not zero. A simple model for this nonzero output impedance is the addition of a resistance, R_o, in the OpAmp macromodel, as shown in Figure 6.2. A more accurate modeling, including the modification of the output impedance with resistance results, if the model uses a structure formed of two resistors, R_o, and two capacitors, C_o, connected as shown in Figure 6.7.

- *Distortion*: W. Sansen defines the distortion of a circuit as the difference in time in the shapes of the output and input signals [9]. Distortion changes the shape of the output

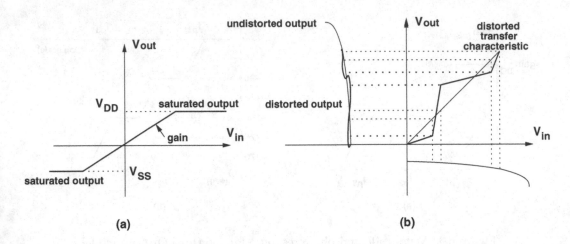

Figure 6.5: Definition of circuit distortion.

signal from that of the input signal. For a linear OpAmp without distortion, its output signal has the same shape as the input signal, even though the latter has been amplified, or attenuated.

Figure 6.5(b) illustrates the effect of circuit distortion on the output signal and shows the circuit's transfer characteristics. The straight line corresponds to an ideal linear circuit, for which $V_{out} = gain\ V_{in}$. The bold line represents the distorted transfer characteristics. The output signal has a different shape from that of the input signal, as shown in bold lines by signal V_{out}.

Nonlinear distortion is produced by a circuit's nonlinear transfer function V_o/V_{in} [9]. Nonlinear distortion can be classified as either (1) weak or (2) hard distortion. From a mathematical point of view, weak distortion produces continuous output signals, whereas hard distortion generates discontinuous output waveforms [9]. In most cases, hard distortions occur because of the improper use of the circuit, and are therefore not taken into account during the analog circuit design process.

Weak distortion can be expressed in terms of a Taylor series expansion as [8, 9]:

$$V_o = a_o + a_1 V_{in} + a_2 V_{in}^2 + a_3 V_{in}^3 + a_4 V_{in}^4 + \cdots \qquad (6.17)$$

where

$$a_i = \frac{1}{i!} \frac{\partial^i V_o}{\partial V_{in}^i} \qquad (6.18)$$

In most cases, it is sufficiently accurate to consider an approximation that includes only the first four terms of this series:

$$V_o = a_o + a_1 V_{in} + a_2 V_{in}^2 + a_3 V_{in}^3 \qquad (6.19)$$

where a_o is the DC component of V_{out}, a_1 is the linear gain of the circuit, and

$$a_2 = \frac{1}{2} \frac{\partial^2 V_{out}}{\partial V_{in}^2} \qquad (6.20)$$

and

$$a_3 = \frac{1}{6}\frac{\partial^3 V_{out}}{\partial V_{in}^3} \tag{6.21}$$

As shown in [9], for an input signal given by:

$$V_{in}(t) = V\cos(\omega t) \tag{6.22}$$

and considering only first- second- and third-order nonlinearities, the output signal can be expressed as

$$V_o(t) = \left[a_0 + V^2\frac{a_2}{2}\right] + \left[a_1 + V^2\frac{3}{4}a_3\right]V\cos(\omega t) \tag{6.23}$$

$$+ \left[V^2\frac{a_2}{2}\right]\cos(2\omega t) + \left[V^3\frac{a_3}{4}\right]\cos(3\omega t) \tag{6.24}$$

Two performance attributes describe the nonlinearity of an OpAmp: second-order harmonic distortion (HD_2), and third-order harmonic distortion (HD_3). For sinusoidal input signals, they are defined as:

$$HD_2 = \frac{1}{2}\frac{a_2}{a_1}V \tag{6.25}$$

$$HD_3 = \frac{1}{4}\frac{a_3}{a_1}V^2 \tag{6.26}$$

In general, HD_n is the ratio of the output component at $n\omega$ and the output component at ω.

The *total harmonic distortion* (THD) of a circuit is defined as

$$THD = \sqrt{HD_2^2 + HD_3^2 + HD_4^2 + ...} \tag{6.27}$$

- *Offset voltage*: The offset voltage, V_{OS}, describes the impact of transistor mismatches on the output voltage. For ideal OpAmps, the output voltage, V_o, is zero when the two differential inputs are connected. However, for real circuits $V_o \neq 0$. By definition, V_{OS} is the input voltage that would return V_o to zero [2]. The offset voltage V_{OS} is shown in the model in Figure 6.7.

- *Common-mode rejection ratio*: The common-mode rejection ratio (CMRR) characterizes the capability of an OpAmp to reject common signals applied at its differential inputs [8]. By definition,

$$CMRR(dB) = 20\log\left|\frac{A_d}{A_c}\right| \tag{6.28}$$

where A_d is the differential gain of the OpAmp, and A_c is the common mode gain. The common mode gain is defined as the ratio of the output signal and the corresponding common input signal applied to the differential inputs. For best performance, CMRR should be as high as possible.

Another definition of CMRR is based on the impact of the common mode voltage V_c on the offset voltage [2]:

$$CMRR(dB) = -20\log\left|\frac{\partial V_{OS}}{\partial V_c}\right| \tag{6.29}$$

- *Power Supply Rejection Ratio*: The power supply rejection ratio (PSRR) describes the capability of an OpAmp to operate correctly, if its power supply lines are noisy and the noise propagates to the OpAmp's output. PSRR is defined as

$$PSRR = \frac{A_d}{circuit\ gain\ from\ power\ supply\ to\ output} \tag{6.30}$$

where A_d is the differential gain and the denominator is the ratio of the output voltage to the corresponding power supply noise.

- *Output saturation*: The output signal of a real OpAmp is limited to the voltage range $(-V_{neg}, +V_{pos})$. This circuit behaves as an amplifier, only if the output voltage is within this range. Otherwise, the distortion of the output signal is very high. Figure 6.5(a) shows an example of output saturation. Output saturation can be modeled using clipping, as shown in Figure 6.7.

- *Slew Rate*: Slew Rate (SR) defines the maximum rate of change of the output signal that can accommodated by the OpAmp without significant distortion of the output [2, 8]:

$$SR = \frac{dV_{out}}{dt} \tag{6.31}$$

If the rate of change is beyond the SR then the output signal experiences a large distortion, for example the output signal may saturate. SR is caused by the inability to charge or discharge the OpAmp's capacitors fast enough, especially the OpAmp compensation capacitor [2].

- *Noise*: Circuit noises are signals produced by complex physical phenomena, such the random movement of electrons in conductors, random trapping and releasing of carriers by dangling bonds, movement of carriers across potential barriers, influence of generation–recombination centers on the diffusion of carriers, trapping and releasing of carriers flowing through discontinuous materials, generation of additional carriers by high velocity carriers, and so on [1, 8, 13]. This can be modeled as random signals with the following characteristics.

The *mean square value* of a noise signals $x(t)$ is defined as

$$X^2 = \frac{1}{T} \int_0^T x^2(t)dt \tag{6.32}$$

where T is a time interval. This mean square value represents the power delivered by signal $x(t)$ in time T to a resistance of $1\ \Omega$. The *Root Mean Square* (RMS) is defined for signal $x(t)$ as

$$X = \sqrt{\frac{1}{T} \int_0^T x^2(t)dt} \tag{6.33}$$

The equivalent RMS value of two uncorrelated noise voltages in series or two uncorrelated noise currents in parallel, is given by

$$X_{equivalent} = \sqrt{X_1^2 + X_2^2} \tag{6.34}$$

where x_1 and x_2 are two uncorrelated noise sources, and X_1 and X_2 are their respective RMS values. In general, for n noise voltages in series, or n noise currents in parallel, the equivalent RMS value is equal to

$$X_{equivalent} = \sqrt{X_1^2 + X_2^2 + \cdots + X_n^2} \tag{6.35}$$

Noise power density and *power spectral density* characterize the way noise power is distributed over a frequency range [2]. The definition of noise power density, depends on whether the noise signal is a voltage or a current. If the noise is a voltage, E, then the noise power density is equal to:

$$e^2(f) = \frac{dE^2}{df} \tag{6.36}$$

And if noise is a current, I, then the noise power density is given by

$$i^2(f) = \frac{dI^2}{df} \tag{6.37}$$

where E^2 and I^2 are the mean square values of the two noise signals and e^2 is expressed in units of V^2/Hz, and i^2 in units of A^2/Hz.

Power spectral density (PSD) is given by

$$e(f) \;=\; \sqrt{\frac{dE^2}{df}} \tag{6.38}$$

if the noise signal is voltage E and

$$i(f) \;=\; \sqrt{\frac{dI^2}{df}} \tag{6.39}$$

if noise is current I. *White noise* has constant power spectral density over the frequency band, $e(f) = n_0$ [2, 8]. Hence, for a frequency range (f_{low}, f_{high}), the RMS value of a white noise signal is expressed as [2, 10]

$$E \;=\; n_0 \sqrt{f_{high} - f_{low}} \tag{6.40}$$

The RMS value increases with the square of the bandwidth $f_{high} - f_{low}$, but does not depend on the actual frequency values, as long as, the difference is constant. If the frequency range increases k times, then the RMS value increases by the factor $\sqrt{k}$.

$\frac{1}{f}$ noise is characterized by a power density that is inversely proportional to the frequency [2]:

$$e^2(f) \;=\; \frac{n_1^2}{f} \tag{6.41}$$

Then, the RMS value of the noise signal for a frequency range (f_{low}, f_{high}) is equal to:

$$E = n_1 \sqrt{\ln\left[\frac{f_{high}}{f_{low}}\right]} \tag{6.42}$$

$\frac{1}{f}$ noise is dominant at low frequencies, whereas white noise is more important at high frequencies. By definition, the *corner frequency* f_c is the frequency at which the power density of white noise is equal to the power density of $\frac{1}{f}$ noise. Considering only white and $\frac{1}{f}$ noise, the overall noise power density in a circuit can be described as [2, 10]

$$e^2 \;=\; n_0^2 \left[1 + \frac{f_c}{f} \right] \tag{6.43}$$

For a frequency range (f_{low}, f_{high}), the mean square value of the overall noise is given by the following:

$$E^2 \;=\; \int_{f_{low}}^{f_{high}} e^2(f) df \tag{6.44}$$

$$=\; n_0^2 \int_{f_{low}}^{f_{high}} \left[1 + \frac{f_c}{f} \right] df \tag{6.45}$$

$$=\; n_0^2 \left(f_{high} - f_{low} + f_c \ln\left[\frac{f_{high}}{f_{low}} \right] \right) \tag{6.46}$$

Thus, the RMS value of the overall noise is equal to

$$E = n_0 \sqrt{ f_{high} - f_{low} + f_c \ln\left[\frac{f_{high}}{f_{low}} \right] } \tag{6.47}$$

Example [10]: For a given constant PSD of the white noise equal to

$$n_0 = 25 \frac{nV}{\sqrt{\text{Hz}}}, \tag{6.48}$$

the corner frequency $f_c = 10KHz$, and the frequency range $(f_{low}, f_{high}) = (60Hz, 20KHz)$, the RMS value of the overall noise is equal to:

$$v = 25 \frac{nV}{\sqrt{\text{Hz}}} \sqrt{ 20K\text{Hz} - 60\text{Hz} + 10K\text{Hz} \ln\left[\frac{20K\text{Hz}}{60\text{Hz}} \right] } = 6.98 \; \mu V \tag{6.49}$$

In a circuit, resistor and MOSFET transistor white noise is due to the thermal noise [2, 8]. Thermal noise originates because of the random motion of carriers in a resistor or a transistor channel. The thermal noise of resistors is modeled as a voltage source in series with the resistor, as shown in Figure 6.6(a). The mean square value of the noise voltage source is:

$$V_{thermal}^2 \;=\; 4\,k\,T\,R \tag{6.50}$$

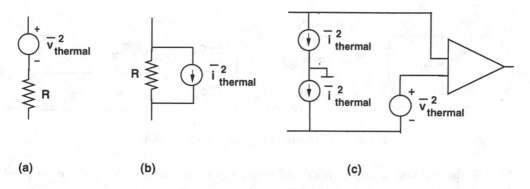

(a) **(b)** **(c)**

Figure 6.6: Noise modeling.

where k is the Boltzmann constant ($k = 1.38 \; 10^{-23} J/K$), T is the absolute temperature in degrees Kelvin, and R is the resistance. Thermal noise can also be modeled as a noise current source in parallel with the resistor, as shown in Figure 6.6(b). Then, the mean square value of the noise current source is expressed as

$$I_{thermal}^2 = \frac{4\,k\,T}{R} \qquad (6.51)$$

The thermal noise in the channel of a MOSFET transistor is modeled as a noise current source connected across the drain and source terminals of the transistor [8]. The RMS value of the current source is equal to

$$I_{thermal}^2 = 4\,k\,T\,\gamma\,g_m \qquad (6.52)$$

where γ is a process dependent constant, and g_m is the transconductance of the transistor,

$$g_m = \frac{\partial I_D}{\partial V_{gs}} \qquad (6.53)$$

for a constant V_{ds}.

In a circuit, $1/f$ noise is due primarily to the flicker noise of the MOSFET transistors [2, 8]. Flicker noise is modeled as a voltage noise source connected to the gate of a transistor. The RMS value of the source is [8]:

$$V_{flicker}^2 = \frac{K}{C_{ox}W\,L}\left(\frac{1}{f}\right) \qquad (6.54)$$

where K is a constant, C_{ox} is the capacitance per unit area, W and L are the width and length of the transistor, and f is the frequency.

All noise sources in a differential amplifier can be lumped into three equivalent noise sources connected at the differential inputs of an OpAmp[2]. Figure 6.6(c) shows the two noise current sources, and the noise voltage source.

As shown throughout this chapter, this OpAmp model is an important way to evaluate the noisiness of an analog block, for example an inverting and noninverting amplifier, by computing the equivalent, input-referred noise of the block. If X_{signal}^2 is the mean square value of the input signal, and X_{noise}^2 is the mean square value of the input-referred noise of the block, then the *signal-to-noise ratio* (SNR) of the circuit is defined as

$$SNR(dB) = 10 \; \log\left[\frac{X_{signal}^2}{X_{noise}^2}\right] \qquad (6.55)$$

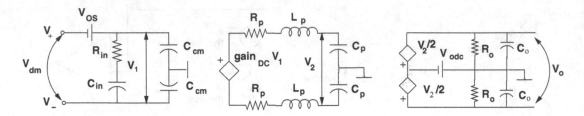

Figure 6.7: Detailed OpAmp macromodel.

SNR is the ratio of the input signal and noise, and should have as high a value as possible to provide good analog processing accuracy.

Most of the circuit characteristics discussed thus far can be incorporated into the more precise macromodel shown in Figure 6.7. The macromodel is important for fast, yet accurate, simulation of analog circuits. This macromodel captures finite gain, one pole and one zero, input impedance, common-mode input impedance, output impedance, offset voltage, common mode gain, and output saturation.

6.1.3 OpAmp Macromodeling[2]

This subsection illustrates a systematic process for developing the macromodel of a single stage differential OpAmp, which can be automated as a computer program. The OpAmp considered in the section is shown in Figure 6.8.

Different methods are available for circuit macromodeling, including circuit analysis, tabular forms and curve fitting and neural networks among others. Many of these methods produce blackbox models, which are mathematical expressions for predicting the performance of circuits. However, these models do not offer insight into circuit design, because most of the physical (i.e., of the performance attributes is lost). In contrast, structural models are networks of current and voltage sources, capacitors, resistors, diodes, and so on. The component parameters directly relate to the transistor parameters in the circuit. Structural models not only predict the circuit's performance, but also offer insight into circuit design.

This modeling technique is detailed in [15] and [16]. Steps 1-6 generate the circuit models in the frequency (AC) domain, step 7 describes circuit noise, step 8 defines circuit clipping, step 9 expresses CMRR, step 10 models the circuit PSRR, and step 11 presents a summary on modeling the harmonic distortion of the circuit. (More details about modeling of harmonic distortion using this method can be found in [16].)

The modeling method consists of the following activities:

Step 1 (Grounding of biasing voltages): First, all biasing voltages are grounded, for example V_{bn} in Figure 6.8. Biasing transistors not only supply the DC current, or voltage, for the circuit,

---

[2]This subsection was co-written by Y. Wei and A. Doboli. This section is based on the papers Y. Wei, A. Doboli, "Systematic Development of Analog Circuit Structural Macromodels through Behavioral Model Decoupling", Proceedings of the Design Automation Conference, 2005, pp. 57-62. Y. Wei, A. Doboli, "Systematic Development of Nonlinear Analog Circuit Macromodels Through Successive Operator Composition and Nonlinear Model Decoupling", Proceedings of the Design Automation Conference, 2006, pp. 1023-1028 and Y. Wei, A. Doboli, "Structural Macromodeling of Analog Circuits Through Model Decoupling and Transformation", IEEE Transactions on Computer-Aided Design of Integrated Circuits and Systems, Vol. 27, No. 4, April 2008, pp. 712–725.

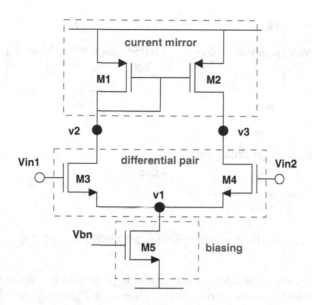

Figure 6.8: Single-stage OpAmp [18] © 2010 IEEE.

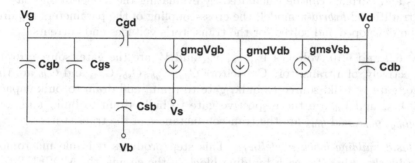

Figure 6.9: Linear *hybrid-π* MOSFET model.

but also affect the performance of the circuit. For example, transistors such as $M5$, that supply the biasing current, also connect other transistors to the power supply, so they influence the power supply rejection ratio (PSRR) of the OpAmp. Also, biasing transistors introduce thermal and other noise into the circuit.

Step 2 (Circuit partitioning): The OpAmp is decomposed into basic building blocks. Next, macromodels for the basic building blocks are developed separately, and put together to form the model of the entire circuit. This step helps reduce the macromodel development effort through reuse of the basic building block models.

Many popular OpAmps are based on fundamental building blocks, including differential pairs, wide-swing current mirrors, cascade current mirrors, cascode stages, active loads, and folded cascode stages [8]. To achieve the required performance, building blocks are connected together directly, or by single transistors. For example, the OpAmp in Figure 6.8 includes one current mirror, one differential pair, and one biasing transistor ($M5$). The circuit partitioning into building blocks is shown in the figure.

Step 3 (MOSFET transistor models): For MOSFET transistors, each of the circuit's MOS-FET transistors is expressed as a small-signal equivalent circuit based on a transformed *hybrid-π*

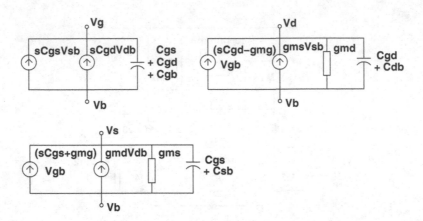

Figure 6.10: Transformed linear *hybrid-π* MOSFET model [18] © 2010 IEEE.

transistor model. The transformed MOSFET model is presented in Figure 6.10. This is different from the popular *hybrid-π* transistor model [14] shown in Figure 6.9 in that the cross-coupled dependencies between the model parameters have been removed by decoupling, so that all transistor voltages and currents can be calculated by evaluating the model in one pass, from left to right. In the traditional *hybrid-π* model, the cross-coupling of the parameters requires that a set of equations be developed and solved for the transistor's voltages and currents.

In Figures 6.9 and 6.10, voltages V_g, V_s, V_d, and V_b are the gate, source, drain, and bulk voltages, respectively, of a transistor. Capacitors C_{gs}, C_{gb}, C_{sb}, C_{gd}, and C_{db} are the respective gate to source, gate to bulk, source to bulk, gate to drain, and drain to bulk capacitances [13]. Voltages V_{gb}, V_{db}, and V_{sb} are the respective gate to bulk, drain to bulk, and source to bulk voltages and g_{mg}, g_{md}, and g_{ms} are the transconductances of the transistor.

Step 4 (Basic building block modeling): This step produces reusable macromodels for the basic building blocks. First, for each building block in the circuit, the MOSFET transistors are replaced by their transformed *hybrid-π* models. The terminal voltages of a building block are chosen as the independent variables in the model of the building block. The current flowing into the terminal is determined by the circuit parameters and the terminal voltages. For example, the differential input, biasing circuit, and current mirror of the single-stage OpAmp are represented as macromodels in Figure 6.11. These models are obtained by replacing the MOSFET transistors with their decoupled models. All voltages in the basic block models are related to the transistor parameters. Note that all nodes in a building block are visible (accessible) from the outside, so that any node can be connected to any other node in the circuit, depending on the building block's connections to other building blocks.

Step 5 (Producing coupled circuit macromodels): The models for all building blocks in a circuit form the circuit macromodel. For example, the two device netlists connected between node v_1 and ground in Figure 6.11 are merged to form the netlist connected between node v_1 and ground in Figure 6.12. The resulting circuit macromodel is coupled because there are cross-coupled dependencies between the model parameters. For the resulting netlist:

$$C_1 = C_{gs3} + C_{sb3} + C_{gs4} + C_{sb4} + C_{gb5} + C_{db5} \tag{6.56}$$

and

$$R_1 = \frac{1}{g_{ms3} + g_{ms4} + g_{md5}} \tag{6.57}$$

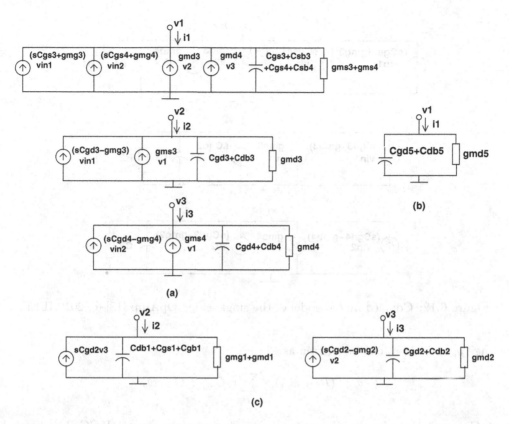

Figure 6.11: Macromodels of the single-stage OpAmp basic building blocks: (a) differential input, (b) biasing circuit, and (c) current mirror.

Similarly, the two netlists connected between node v_2 and ground in Figure 6.11 are merged to build the netlist in Figure 6.12 connected between v_2 and ground. For this netlist

$$C_2 = C_{db1} + C_{gs1} + C_{gb1} + C_{gd3} + C_{db3} \tag{6.58}$$

$$\tag{6.59}$$

and

$$R_2 = \frac{1}{g_{md1} + g_{mg1} + g_{md3}} \tag{6.60}$$

Finally, the two netlists connected between node v_3 and ground are merged, as shown in Figure 6.11, to build the netlist in Figure 6.12 connected between v_3 and ground, so that

$$C_3 = C_{db2} + C_{gd2} + C_{gd4} + C_{db4} \tag{6.61}$$

and

$$R_3 = \frac{1}{g_{md2} + g_{md4}} \tag{6.62}$$

In general, each node's voltage depends only on the voltages of its adjacent nodes. Node n_j is said to be *adjacent* to node n_i, if the two nodes are the terminals of the same block. Hence,

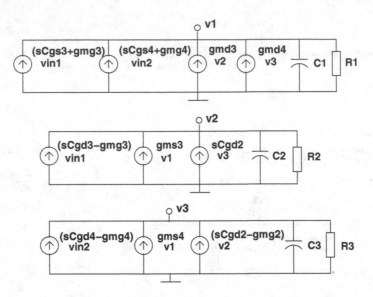

Figure 6.12: Coupled macromodel of the single-stage OpAmp [18] © 2010 IEEE.

the voltage v_i of node n_i can be written as

$$v_i = (R_i + sC_i) \times \sum_j (sC_{m_j} \pm G_{m_j})v_j \qquad (6.63)$$

where $(sC_{m_j} \pm G_{m_j})v_j$ denotes the voltage controlled current source $(VCCS)$ in block i. R_i and C_i are the total resistance and capacitance associated with node n_i. The complete coupled macromodel of the single-stage OpAmp in Figure 6.8 is shown in Figure 6.12.

The advantage of the structural macromodel is that it shows how circuit nodes relate to the circuit parameters and there is a one-to-one mapping from the parameters of the original circuit to the macromodel parameters. For example, in order to see how transistor $M3$ is connected in the circuit, one would find all the elements in the model that depend on the parameters of transistor $M3$. It turns out that the first and third voltage-controlled current sources of block v_1, and the first two voltage controlled current sources of block v_2 depend on the parameters of transistor $M3$. A comparison of these current sources with the decoupled linear MOS model, shows that the gate, drain, and source of transistor $M3$ are connected to v_{in1}, v_2 and v_1.

Pole and zero modeling. There is a pole in the model at each internal node formed by resistor R_i and capacitor C_i. Zeros exist for some voltage controlled current sources, for example the $(sC_{gd3} - g_{mg3})v_{in1}$ current source in the block of v_2. All zeros and poles are clearly shown in the structural macromodel. However, their influence on the transfer function cannot be derived directly due to the coupling between blocks. In the model in Figure 6.12, voltage v_1 is expressed as a function of voltages v_2 and v_3, voltage v_2 depends on voltages v_1 and v_3, and voltage v_3 on voltages v_1 and v_2. To address this issue, the circuit macromodel has to be further refined by uncoupling the cross-coupled dependencies in the model. The *uncoupled* circuit macromodel results from removing all cross-coupled dependencies in the coupled model.

Step 6 (Uncoupled circuit macromodels): The model uncoupling step modifies the coupled macromodel, in that all of the voltage dependencies are replaced by their equivalent functions of the inputs. The decoupling step defines the decoupling sequence by signal path tracing so that

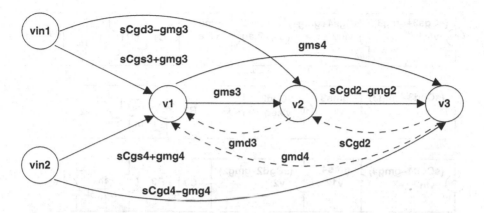

Figure 6.13: Signal path of the single-stage OpAmp [18] © 2010 IEEE.

the symbolic expression for the voltage that needs to be decoupled is derived starting from the lower order to the higher order.

For uncoupling, it is necessary to first determine which voltage dependencies should be decoupled. For example, in Figure 6.12, voltage v_1 depends on voltages v_{in1}, v_{in2}, v_2, and v_3. and voltage v_2 depends on voltages v_{in1}, v_1, and v_3. Obviously, there is a coupling, i.e., a cross-dependency, between voltages v_1 and v_2. Hence, the voltage that be should be solved first must be determined.

The signal-path tracing algorithm of Huang, Gathercole, and Mantooth [5] can be used to find the sequence of the nodes in the circuit. The signal path for the OpAmp in Figure 6.8 is shown in Figure 6.13. The dotted line represents the feedback voltage dependency, which needs to be decoupled. The modeling sequence follows the signal path from the input to the output as $v_{in1,2} \rightarrow v_1 \rightarrow v_2 \rightarrow v_3$.

Then, the coupled model is transformed using the uncoupling sequence. The feedback dependent voltage v_i is replaced by its equivalent voltage $v_{i,eq}$, as shown in Figure 6.14. This corresponds to replacing each dotted feedback path in the signal path by the feedforward paths from the inputs in Figure 6.13.

In the frequency domain, the equivalent voltages $v_{i,eq}$ have the general form [15]:

$$v_{i,eq} = \sum_{p=1}^{P} \frac{a_{i,0} + a_{i,1}s + \cdots + a_{i,n-1}s^n}{b_{i,0} + b_{i,1}s + \cdots + b_{i,n-1}s^n} v_{in,p} \tag{6.64}$$

where P is the number of inputs. The coefficients of $\mathbf{s^i}$ are given by

$$a_{i,j} = \sum_{t} \left((\pm 1) \prod_{k_i k_j} G_{m,k_i k_j}^{\alpha_k} \prod_{l_i l_j} C_{m,l_i l_j}^{\beta_l} \right) \tag{6.65}$$

where $\alpha_{\mathbf{k}}$ and $\beta_{\mathbf{l}}$ are 0 or 1, $G_{m,k_i k_j}$ is the transconductance between node k_i and k_j, and $C_{m,l_i l_j}$ is the transcapacitance between node l_i and l_j. If there are N nodes in the coupled model, then

$$\sum_{k=1}^{K} \alpha_k = N - j, \sum_{l=1}^{L} \beta_l = j \tag{6.66}$$

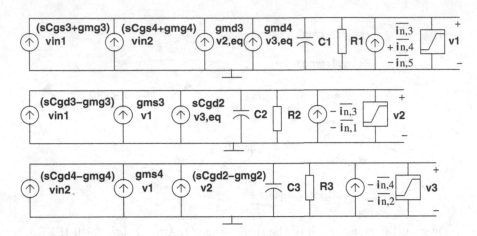

Figure 6.14: Uncoupled macromodel for single-stage OpAmp [18] © 2010 IEEE.

where K and L are the total number of G_{m,k_ik_j} and C_{m,l_il_j} in the model.

Next, the symbolic expressions for all coefficients are found by solving a set of equations [15]. Feedback dependencies are substituted by their equivalent symbolic functions and the equivalent voltage-controlled current sources are functions of the input frequency, as shown in Figure 6.14.

Step 7 (Noise modeling): Thermal noise and $1/f$ noise are the dominant noise sources in the MOS transistor. The noise current density can be approximated as Equation (6.67) for thermal noise and Equation (6.68) flicker noise [8].

$$\overline{i^2_{n,th}} = \frac{8kT}{3}g_{ms} \tag{6.67}$$

$$\overline{i^2_{n,1/f}} = \frac{K}{C_{ox}WL}\left(\frac{1}{f}\right)g^2_{mg} \tag{6.68}$$

where, as discussed previously, γ is technology dependent and $\approx 2/3$ for long channel devices, and 2.5 for submicron MOS devices [8].

Noise current sources are added to each circuit node, as shown in Figure 6.14. To calculate the total output noise, all input sources are removed, and then the symbolic expression for the output noise is derived directly from Figure 6.14, without solving any equations. The noise current sources in Figure 6.14 have the general form presented in expressions (6.69) and (6.70):

$$\overline{i^2_{n,i}} = \frac{8kT}{3}g_{ms,i} + \frac{K}{C_{ox}W_iL_i}\left(\frac{1}{f}\right)g^2_{mg,i} \tag{6.69}$$

and

$$\overline{i^2_{n,R}} = \frac{4kT}{R} \tag{6.70}$$

The total output noise in the band from f_l to f_h is the integral of the output noise density over the given bandwidth, which can be approximated as

$$V_{n,tot} = \sum_{k=1}^{k_0} V_{n,out}(f_k)\left[\frac{f_h - f_l}{k_0}\right] \tag{6.71}$$

where $V_n(f_k)$ is the output noise at frequency f_k, and

$$f_k = f_l + (f_h - f_l)\left(\frac{k}{k_0}\right) \tag{6.72}$$

Step 8 (Modeling of clipping): The transistors connected to the node that have the largest swing dominate the distortion of the circuit. If the swing is too large, it may drive the transistor into a linear, or cutoff region, resulting in voltage clipping. Clipping can be modeled by calculating the maximum swing a node can tolerate, which is set by the DC operating point. In other words, all the transistors should be in the correct operating region in order to give the proper performance. The maximum swing the node can tolerate is equal to $min(V_{max} - V_0, V_0 - V_{min})$, where V_{max} and V_{min} are the highest and lowest voltages, respectively, that the node can reach without clipping, and V_0 is the DC operating voltage of the node. The maximum swing for Figure 6.14 can be expressed as:

$$\begin{cases} V_B - V_{Tn5} < V_1 < V_{in1,2} - V_{Tn3,4} \\ V_{in1} - V_{Tn3} < V_2 < V_{DD} - |V_{Tp1,2}| \\ V_{in2} - V_{Tn2} < V_3 < V_2 - |V_{Tp2}| \end{cases} \tag{6.73}$$

Step 9 (Modeling of common-mode rejection ratio): The differential-mode gain A_{dm} is calculated from the macromodel by setting $v_{in1} = v_{id}/2$ and $v_{in2} = -v_{id}/2$, and the common-mode gain A_{cm} is calculated from the macromodel by setting $v_{in1} = v_{ic}$ and $v_{in2} = v_{ic}$. CMRR is the ratio of the two gains.

Step 10 (Modeling of power-supply rejection ratio): Variations in the power-supply voltages are propagated to the OpAmp output. If the small signal variation in the positive (V_{DD}) and negative (V_{SS}) power supply voltages are denoted by v_{dd} and v_{ss}, respectively, then the small signal, OpAmp output voltage is:

$$v_o = A_{dm}v_{id} + A^+v_{dd} + A^-v_{ss} \tag{6.74}$$

where A^+ and A^- are the small signal gains from the positive and negative power supplies to the output, respectively. The positive and negative power supply rejection ratio are defined as [3]

$$PSRR^+ = \frac{A_{dm}}{A^+} \tag{6.75}$$

and,

$$PSRR^- = \frac{A_{dm}}{A^-} \tag{6.76}$$

$PSRR^+$ and $PSRR^-$ can be calculated from the macromodel excited by the variation (v_{dd} and v_{ss}) on the power-supplies, as shown in Figure 6.15.

Step 11 (Modeling of harmonic distortion): Circuit nonlinearity is modeled by including nonlinear current sources at each voltage node, as shown in Figure 6.16. Nonlinear current sources are symbolic functions of the nonlinearity coefficients [14] and small signal parameters of the MOSFET transistor, which is described in detail in [16].

For example, the equivalent nonlinear current source for the drain current of transistor $M3$ is

$$i_{d3}^{n2} = K_{2_{g3}}v_{in1}^2 + (K_{2_{gd3}} + K_{2_{d3}})TF_{v_2}v_2 + (K_{2_{gs3}} + K_{2_{d3}} - K_{2_{s3}})TF_{v_1}v_1 \tag{6.77}$$

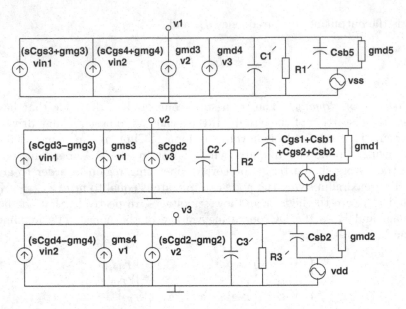

Figure 6.15: Macromodel for single-stage OpAmp with power-supply variation.

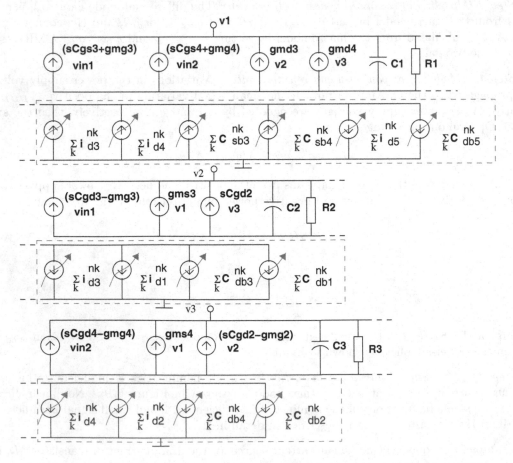

Figure 6.16: Nonlinear macromodel for single stage OpAmp [18] © 2010 IEEE.

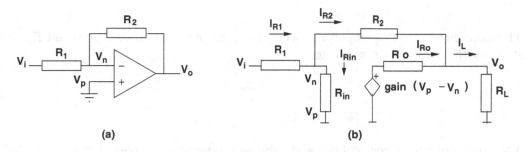

Figure 6.17: Inverting amplifier.

where,

$$
\begin{aligned}
TF_{v_1} = {} & (C_3C_2 + C_3G_2 + G_3C_2 + G_3G_2 + s^2 C_{gd2}C_{gd3} + g_{mg2}sC_{gd3})((-g_{mg3} - sC_{gs3})v_{in1} \\
& + (-g_{mg4} - sC_{gs4})v_{in2}) + (g_{md3}G_3 + g_{md4}g_{mg2} + g_{md4}sC_{gd2} + g_{md3}C_3) \\
& (g_{mg3} - sC_{gd3})v_{in1} + (g_{md4}G_2 + g_{md4}C_2 + g_{md3}sC_{gd3})(g_{mg4} - sC_{gd4})v_{in2} \\
TF_{v_2} = {} & (g_{ms3}C_3 + g_{ms3}G_3 + Cgd3sg_{ms4})((-g_{mg3} - sC_{gs3})v_{in1} + (-g_{mg4} - sC_{gs4})v_{in2}) \\
& + (G_1G_3 + G_1C_3 + C_3sC_1 + G_3sC_1 + g_{md4}g_{ms})v_{in1} + (sC_{gd3}G_1 + s^2 C_{gd3}C_1) \\
& + g_{ms3}g_{md4})(g_{mg4} - sC_{gd4})v_{in2}
\end{aligned}
$$

$$(6.78)$$

6.2 Continuous-Time Analog Building Blocks

This section provides a brief discussion of the basic analog circuits, inverting amplifiers, noninverting amplifiers, summing amplifiers, difference amplifiers, integrators, and comparators. For each analog block, the functioning of the circuit and the impact of nonidealities on the circuit operation are also discussed.

6.2.1 Inverting Amplifiers

The behavior of an ideal inverting amplifier, shown in Figure 6.17(a), is described by the following:

$$V_o \quad = \quad gain \, V_i \qquad (6.79)$$

where $gain \in R$ and $gain < 0$. Note that this behavior does not depend on frequency, or input signal magnitude.

The inverter gain depends only on the values of the two resistors R_1 and R_2. For an ideal OpAmp, voltages $V_n = V_p = 0$, and there is no current flowing into the differential OpAmp inputs. Therefore, the currents i_{R_1} and i_{R_2} flowing through the resistors R_1 and R_2 are equal.

$$i_{R_1} = \frac{V_i}{R_1} = i_{R_2} = -\frac{V_o}{R_2} \qquad (6.80)$$

and therefore,

$$V_o = gain \, V_i = -\frac{R_2}{R_1} \, V_i \qquad (6.81)$$

The gain of the inverting amplifier is set by the ratio of the two resistors R_1 and R_2, and is constant for the entire frequency range,

$$gain = -\frac{R_2}{R_1} \tag{6.82}$$

The impact of circuit nonidealities and noise on the inverter behavior, for example OpAmp nonidealities, are treated next.

Impact of finite OpAmp gain, finite input resistance, and nonzero output resistance. Real OpAmps have finite DC gain, finite input impedance, and nonzero output impedance. To characterize the effect of OpAmp nonidealities on the noninverting amplifier's behavior, the OpAmp macromodel, shown in Figure 6.17(b), is used. Resistance R_{in} is the input resistance of the OpAmp, R_o the output resistance, and $gain_o$ is the finite DC gain. The following set of equations can be derived for this circuit:

$$
\begin{array}{rclcrcl}
I_{R1} & = & I_{R_{in}} + I_{R_2} & \qquad & V_n - V_o & = & I_{R_2}\, R_2 \\
V_n & = & I_{R_{in}}\, R_{in} & & V_o & = & I_L\, R_L \\
V_i - V_n & = & I_{R_1}\, R_1 & & I_{R_2} + I_{R_o} & = & I_L \\
gain_0\, (V_p - V_n) & = & I_{R_o}\, R_o + V_o & & V_p & = & 0
\end{array}
$$

where currents I_{R_1}, I_{R_2}, $I_{R_{in}}$, I_{R_o}, and I_L are the currents through the resistors R_1, R_2, R_{in}, R_o, and R_L. The unknowns are currents I_{R_1}, I_{R_2}, $I_{R_{in}}$, I_{R_o}, and I_L, and voltages V_o and V_n.

The gain of the inverting amplifiers is equal to:

$$gain = -\frac{(gain_0\, R_2 - R_o)R_{in}R_L}{R_1((gain_0+1)R_{in}+R_2)R_L+R_{in}R_LR_o+R_1(R_2+R_{in}+R_L)R_o+R_2R_{in}(R_L+R_o)} \tag{6.83}$$

The input resistance of the inverting amplifier is:

$$
\begin{aligned}
R_{in} & = \frac{V_i}{I_{R_1}} \\[2mm]
& = \frac{R_1\left[(R_{in}(gain_0+1)+R_2)R_L \;+\; (R_{in}+R_2+R_L)R_o\right]+R_{in}(R_oR_L+R_2(R_o+R_L))}{(R_2+R_{in}+gain_0\, R_{in})R_L+(R_2+R_{in}+R_L)R_o}
\end{aligned}
\tag{6.84}
$$

The output resistance of the inverting amplifier is:

$$R_{out} = \frac{V_o}{I_L} = \frac{gain_0\, R_2 \;-\; R_o}{1 \;+\; \dfrac{gain_0}{1+\frac{R_2}{R_L}}} \tag{6.85}$$

Because R_{in}, R_L are much larger than R_1, and R_2 is much larger than R_o, the above expressions become

$$gain \approx -\frac{R_2}{\left[R_1 + \frac{R_1+R_2}{gain_0}\right]} = gain_{ideal}\frac{1}{1 + \left[\frac{1-gain_{ideal}}{gain_0}\right]} \tag{6.86}$$

$$R_{input} \approx R_1 + \frac{R_2}{1 + gain_0} \tag{6.87}$$

$$R_{output} \approx \frac{R_2}{1 + \frac{1}{gain_0}} \tag{6.88}$$

Table 6.1: Design tradeoffs for noninverting amplifier

OpAmp gain	$\epsilon = 0.9$	$\epsilon = 0.99$	$\epsilon = 0.999$
$gain_0 = 100$	$\frac{R_2}{R_1} < 10$	$\frac{R_2}{R_1} < 0.01$	not possible
$gain_0 = 1,000$	$\frac{R_2}{R_1} < 110$	$\frac{R_2}{R_1} < 9.1$	$\frac{R_2}{R_1} < 0.001$
$gain_0 = 10,000$	$\frac{R_2}{R_1} < 1,110$	$\frac{R_2}{R_1} < 100$	$\frac{R_2}{R_1} < 9$

Table 6.1 summarizes the design tradeoffs existing between (i) the accuracy of achieving a specified gain for the inverting amplifier, (ii) the DC gain of the OpAmp, and (iii) the gain value of the inverting amplifier. The accuracy is described by the value ϵ, such that

$$\epsilon < \frac{gain_{Inv\ Amp}}{gain_{ideal}} < 1 \tag{6.89}$$

Note that the ratio:

$$\frac{gain_{Inv\ Amp}}{gain_{ideal}} < 1 \tag{6.90}$$

is always less than one. For very high accuracy, for example $\epsilon = 0.999$, it is required that the OpAmp has a large DC gain, for example $gain_0 = 10,000$. Also, the maximum amplifier gain that can be achieved is less than 9. If $gain_0$ is small, the required accuracy cannot be achieved, or is achieved at the expense of an extremely low gain (attenuation) for the inverting amplifier, as shown in the table for $gain_0$ values of 100 and 1,000. If the accepted accuracy can be less, for example $\epsilon = 0.9$, then OpAmps with smaller DC gain can be used, e.g., $gain_0 = 100$, and higher amplifier gains can be achieved by the inverting amplifier, as is illustrated by columns 3 and 4 in the table.

Impact of OpAmp poles. OpAmp poles have a significant impact on the frequency behavior of the inverting amplifier. The following analysis assumes that the OpAmp behavior is modeled using a single (dominant) pole. Figure 6.18 presents the macromodel for the single pole model. The analysis presents the magnitude and phase response of the inverting amplifier in the frequency domain.

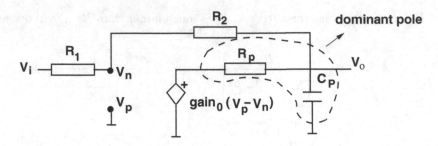

Figure 6.18: Impact of the OpAmp pole on the amplifier behavior.

Single pole model. Following an analysis similar to the previous case, the gain of an inverting amplifier is

$$gain(f) = -\frac{gain_0 \, R_2 - R_p}{R_1(gain_0 + 1 + \frac{R_2}{R_1} + \frac{R_p}{R_L}\frac{R_1+R_2+R_L}{R_1} + \frac{R_p}{Xc(f)}\frac{R_1+R_2}{R_1})} \tag{6.91}$$

where

$$X_c(f) = \frac{1}{j \, \omega \, C_p} = \frac{1}{j \, 2\pi \, f \, C_p} \tag{6.92}$$

is the impedance of capacitor C_p.

The inverting amplifier gain can be approximated by

$$gain(f) \approx -\frac{gain_{ideal}}{(1 + \frac{R_1+R_2}{gain_0 R_1} + \frac{R_1+R_2}{gain_0 R_1}\frac{R_p}{Xc})} \tag{6.93}$$

Or, because $gain_0 \, R_1 >> R_1 + R_2$

$$gain(f) \approx -\frac{gain_{ideal}}{(1 + \left[\frac{R_1+R_2}{gain_0 R_1}\right]\frac{R_p}{Xc})} = -\frac{gain_{ideal}}{1 + 2\pi R_p C_p f \left[\frac{R_1+R_2}{gain_0 R_1}\right] j} \tag{6.94}$$

The magnitude response in the frequency domain is equal to

$$|gain(f)| \approx -\frac{gain_{ideal}}{\sqrt{1 + (2\pi R_p C_p f \left[\frac{R_1+R_2}{gain_0 R_1}\right])^2}} \tag{6.95}$$

The phase response is expressed as

$$\tan \varphi(f) \approx -2\pi R_p C_p f \left[\frac{R_1 + R_2}{gain_0 \, R_1}\right] \tag{6.96}$$

The 3 dB frequency f_{3dB} of the inverting amplifier is given by

$$f_{3dB} = \frac{1}{2\pi R_p C_p}\left[\frac{gain_0 \, R_1}{R_1 + R_2}\right] = f_{3dB}^{OpAmp}\left[\frac{gain_0}{1 + |gain_{ideal}|}\right] \tag{6.97}$$

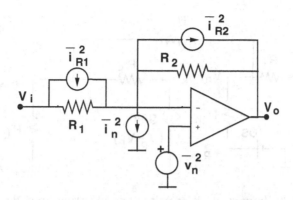

Figure 6.19: Noise analysis for inverting amplifier.

Equation shows that inverting amplifiers are characterized by gain-bandwidth tradeoffs. Increasing the gain of the amplifier decreases the bandwidth of the amplifier, and decreasing the gain increases the bandwidth. For example, the performance of a very good OpAmp is about $gain_0 = 10^4 \frac{V}{V}$, and its 3dB frequency is $\approx f_{3dB} = 10KHz$. For these values,

$$f_{3dB} = \frac{10^5}{1 + |gain_{ideal}|} kHz \tag{6.98}$$

and, $f_{3dB} = 50MHz$ for $gain_{ideal} = 1$, and $f_{3dB} = 9MHz$ for $gain_{ideal} = 10$. For a more common OpAmp, e.g., the 741 OpAmp, $f_{3dB} = 1KHz$ and $gain_0 = 10^3 \frac{V}{V}$ [2]:

$$f_{3dB} = \frac{10^3}{1 + |gain_{ideal}|} kHz \tag{6.99}$$

The 3 dB frequency of the inverting amplifier is $f_{3dB} = 500KHz$ for $gain_{ideal} = 1$, and $f_{3dB} = 90KHz$ for $gain_{ideal} = 10$.

Impact of noise. Figure 6.19 shows the primary noise sources in an inverting amplifier. Resistor R_1 is in parallel with a noise current source with power density $\overline{i}^2_{R1}$. Resistor R_2 has in parallel a noise current source with the power density $\overline{i}^2_{R1}$, and the OpAmp introduces two noise sources, the noise voltage source $\overline{v}^2_n$ and the noise current source $\overline{i}^2_n$. As previously discussed, the noise of the two resistors is thermal noise and can be described as $\overline{i}^2_{R1} = 4\,k\,T/R_1$ and $\overline{i}^2_{R2} = 4\,k\,T/R_2$, respectively.

Applying the superposition principle, the overall power density at the output of the inverting amplifier is expressed as

$$\overline{v}^2_o = \overline{v}^2_n \left(1 + \frac{R_2}{R_1}\right)^2 + \overline{i}^2_n\, R_2^2 + \overline{i}^2_{R1}\, R_2^2 + \overline{i}^2_{R2}\, R_2^2 \tag{6.100}$$

or

$$\overline{v}^2_o = \overline{v}^2_n \left(1 + \frac{R_2}{R_1}\right)^2 + \overline{i}^2_n\, R_2^2 + \left(\frac{4\,k\,T}{R_1}\right) R_2^2 + \left(\frac{4\,k\,T}{R_2}\right) R_2^2 \tag{6.101}$$

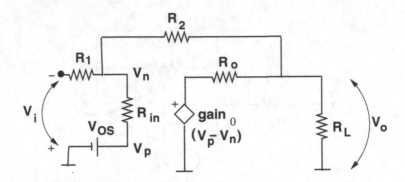

Figure 6.20: Impact of offset voltage on inverting amplifier output.

Assuming an ideal OpAmp, the input-referred power density of the inverting amplifier is equal to

$$\overline{v}^2_{input-referred} = \frac{\overline{v}^2_n \left(1 + \frac{R_2}{R_1}\right)^2 + \overline{i}^2_n R_2^2 + \left(\frac{4\,k\,T}{R_1}\right) R_2^2 + \left(\frac{4\,k\,T}{R_2}\right) R_2^2}{\left(\frac{R_2}{R_1}\right)^2} \tag{6.102}$$

The amplifier SNR is

$$SNR = 10 \log \left[\frac{v_i^2}{\overline{v}^2_{input-referred}}\right] \tag{6.103}$$

$$= 10 \log \left[\frac{v_i^2 \left(\frac{R_2}{R_1}\right)^2}{\overline{v}^2_n \left(1 + \frac{R_2}{R_1}\right)^2 + \overline{i}^2_n R_2^2 + \frac{4\,k\,T}{R_1} R_2^2 + \frac{4\,k\,T}{R_2} R_2^2}\right] \tag{6.104}$$

The above expressions show the existence of a tradeoff involving the amplifier noise, gain, and SNR. The overall power density at the output of the inverting amplifier is reduced by lowering the value of resistor R_2 and increasing the value of resistor R_1. However, this results in low amplifier gains. If the amplifier gain has to be large then the overall noise at the output also increases. In addition, increasing R_1 and/or decreasing R_2, reduces the SNR.

Impact of offset voltage. Figure 6.20 shows the circuit used to analyze the impact of the offset voltage V_{OS} on the output voltage V_o of the inverting amplifier. The output voltage of the inverting amplifier is

$$V_o = -\frac{A(R_2 V_{in} + R_1 V_{OS} + R_2 V_{OS}) - R_o V_{in} + \frac{R_1 R_o}{R_{in}} V_{OS}}{R_1 (gain_0 + 1 + \frac{R_2}{R_{in}}) + R_o + \frac{R_1 R_o}{R_{in}}(1 + \frac{R_{in} + R_2}{R_L}) + R_2(1 + \frac{R_o}{R_L})} \tag{6.105}$$

This expression can be approximated by:

$$V_o \approx \frac{R_2 V_{in} + (R_1 + R_2) V_{OS}}{R_1 + \frac{R_2}{A} + \frac{R1 R_o}{A}(\frac{1}{R_{in}} + \frac{1}{R_L} + \frac{R_2}{R_L R_{in}})} \tag{6.106}$$

$$\approx gain_{ideal} V_{in} - (1 + |gain_{ideal}|) V_{OS} \tag{6.107}$$

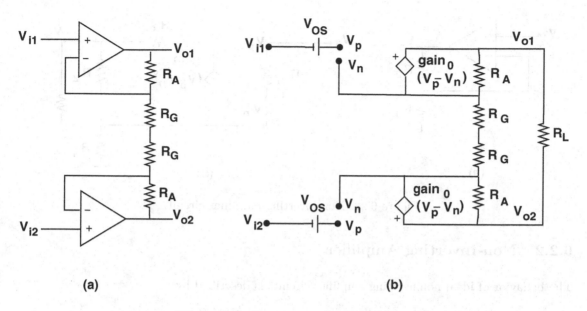

Figure 6.21: Instrumentation amplifier.

An error voltage at the amplifier output that increases with the gain of the amplifier is introduced by the offset voltage. This error is greater for larger amplifier gains and therefore the signal processing is less precise for large gains.

Instrumentation amplifiers are used for precise amplification of the difference of two input voltages [2]. Figure 6.21(a) shows the schematics of an instrumentation amplifier based on the circuit described in [11]. Figure 6.21(b) shows the macromodel for this circuit. The models used for the two OpAmps describe only the offset voltage V_{OS} and the finite gain, but, for simplicity do not include finite input and nonzero output impedances, poles, distortion, and so on. Also, it is assumed that the two OpAmps have similar characteristics, and hence their offset voltages and gains are equal.

The output voltage of the instrumentation amplifier is expressed by

$$V_o = \frac{R_A + R_G}{R_G + \frac{R_1 + R_A + R_G}{gain_0} + \frac{2\,R_1(R_A + R_G)}{gain_0 R_L}}(V_{i1} - V_{i2}) \tag{6.108}$$

$$\approx \left(1 + \frac{R_A}{R_G}\right)(V_{i1} - V_{i2}) \tag{6.109}$$

This shows that the output voltage of the instrumentation amplifier does not depend on the offset voltages. In reality, the effect of the offset voltage cannot be entirely removed due to the mismatches existing between the characteristics of the two OpAmps and the values of the external components. The gain of the circuit is fixed by the ratio $\frac{R_A}{R_G}$.

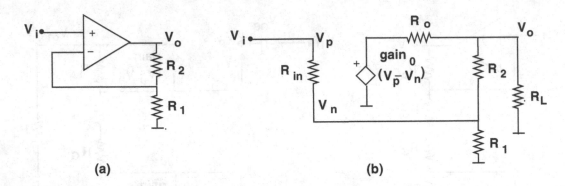

Figure 6.22: Non-inverting amplifier circuit.

6.2.2 Non-Inverting Amplifier

The behavior of ideal noninverting amplifier circuits is described by

$$V_o = gain_{ideal}\, V_i \tag{6.110}$$

where $gain_{ideal}$ is the gain of the noninverting amplifier, and $gain_{ideal} \in R$ and $gain_{ideal} > 0$.

Figure 6.22(a) shows the schematics of noninverting amplifiers. For ideal OpAmps, the output voltage is equal to the value:

$$V_o = \left(\frac{R_1 + R_2}{R_1}\right) V_i = \left(1 + \frac{R_2}{R_1}\right) V_i \tag{6.111}$$

and

$$gain_{ideal} = 1 + \frac{R_2}{R_1} \tag{6.112}$$

Impact of finite OpAmp gain, finite input resistance, and nonzero output resistance.

Figure 6.22(b) shows the circuit used to study the impact of these nonidealities on amplifier gain which is given by:

$$gain = \frac{[gain_0(R_1+R_2)R_{in}+R_1R_o]R_L}{R_1(R_2+R_{in}+AR_{in})R_L+R_{in}R_LR_o+R_1(R_2+R_{in}+R_L)R_o+R_2R_{in}(R_L+R_o)} \tag{6.113}$$

and can be approximated by:

$$gain \approx gain_{ideal}\, \frac{1}{1 + \frac{1}{gain_0}gain_{ideal}} \tag{6.114}$$

assuming that $R_{in}, R_L \gg R_1, R_2$, and $R_o \ll gain_0\, R_1$.

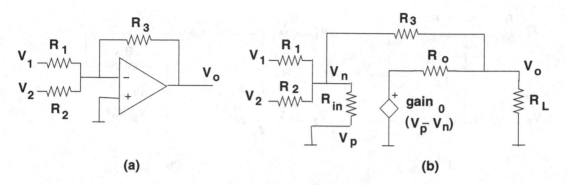

Figure 6.23: Summing amplifier circuit.

6.2.3 Summing Amplifier

The behavior of the summing amplifier is described by the expression

$$V_o = gain_1 \ V_{i1} + gain_2 \ V_{i2} \tag{6.115}$$

Thus, the output voltage V_o is the weighted sum of the two voltage inputs, V_{i1} and V_{i2}. Figure 6.23(a) shows the circuit of a summing amplifier. Following the same reasoning as for the inverting amplifier circuit, the output voltage of the circuit is described by:

$$V_o = -\left(\frac{R_3}{R_1}V_{i1} + \frac{R_3}{R_2} \ V_{i2}\right) \tag{6.116}$$

where $gain_1 = -\frac{R_3}{R_1}$, and $gain_2 = -\frac{R_3}{R_2}$. If $R_1 = R_2 = R$ then $gain = gain_1 = gain_2 = -\frac{R_3}{R}$, and $V_o = -\frac{R_3}{R}(V_{i1} + V_{i2})$.

As with other amplifier types, the behavior of the ideal summing amplifier is not influenced by frequency and the OpAmp nonidealities, e.g., finite gain and input impedance, nonzero output impedance, offset voltage, distortion, and noise. However, the behavior of real circuits is influenced by all of these factors. The impact of OpAmp nonidealities on the behavior of the summing amplifier is discussed next.

Impact of finite OpAmp gain, finite input resistance, and nonzero output resistance. Figure 6.23(b) shows the circuit for finding the influence of these nonidealities on the summing amplifier output. The output voltage is given by:

$$V_o = -\frac{[\frac{R_3}{R_1} \ V_{i1} + \frac{R_3}{R_2} \ V_{i2}][1 - \frac{R_o}{gain_0 R_3}]}{1 + \frac{1}{gain_0} + \frac{R_3}{gain_0 R_{in}} + \frac{R_3 + R_o(1 + \frac{R_3 + R_2}{R_L} + \frac{R_2}{R_{in}} + \frac{R_2 R_3}{R_{in}R_L})}{gain_0 R_2} + \frac{R_o + R_3(1 + \frac{R_o}{R_L})}{gain_0 R_1}} \tag{6.117}$$

If $R_{in} >> R_2, R_3$; $R_L >> R_o, R_2, R_3$; $gain_0 R_1 >> R_o$; $gain_0 R_2 >> R_o$; and $gain_0 R_3 >> R_o$ then Equation (6.117) can be reduced to:

$$V_o = -\left[\frac{R_3}{R_1}V_{i1} + \frac{R_3}{R_2}V_{i2}\right]\left[1 + \frac{1}{gain_0}(1 + \frac{R_3}{R_2} + \frac{R_3}{R_1})\right]^{-1} \tag{6.118}$$

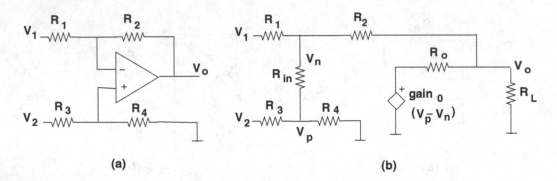

(a)   (b)

Figure 6.24: Difference amplifier circuit.

6.2.4   Difference Amplifier

The output voltage $\mathbf{V_o}$ of the difference amplifier is the weighted difference of its two input voltages. The behavior of the amplifier is specified by:

$$V_o = gain_2 \; V_{i2} \; - \; gain_1 \; V_{i1} \tag{6.119}$$

where $gain_1 > 0$ and $gain_2 > 0$. If the two gains are equal then the difference amplifier is called a differential amplifier.

Figure 6.24(a) shows the difference amplifier circuit. For an ideal OpAmp,

$$V_o = \left[\frac{1 \; + \; \frac{R_2}{R_1}}{1 \; + \; \frac{R_3}{R_4}} \right] V_{i2} \; - \; \left[\frac{R_2}{R_1} \right] V_{i1} \tag{6.120}$$

with

$$gain_1 = \frac{R_2}{R_1} \tag{6.121}$$

and,

$$gain_2 = \frac{1 + \frac{R_2}{R_1}}{1 + \frac{R_3}{R_4}} \tag{6.122}$$

The impact of nonidealities can be computed by following a procedure similar to the noninverting amplifier, and using the model in Figure 6.24(b).

6.2.5   Integrator

The output voltage of an integrator is the integral, over time, of the input voltage. The circuit's behavior, as a function of time, is given by:

$$V_o(t) = K \int V_i(t) \mathrm{d}t \tag{6.123}$$

where K is a constant.

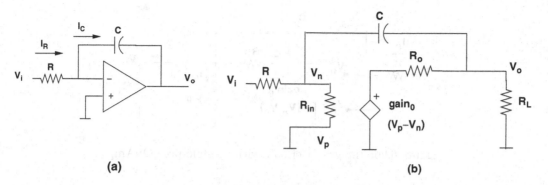

Figure 6.25: Integrator circuit.

Figure 6.25(a) shows an integrator circuit. The behavior of the ideal integrator circuit is characterized by the following set of equations:

$$V_i(t) = I_R(t) \, R \tag{6.124}$$

$$I_C(t) = \frac{d \, Q(t)}{dt} \tag{6.125}$$

$$= -C \frac{d \, V_o(t)}{dt} \tag{6.126}$$

$$I_R(t) = I_C(t) \tag{6.127}$$

or

$$\frac{d \, V_o(t)}{dt} = -\frac{1}{R \, C} \, V_i(t) \tag{6.128}$$

Thus,

$$V_o(t) = -\frac{1}{R \, C} \int V_i(t) dt \tag{6.129}$$

Impact of OpAmp poles. Figure 6.26 shows the circuit used to study the impact of the OpAmp pole on the integrator's behavior. The OpAmp pole is represented by resistor R_p and capacitor C_p, and the finite OpAmp gain is $gain_0$. For this analysis, the input resistance R_{in} of the OpAmp is assumed to be infinite, and the output resistance R_o to be zero.

The following set of equations describes the behavior of the single-pole integrator model in the frequency domain

$$
\begin{aligned}
V_i - V_n &= I_R \, R \\
V_n - V_o &= I_C \, X_c \\
I_R &= I_C \\
-gain_0 \, V_n &= I_{R_p} \, R_p + V_o \\
I_C + I_{R_p} &= I_{c_p} + I_L \\
V_o &= I_{C_p} \, X_{c_p} \\
V_o &= I_L \, R_L
\end{aligned}
$$

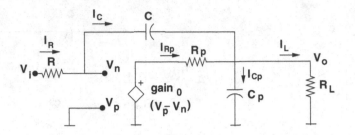

Figure 6.26: Integrator circuit with a single-pole OpAmp.

After simplifications the integrator output is given by

$$V_o = -\frac{gain_0(1 - \frac{R_p}{gain_0\ X_c})X_cX_{cp}R_L\ V_i}{R_L\ R_p(R+X_c)+R\ X_{cp}\ (gain_0R_L+R_L+R_p)+R_p\ X_c\ X_{cp}+R_L\ X_{cp}\ (R_p+X_c)} \tag{6.130}$$

If $R_L \gg R_p$ and $gain_0\ R \gg R_p$, then

$$V_o \approx -gain_{DC}\ \frac{(1 - \frac{R_p}{gain_0\ X_c})}{1 + gain_0\frac{R}{X_c} + \frac{R_p}{X_{cp}} + \frac{R}{X_c}\frac{R_p}{X_{cp}}}\ V_i \tag{6.131}$$

$$\approx -gain_{DC}\ \frac{1}{1 + gain_0\frac{R}{X_c} + \frac{R_p}{X_{cp}} + \frac{R}{X_c}\frac{R_p}{X_{cp}}}\ V_i \tag{6.132}$$

and the integrator's transfer function becomes:

$$H(f) \approx -\frac{gain_0}{1 + gain_0\frac{R}{X_c} + \frac{R_p}{X_{cp}} + \frac{R}{X_c}\frac{R_p}{X_{cp}}} \tag{6.133}$$

Note that the denominator of the transfer function $H(f)$ is a second-order polynomial of frequency f.

Example: If $R = 250k\Omega$, $C = 2nF$, $gain_0 = 10^4\frac{V}{V}$, and $k = R_pC_p = 0.025$, then the OpAmp dominant pole is near $10\,\text{Hz}$. The magnitude of the transfer function is:

$$|H(f)| \approx \frac{10^4}{\sqrt{(1 - 0.493\ 10^{-3}\ Hz^{-2}f^2)^2 + 0.986\ f^2\ 10^3}} \tag{6.134}$$

and the phase is

$$\varphi(f) \approx \arctan\frac{0.3141\ 10^3\ f}{1 - 0.493\ 10^{-3}Hz^{-2}\ f^2} \tag{6.135}$$

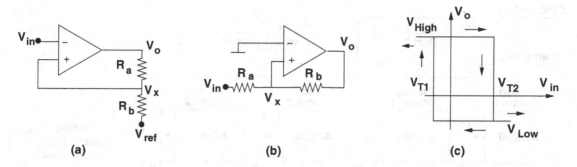

Figure 6.27: Schmitt trigger: (a) inverting circuit, (b) noninverting circuit, and (c) transfer characteristics.

6.2.6   Comparator

The behavior of a comparator is described by:

$$V_o = V_{high}, \; if \; V_{in} > V_T \tag{6.136}$$
$$V_o = V_{low}, \; if \; V_{low} < V_T \tag{6.137}$$

where V_{in} and V_o are the input and output voltages, V_{high} is the high reference voltage, V_{low} is the low reference voltage and V_T is the threshold voltage. If the input voltage is larger than the threshold voltage then the comparator output is the high reference voltage, otherwise the comparator output is the low reference voltage.

Schmitt triggers are comparators that switch, depending on different threshold voltages, to the high and low reference voltages. The behavior of a Schmitt trigger is determined by

$$V_o = V_{high}, \; if \; V_{in} > V_{T1}, \tag{6.138}$$
$$V_o = V_{low}, \; if \; V_{in} < V_{T2} \tag{6.139}$$

Having different threshold voltages for switching to the two reference voltages is useful in avoiding repeated switching of the circuit due to fluctuations of the input voltage. Figure 6.27(a) presents an inverting Schmitt trigger, and Figure 6.27(b) shows a noninverting Schmitt trigger.

Figure 6.27(c) summarizes the behavior of the inverting Schmitt trigger. If the input voltage exceeds the threshold voltage V_{T1} then the circuit output switches to V_{high}. As long as $V_{in} < V_{T2}$ the circuit output remains V_{high}. If the input voltage exceeds the threshold value V_{T2} then the output switches to the low reference value V_{low}. If the input voltage is then lowered, the output signal remains V_{low} until the input voltage falls below the value V_{T1}, at which point the output swings back to V_{high}.

The equations for the two threshold values are:

$$V_{T1} = V_{low}\left[\frac{R_B}{R_A + R_B}\right] + V_{ref}\left[\frac{R_A}{R_A + R_B}\right] \tag{6.140}$$
$$V_{T2} = V_{high}\left[\frac{R_B}{R_A + R_B}\right] + V_{ref}\left[\frac{R_A}{R_A + R_B}\right] \tag{6.141}$$

The width of the hysteresis function is given by:

$$V_{T2} - V_{T1} = (V_{high} - V_{low})\frac{R_B}{R_A + R_B} \tag{6.142}$$

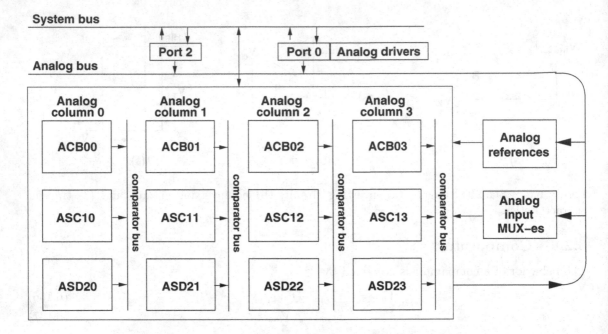

Figure 6.28: Analog section of the PSoC architecture [17].

6.3 Reconfigurable Continuous-Time Analog Blocks

PSoC's analog section includes an analog array with up to 12 analog blocks as shown in Figure 6.28. The analog blocks are grouped in columns, with three blocks in each column. Each column includes continuous-time analog blocks (blocks ACB00 to ACB03 in the figure), and switched-capacitor blocks (blocks ASC10, ASC12, ASC21, and ASC23, and blocks ASD20, ASD11, ASD22, and ASD13 as shown in the figure).

The functionality and interconnection of each analog block can be programmed specifically for each application. Each of the analog blocks can be connected to a variety of signals, e.g., different reference voltages, analog bus (ABUS), ground (AGND), neighboring analog blocks, and input ports. A block's output can be routed to different destinations, such as to the analog bus and neighboring blocks. The nature of the input and output signals supplied to each of the blocks in an analog array is controlled by a set of programmable control registers.This subsection details the continuous-time analog processing blocks. (The switched capacitor blocks are discussed in the following chapter.) Figure 6.29 shows the topology of a reconfigurable continuous-time analog processing block which includes both programmable amplifier gain and selection of the input and output signals.

The following are the main parts of a continuous-time block:

- *Control registers*: The functionality and interfacing of each continuous-time analog block is controlled by programming four registers called *ACBxxCR0*, *ACBxxCR1*, *ACBxxCR2*, and *ACBxxCR3* (*xx* indicates the analog block *ACBxx*). The *ACBxxCR0* register programs the resistor matrix, the *ACBxxCR1* register selects the input signals and outputs of the block, and the *ACBxxCR2* register determines the block's functionality with respect to comparator mode, testing, and power levels. The *ACBxxCR3* register determines the block's functionality (e.g., amplifier or comparator) configuration as an instrumentation

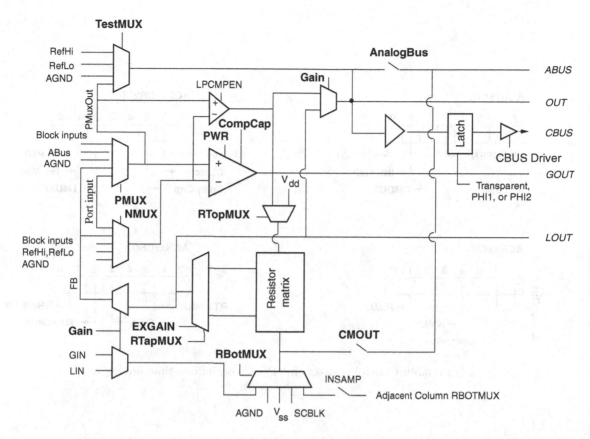

Figure 6.29: Reconfigurable continuous-time analog block [17].

amplifier, etc. Figure 6.30 shows the structure of the four control registers associated with each continuous-time, analog block.

The *ACR00CR1* register at address 72H controls Block *ACB00*; the *ACB01CR1* register at 76H, Block *ACB01*; the *ACB02CR1* register at 7AH Block *ACB02*; and the *ACB03CR1* register at 7EH block *ACB03*.

- *OpAmp*: Each block includes a differential OpAmp for implementing the signal processing functionality. The OpAmp mode is selected by setting bit *CompCap* - bit 5 of the *ACBxxCR2* register.

- *Comparator*: Each continuous-time analog block has a comparator that operates only if the OpAmp is not enabled, and vice versa. The comparator is enabled by bit *LPCMPEN*, which is bit 3 of the *ACBxxCR3* register. In the comparator mode bit *CompCap* is reset, which disables the OpAmp. Bit *CompCap* corresponds to bit 5 in the *ACBxxCR2* register.

- *Resistor matrix*: The resistor matrix determines the programmable gain of an analog block. Bits *RTapMUX* are used to select the needed gain value from a set of 18 possible gain values, or 18 different attenuation values. Bit *EXGAIN* decides between two gain domains: *standard gain* which includes 16 of the 18 gain values, and *high gain* which has 2 possible gains. Finally, bit *Gain* decides whether the analog block operates as an amplification block with a gain magnitude larger than one, or as an attenuation block with a gain magnitude

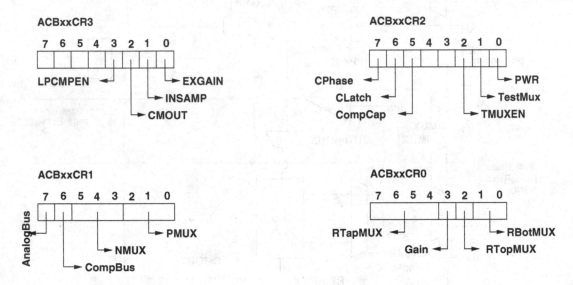

Figure 6.30: Control registers for PSoC continuous-time analog block.

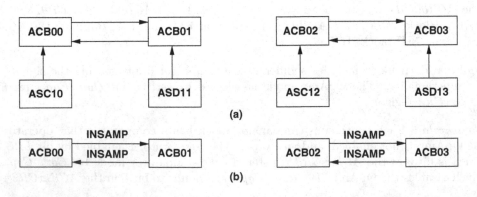

Figure 6.31: Possible connections to the resistor matrix bottom node in CT blocks [17].

Table 6.2: Programmable connections to the resistor matrix bottom node [17].

	bit **INSAMP**	bits **RbotM** (2 bits)	**ACB00**	**ACB01**	**ACB02**	**ACB03**
1	0	00	ACB01	ACB00	ACB03	ACB02
2	0	01	AGND	AGND	AGND	AGND
3	0	10	Vss	Vss	Vss	Vss
4	0	11	ASC10	ASD11	ASC12	ASD13

Table 6.3: Programmable gains of a continuous-time analog block [17].

	RTap bits (4 bits)	**EXGain** bit	Loss	Gain	Domain
1	0000	1	0.0208	48.000	High gain
2	0001	1	0.0417	24.000	High gain
3	0000	0	0.0625	16.000	Standard gain
4	0001	0	0.1250	8.000	Standard gain
5	0010	0	0.1875	5.333	Standard gain
6	0011	0	0.2500	4.000	Standard gain
7	0100	0	0.3125	3.200	Standard gain
8	0101	0	0.3750	2.667	Standard gain
9	0110	0	0.4375	2.286	Standard gain
10	0111	0	0.5000	2.000	Standard gain
11	1000	0	0.5625	1.778	Standard gain
12	1001	0	0.6250	1.600	Standard gain
13	1010	0	0.6875	1.455	Standard gain
14	1011	0	0.7500	1.333	Standard gain
15	1100	0	0.8125	1.231	Standard gain
16	1101	0	0.8750	1.143	Standard gain
17	1110	0	0.9375	1.067	Standard gain
18	1111	0	1.0000	1.000	Standard gain

less than one. Table 6.3 lists the possible amplification and attenuation values for an analog block, and the corresponding configuration of bits $RTapMUX$ and $EXTgain$. Bits $RTapMUX$ are the four bits 7-4 in the $ACBxxCR0$ register. Signal $EXgain$ is bit 0 of the $ACBxxCR3$ register. Finally, bit $Gain$ is bit 3 in the $ACBxxCR0$ register.

In addition, different feedback structures can be programmed by selecting different top and bottom nodes for the resistor matrix. The top node of the resistor matrix can be either the supply voltage V_{dd} or the output voltage of the OpAmp. This selection is determined by bit $RTopMUX$.

The bottom node of the resistor matrix is determined by two bits $RBotMUX$, which selects one of five different possibilities: analog ground ($AGND$), V_{SS}, a neighboring switched-capacitor block ($SCBLK$), the resistor matrix of a neighboring continuous-time analog block ($RBotMUX$), and (GIN (LIN)). Table 6.2 shows the control bit values that select

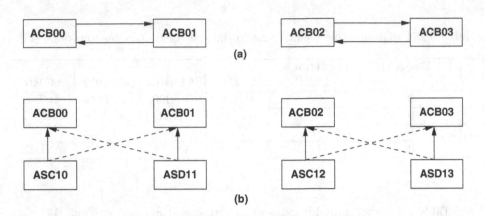

Figure 6.32: Types of links between continuous-time and switched capacitor analog blocks.

one of four different options. Note that in each of the four cases bit *INSAMP* is reset. Figure 6.31 shows the connection styles presented in Table 6.2.

The resistor matrices of neighboring blocks can be connected to each other to build an instrumentation amplifier. Bit *INSAMP* has to be set for this purpose. Figure 6.31(a) shows the continuous-time analog blocks involved in the connections for this case.

Bit *RTopMUX* is bit 2 of register *ACBxxCR0*. Signals *RBotMux* are implemented using bits 1-0 of register *ACBxxCR0*. Bit *INSAMP* corresponds to bit 1 of the *ACBxxCR3* register.

- *Reconfigurable input connections to the inverting OpAmp*: An analog block's functional flexibility is provided by the ability to select the kind of signal applied to the inverting input of the OpAmp. The signal selection is controlled by the three *NMUX* bits and an 8-1 multiplexer, also referred to as *NMux* [17], at the OpAmp input.[3] The three *NMUX* bits represent bits 5-3 of the *ACRxxCR1* register corresponding to the analog block *xx*.

Table 6.4: Programmable inverting input connections of a CT analog block [17].

	NMUX bits (3 bits)	**ACB00**	**ACB01**	**ACB02**	**ACB03**
1	000	ACB01	ACB00	ACB03	ACB02
2	001	AGND	AGND	AGND	AGND
3	010	RefLo	RefLo	RefLo	RefLo
4	011	RefHi	RefHi	RefHi	RefHi
5	100	Feedback	Feedback	Feedback	Feedback
6	101	ASC10	ASD11	ASC12	ASD13
7	110	ASD11	ASC10	ASD13	ASC12
8	111	Port input	Port input	Port input	Port input

Table 6.4 summarizes the types of input signals for different bit values for the *NMUX* bits. Different kinds of inputs can be provided:

- Feedback connection coming from the resistor matrix.

---

[3]Refer to Chapter 3 for additional details on the interconnect for analog blocks.

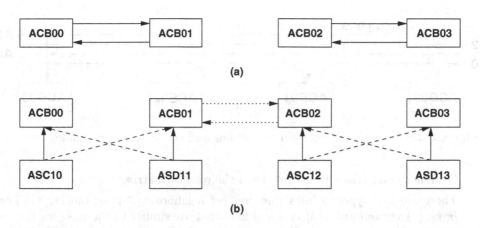

Figure 6.33: Types of links between continuous-time analog blocks.

- Connections to reference voltages, reference low (*RefLo*) and reference high (*RefHi*), and to the analog ground (*AGND*).

- Signal connection to PSoC's input ports.

- Connections to neighboring continuous-time and switched-capacitor, analog blocks as shown in Figure 6.32. These connections are of three types: (i) *horizontal links* to neighboring analog blocks on the same row as illustrated in Figure 6.32(a), (ii) *vertical links* coming from blocks on the same column but on the next row. These links are illustrated in Figure 6.32(b), and (iii) *crossing connections* that come from neighboring blocks positioned on the diagonal, as shown by the dashed lines in Figure 6.32(b). Other available connections are also shown.

Table 6.5: Programmable, noninverting, input connections of a analog block [17].

	PMUX bits (3 bits)	ACB00	ACB01	ACB02	ACB03
1	000	RefLo	ACB02	ACB01	RefLo
2	001	Port input	Port input	Port input	Port input
3	010	ACB01	ACB00	ACB03	ACB02
4	011	AGND	AGND	AGND	AGND
5	100	ASC10	ASD11	ASC12	ASD13
6	101	ASD11	ASC10	ASD13	ASC12
7	110	ABUS0	ABUS1	ABUS2	ABUS3
8	111	Feedback	Feedback	Feedback	Feedback

- *Reconfigurable input connections to the noninverting OpAmp input*: There is a similar capability to program the nature of the signals applied to the positive input of the OpAmp. Signal selection is controlled by the three bits *PMUX* applied to an 8–1 multiplexer called *PMUX* [17]. Bits *PMUX* are bits 2-0 of the *ACBxxCR1* registers, that also control the reconfigurable connections to the inverting inputs.

Table 6.5 lists the input signals that can be applied to the noninverting input of the OpAmp. These signals are of various types:

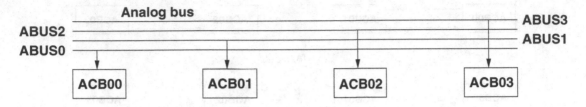

Figure 6.34: Connections between analog bus and continuous-time analog blocks.

- Feedback connections that come from the resistor matrix.

- There are four types of links that connect neighboring analog blocks: (1) *horizontal links*, (2) *vertical links*, (3) *crossing links* that are similar to the links for the inverting OpAmp input as shown in Figure 6.33(a) and (b), and (4) *extension links*, shown by dotted lines in Figure 6.33(b), that connect analog blocks *ACB01* and *ACB02*. These links do not exist for the inverting inputs.

- Reference low, *RefLo*, signal and analog ground, *AGND*, that are applied to the OpAmp noninverting input.

- Input signals that come from the input ports.

- Signals that come from the analog bus (cf. Figure 6.28). The analog Block *ACB00* is connected to bit 0 (*Bit ABUS0*) of the analog bus, Block *ACB01* to bit 1 (*Bit ABUS1*) of the analog bus, Block ACB02 to bit 2 (*Bit ABUS2*) of the analog bus, and Block *ACB03* to bit 3 (*Bit ABUS3*) of the analog bus. Figure 6.34 illustrates these connections.

- *Programmable outputs*: The output signal of a continuous-time block is produced at the output of the enabled circuit, which is either an OpAmp or a comparator, but not both. The output signal can be routed to several destinations as follows.

  - *Analog bus*: The output signal is routed to PSoC's analog bus *ABUS*. This is controlled *Bit AnalogBus*, which is bit 7 of the *ACBxxCR1* register[4].

  - *Comparator bus*: The output signal is applied to PSoC's comparator bus *CBUS*. This is programmed by setting bit *CompBus*, bit 6 of the *ACBxxCR1* register. The connection to the bus *CBUS* includes a latch circuit, which can be programmed to store the information, or pass it through transparently. This functionality is selected by bit *CLatch*, which is bit 6 of the *ACBxxCR2* control register. In addition, this latch can be programmed to be transparent, either for clock *PHI1* (if bit *CPhase* is reset), and for clock *PHI2* (if bit *CPhase* is set). Bit *CPhase* corresponds to bit 7 in the *ACBxxCR2* register.

- *Test mode*: For testing purposes, some of the input signals applied to a continuous-time block can be routed to the analog bus, and then to the chip pins, so that they can be measured externally. Bit *TMUXEN* enables the test multiplexer. The two *TestMUX* bits determine which of the block inputs is routed to the analog bus *ABUS*. The following signals are selected for the four values of the *TestMUX* bits: (i) positive input for value "00", (ii) analog ground (*AGND*) for 01, Reference Low (*RefLo*) for value "10", and Reference High (*RefHi*) for "11". The selected signal is routed to bit *ABUS0* of the analog block for Block *ACB00*, to bit *ABUS1* for Block *ACB01*, to bit *ABUS2* for Block *ACB02*, and to

---

[4]Chapter 3 provides additional details on the analog bus.

Continuous time analog block ACB00

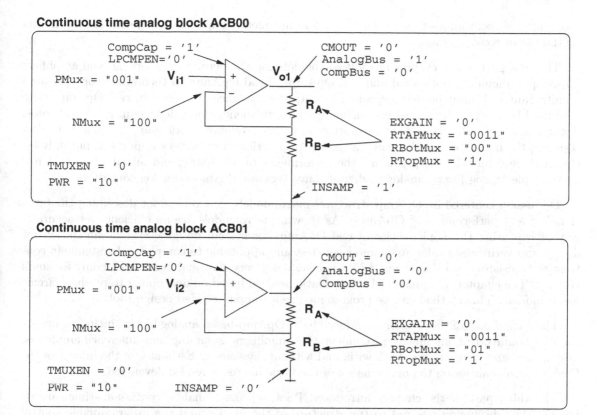

Figure 6.35: Instrumentation amplifier implemented using PSoC [17]. Courtesy of Cypress Semiconductor Corporation.

Bit *ABUS3* for Block *ACB03*. In addition, bit *CMOUT* connects the common mode input of the OpAmp to the corresponding analog bus bit.

Bit *TMUXEN* corresponds to bit 4 of the *ACBxxCR2* register. Bits *TestMux* are bits 3-2 of the same register. Bit *CMOUT* is bit 2 of the *ACBxxCR3* register.

- *Power mode*: The power mode of the OpAmp is determined by the two *PWR* bits. These bits are bits 1-0 of the register. The following power modes are selected by the two bits: (i) power off, if the bits are "00", (ii) low power, if the bits are "01", (iii) medium power for bits "10", and (iv) high power for bits "11". These power levels are doubled in the high bias mode [17].

Figure 6.35 shows the implementation of an instrumentation amplifier using two continuous-time analog blocks and the programming of the four control registers for each of the two blocks [17].

6.4 Conclusions

The chapter has presented the basic concepts of continuous-time analog circuits, including circuit operation and performance characterization in the presence of nonidealities and discussed PSoC's programmable continuous-time analog blocks for implementing application-specific analog circuits by the use of reconfiguration.

The first part of the chapter introduced the defining characteristics of operational amplifiers (OpAmps), including both ideal and real OpAmps. Ideal OpAmps, a theoretical concept, have infinite gain and input impedance, and zero output impedance. In contrast, real OpAmps were described by a set of nonlinearities, including finite, frequency-dependent gain, poles and zeros, distortion, offset voltages, output saturation, slew rate, common mode gain, and noise. Understanding the impact of nonidealities on the OpAmp performance is very important, not only for circuit design, but also for predicting the performance of the analog and mixed-signal systems, for example analog filters, analog-to-digital converters and digital-to-analog converters.

Discussion centered on OpAmp structural macromodels as a vehicle for describing the functionality and performance of OpAmps. As shown, macromodels are an efficient, yet accurate, way of capturing the nonidealities of real OpAmps for system design. System simulation for design and verification using macromodels is fast and applicable to complex subsystems, in contrast to transistor-level simulation, which is precise but very slow and hence useful only for small circuits. The chapter also presented a systematic method based on decoupling to produce circuit macromodels. This method can be programmed as a computer-aided design tool.

The second part of the chapter introduced basic OpAmp-based analog blocks, including inverting and noninverting amplifiers, instrumentation amplifiers, summing and difference amplifiers, integrators, and comparators, both with, and without, hysteresis. Estimates of the impact of the OpAmp nonidealities on the performance of the basic blocks were also developed.

The third part of the chapter introduced PSoC's programmable, continuous-time, analog blocks. The discussion referred to the structure of the blocks, and the programmable control registers for configuring an analog block. Each block can be configured with respect to its functionality, attributes (e.g., gain), inputs, outputs, and power level. Each block also incorporates a comparator. The operation of each block is controlled through four registers called *ACBxxCR0*, *ACBxxCR1*, *ACBxxCR2*, and *ACBxxCR3*. The programmable, switched, and continuous-time capacitor blocks can be connected with each other to form larger networks, for example filters and analog-to-digital converters.

6.5 Recommended Exercises

1. Starting from Equation (6.2), find the OpAmp transfer function. Find the analytical expressions of the magnitude and phase response for the two-pole model, and expressions for the 3 dB frequency and unit gain frequency points.

2. Graph the magnitude and phase response using the expressions derived for Exercise 1. Assume that the dominant pole is at 1 KHz, the secondary pole at 100 KHz and that the DC gain is 40 dB. Identify the position of the 3 dB and unity gain frequency points on your graphs.

3. For the OpAmp model in Exercise 2, assume that nonlinear distortion is approximated by the first three terms in Equation (6.15). Plot the output voltage V_o of the OpAmp model for different values of the nonlinearity coefficient a_2.

4. Find the transfer function for the noninverting amplifier assuming a single-pole model for the OpAmp. Study the effect of the pole position on the gain accuracy of the circuit.

5. For the summing amplifier, derive an analytic expression for the output voltage assuming a one-pole model the OpAmp. Study the impact of the pole position on the amplifier's frequency response.

6. Derive an analytic expression, in the frequency domain, for the voltage $V_{o1} - V_{o2}$ for the instrumentation amplifier illustrated in Figure 6.21(a). Assume a single-pole model for the two OpAmps. Find the conditions that minimize the impact of the two poles on the difference voltage.

7. Propose and implement a PSoC-based solution for offset compensation for high gain amplifier circuits (gain > 100). (Additional information is available in the Application Note by V. Grygorenko, Offset Compensation for High Gain Amplifiers, AN2320, Cypress Semiconductor Corporation, October 28 2005.)

8. Estimate the impact of noise on a noninverting amplifier. Analyze the resulting SNR for different noise power levels and amplifier gains.

9. Study the nonidealities caused by the finite gain of the inverting and noninverting amplifiers. For the inverting amplifier shown in Figure 6.17, plot the relative gain error versus the ratio R_2/R_1 for gains equal to 10, 100, 1,000, and 10,000. Explain your results. Repeat the exercise for the noninverting amplifier shown in Figure 6.22. For the noninverting amplifier, plot the relative gain error versus $gain_0$ for R_2/R_1 equal to 1, 10, 100, and 1000. Compare your results with the results in Table 6.1, and summarize the tradeoffs that exist between gain error, the values of the resistances, and the $gain_0$.

10. For the difference amplifier in Figure 6.24, determine the CMRR for ideal OpAmps and then calculate the CMRR for the following OpAmp nonidealities: finite gain, offset voltage, noise, and poles. Compare the two CMRR expressions. Discuss any tradeoffs.

11. Identify the different discrete outputs from the continuous-time, analog PSoC blocks, and the destinations to which each of the outputs can be routed.

12. Explain the use of *INSAMP* (bit 1 of the *ACBxxCR3* register).

13. Estimate the input-referred noise power density of the circuit in Exercise 1, assuming that the OpAmp is ideal and noiseless, and using only dominant noise sources.

14. Find the analytical expressions for the magnitude and phase response of the circuit in Exercise 1, if the OpAmp has a DC gain of 40 dB and the dominant pole is located at $1 KHz$.

15. A first-order low-pass filter can be created by connecting capacitor C_2 in parallel with resistor R_2, of the inverting amplifier circuit in Figure 6.17. Find the analytical expressions for the magnitude and phase response, assuming that the OpAmp is ideal.

Bibliography

[1] R. J. Baker, CMOS Circuit Design, Layout, and Simulation, second edition, Hoboken, NJ: J. Wiley Interscience, 2005.

[2] S. Franco, Design with Operational Amplifiers and Analog Integrated Circuits, New York: McGraw Hill, 2002.

[3] P. Gray, P. Hurst, S. Lewis, R. Meyer, "Analysis and Design of Analog Integrated Circuits", fourth edition, Hoboken, NJ: J. Wiley & Sons, 2001.

[4] R. Gregorian, G. Temes, "Analog MOS Integrated Circuits for Signal Processing", J. Wiley & Sons, 1986.

[5] X. Huang, C. Gathercole, H. Mantooth, Modeling nonlinear dynamics in analog circuits via root localization, *IEEE Transactions on Computer-Aided Design of Integrated Circuits and Systems*, 50, (7), pp. 895–907, July 2003.

[6] C. Motchenbacher, J. Connelly, *Low Noise Electronic System Design*, Hoboken, NJ: J. Wiley, 2000.

[7] G. Palumbo, S. Pennisi, High-Frequency harmonic distortion in feedback amplifiers: Analysis and Applications, *IEEE Transactions on Circuits and Systems - I: Fundamental Theory and Applications*, 50, (3), pp. 328–340, March 2003.

[8] B. Razavi, *Design of Analog CMOS Integrated Circuits*, New York: 2001.

[9] W. Sansen, "Distortion in elementary transistor circuits", *IEEE Transactions on Circuits and Systems - II: Analog and Digital Signal Processing*, 46, (3), pp. 315–325, March 1999.

[10] D. Seguine, Lower Noise Continuous Time Signal Processing with PSoC, Application Note AN224, *Cypress*, October 12 2004.

[11] D. Van Ess, Differential Amplifier, Application Note AN2367, *Cypress*, April 14 2006.

[12] D. Van Ess, Understanding Switched Capacitor Analog Blocks, Application Note AN2041, *Cypress*, March 30 2004.

[13] S. M. Sze, *Physics of Semiconductor Devices*, Hoboken, NJ: J. Wiley & Sons, 1981.

[14] P. Wambacq, W. Sansen, *Distortion Analysis of Analog Integrated Circuits*, Boston: Kluwer, 1998.

[15] Y. Wei, A. Doboli, Systematic Development of analog circuit structural macromodels through behavioral model decoupling, *Proceedings of the Design Automation Conference (DAC)*, pp. 57–62, 2005.

[16] Y. Wei, A. Doboli, Systematic development of nonlinear analog circuit macromodels through successive operator composition and nonlinear model ecoupling, *Proceedings of the Design Automation Conference (DAC)*, 2006, pp. 1023-1028.

[17] PSoC Mixed Signal Array, Technical Reference Manual, *Document No. PSoC TRM 1.21*, Cypress Semiconductor Corporation, 2005.

[18] Y. Wei, A. Doboli, Structural Modeling of Analog Circuits Through Model Decoupling and Transformation, *IEEE Transactions on Computer-Aided Design of Integrated Circuits and Systems*, 27, (4), pp. 712-725, April 2008.

Chapter 7

Switched-Capacitor Blocks

This chapter[1] presents an introduction to switched capacitor (SC) circuits, and PSoC's programmable SC blocks.

The principle of operation of SC circuits is based on the fact that the movement of charge stored on a capacitor can approximate the average current through a resistor. The behavior of these circuits is influenced by a number of circuit nonidealities, such as the nonzero resistance of "on" switches, channel charge injection, and clock feedthrough.

Basic SC circuits can be employed in a wide variety of circuits, including fixed gain amplifiers, selectable gain amplifiers, comparators, integrators, and differentiators. In addition, the concept of auto-zeroing is explained for SC amplifiers and the effect of the OpAmp finite gain and nonzero switch resistance on the gain value is also discussed.

The second part of this chapter describes PSoC's two types of programmable switched capacitance circuits (viz., type C and type D). Chapter 3 detailed the interconnect structure for SC blocks, which can be connected to neighboring continuous-time and SC blocks to form more complex networks (e.g., filters and ADCs). This chapter presents the structure and the control registers of the programmable SC blocks. The functionality of each SC block can be programmed as inverting/noninverting amplifiers, integrators, and comparators. Four programmable capacitor arrays can be programmed to set the gain value of these blocks. The inputs, outputs, and clocks of a SC block are also configurable.

Chapters 8 and 9 illustrate the use of PSoC's programmable SC blocks to build analog filters and $\Delta\Sigma$ analog-to-digital converters:

This chapter has the following structure.

- Section 1 provides an introduction into SC techniques.

- Section 2 describes the basic SC circuits.

- Section 3 describes PSoC's programmable SC blocks.

- Section 4 provides concluding remarks.

---

[1]This chapter is based in part on material found in reference [4]. Other sections of this chapter are reproduced from [7]. Courtesy of Cypress Semiconductor Corporation.

A. Doboli, E.H. Currie, *Introduction to Mixed-Signal, Embedded Design*,
DOI 10.1007/978-1-4419-7446-4_7, © Springer Science+Business Media, LLC 2011

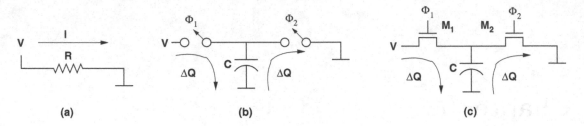

Figure 7.1: Moving charge with current stimulus.

7.1   Introduction To Switched Capacitor Techniques

Analog circuit design typically requires the use of resistors, capacitors, and integrated active devices. Inductors have proven to be of less importance and interest because of size, cost and other issues. It is the nature of integrated circuitry that small precision resistors are more difficult to manufacture and consequently more expensive than capacitors [4]. Given that making capacitors is easier and cheaper, it follows that techniques were developed to use capacitors instead of resistors in high-quality analog circuits [1, 4, 5]. These techniques led to the development of switched capacitor (SC) architectures that utilize capacitors and precise timing of switches to control charge flow both to and from these capacitors to replace many of the resistors that would otherwise be required.

Analog circuit design is largely about controlling the movement of charge between voltage nodes. Figure 7.1 shows charge movement through a resistor and through a switched capacitor. Equation (7.1) represents the current flow from a voltage potential to ground through the resistor shown in Figure 7.1(a):

$$i \; = \; \frac{V}{R} \tag{7.1}$$

Note that this current is a linear continuous movement of charge.

In Figure 7.1(b), when the ϕ_1 switch is closed and the ϕ_2 switch is open, the capacitor charges to the full potential V. The equation for the amount of stored charge is:

$$Q \; = \; C\,V \tag{7.2}$$

where the capacitance C is a constant. Thus for a given capacitor, the stored charge Q is proportional to V, the applied potential.

When the switch ϕ_1 is opened and switch ϕ_2 is closed, the stored charge moves to ground. For each sequential pair of switch closures, a quantum of charge is moved. If the switches are opened and closed at the rate of f_s, the charge quanta also move at this rate. Equation (7.3) shows that the repetitive movement of charge is equivalent to an average current given by:

$$I_{average} \; = \; \frac{Q}{T} \; = \; Q\,f_s \; = \; C\,V\,f_s \tag{7.3}$$

Unlike the resistor, this current is not actually a continuous movement of charge. The charge move in quanta. However, this is not a problem for a system for which the signal is sampled at the end of each switching cycle. Equation (7.4) shows that a switched capacitor is equivalent to a resistor when they have equivalent ratios of voltage drop versus delivered current:

$$R_{equivalent} \;=\; \frac{V}{I_{average}} \;=\; \frac{1}{C\,f_s} \tag{7.4}$$

Hence, the equivalent resistance is inversely proportional to the capacitance and the switching frequency, and can be altered, merely by changing the switching frequency. The larger the capacitance C, the larger the charge quanta, resulting in more current for the same value of f_s, and thus a smaller equivalent resistance. Increasing frequency f_s causes more quanta to be transferred per unit of time, resulting in a higher current and lower effective resistance.

Figure 7.1(c) presents a switched-capacitor circuit that uses the transistors M_1 and M_2 to implement the switches ϕ_1 and ϕ_2. If the clock signal ϕ_1 is high, then the gate to source voltage, V_{gs}, of transistor M_1 is larger than the threshold voltage of the transistor, therefore the transistor is conducting, that is the switch is closed. During this time, clock ϕ_2 is low, hence the gate to source voltage V_{gs} of transistor M_2 is lower then its threshold voltage, and the transistor is off, that is the switch is open. The circuit operation is similar for clock ϕ_1 low and clock ϕ_2 high, but now transistor M_1 is off, and transistor M_2 is on.

Figure 7.2 is a graph of the voltage V_c across the capacitor as a function of time, and for different capacitance values and transistor widths. Note that the rise time of the voltage increases with the capacitance value C, and decreases with the width W of transistor M_1.

In addition, the following timing requirements must be satisfied for switches ϕ_1 and ϕ_2 for the two circuits in the figure to exhibit similar behavior.

- Switches ϕ_1 and ϕ_2 can never be closed at the same time.

- Switch ϕ_1 must have time to open before switch ϕ_2 is closed.

- Switch ϕ_2 must have time to open before switch ϕ_1 is closed.

- The switching rate f_s must allow enough time for the circuitry to fully charge and discharge in the allotted phase cycle.

7.1.1   Nonidealities in Switched Capacitor Circuits

The main nonidealities that influence the behavior of switched capacitor circuits are (a) the nonzero resistance of conducting MOSFETs, (b) channel charge injection, and (c) clock feedthrough [1, 4, 5]. Other nonidealities (e.g., the impact of OpAmp nonidealities on the switched capacitor circuit behavior) are discussed later in this chapter.

A. Nonzero Resistance of Conducting MOSFETs

Equations (7.1)–(7.4) characterize the average current through the conducting transistors, and the final value of the voltage V_c across the capacitor. However, it is more difficult to find the variations in time of the current conducted by the "on" transistor and the voltage across the capacitor. Figure 7.2 shows the current I_D and the voltage V_c as functions of time for different values of the capacitor C and input voltage V_{in}. This section computes the analytical expression for the voltage V_c as a function of time. The analysis refers to the Figure 7.1(c), in which transistor M_1 is conducting (clock ϕ_1 is one), and transistor M_2 is off (clock ϕ_2 is zero).

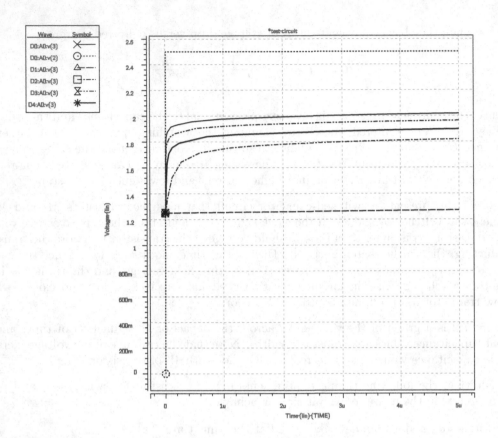

Figure 7.2: Time behavior of switched capacitor circuits.

Expressed analytically, the voltage V_c is given by the solution of the following differential equation:

$$\frac{d}{dt}[V_c(t)] = \frac{I_D(t)}{C} \tag{7.5}$$

The current I_D through the transistor M_1 depends on the operation region of the transistor, which is either linear mode or saturation.

If transistor M_1 is in the linear operation mode (i.e., $V_{gs} - V_{th} > V_{ds}$ then

$$\frac{d}{dt}[V_c(t)] = \frac{\mu_n\, C_{ox}}{2LC}\frac{W}{}\left[(V_{DD} - V_c(t) - V_{th})(V_{in} - V_c(t)) - \frac{(V_{in} - V_c(t))^2}{2}\right] \tag{7.6}$$

After solving the differential equation, the voltage V_c, as a function of time, is given by:

$$V_c(t) = \frac{2\,K\,\exp(AV_{in}\,t) - A\,\exp(A\,V_{in}t) + \exp(K\,t + K[1])\,V_{in}}{A\,\exp(A\,V_{in}\,t) + \exp(K\,t + K[1])} \tag{7.7}$$

where

$$A = \frac{\mu_n\,C_{ox}W}{2\,LC} \tag{7.8}$$

and

$$K = A \ (V_{DD} - V_{th}).$$ (7.9)

The constant $K[1]$ is determined by the initial value of the voltage V_c (the voltage at time 0).

If transistor M_1 is in saturation mode, that is $V_{gs} - V_{th} < V_{ds}$, and λ is neglected (channel length modulation is the influence of the voltage V_{ds} on the properties of the channel), then

$$\frac{d}{dt}[V_c(t)] = \frac{\mu_n \ C_{ox} \ W}{2L \ C}(V_{DD} - V_c(t) - V_{th})^2$$ (7.10)

After solving the differential equation, the voltage V_c, as a function of time, becomes:

$$V_c(t) = \frac{(A \ t - C[1]) \ (V_{DD} - V_{th}) - 1}{A \ t - C[1]}$$ (7.11)

where

$$A = \frac{\mu_n \ C_{oc} \ W}{2L \ C}$$ (7.12)

and the constant $C[1]$ depends on the initial value of the output voltage V_c, which as a function of time, can be represented using Equations (7.7) and (7.11).

If $V_{in} = V_{DD}$, then after the transistor turns on ($V_{\phi_1} \leq V_{th}$), it enters the saturation mode. Voltage V_c is described by Equation (7.11). Imposing the requirement that the voltage V_c is zero when the transistor starts to conduct provides the condition for finding the constant $C[1]$:

$$C[1] = -\frac{1}{V_{DD} - V_{th}}$$ (7.13)

Note that in this case the transistor does not enter the linear operation mode. The condition for linear mode imposes that $V_{gs} - V_{th} > V_{ds}$, which requires that $V_{\phi 1} \geq V_{DD} + V_{th}$. This condition is never satisfied for a clock signal bounded by the supply voltage. The voltage V_c is described by

$$V_c(t) = V_{DD} - V_{th} - \frac{1}{A \ t - C[1]}$$ (7.14)

This equation shows that the constant A must be large for the voltage $V_c(t)$ to increase quickly to the value $V_{DD} - V_{th}$. This requires transistors with large width W. Also, note that the voltage $V_c(t)$ is bound by the value $V_{DD} - V_{th}$.

B. Channel Charge Injection

In Figure 7.1, the charge transferred by transistor M_1 to capacitor C is slightly different from the value expressed in Equation (7.1). Channel charge injection is one of the main reasons for the difference. Figure 7.3(a) indicates that when transistor M_1 conducts a current, the channel stores a charge equal to

$$Q_{channel} = W \ L \ C_{ox} \ (V_{DD} - V_{in} - V_{th}) \ [5]$$ (7.15)

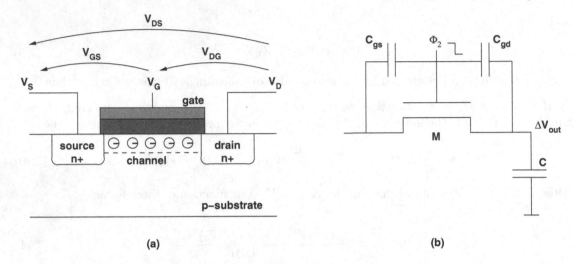

(a)   **(b)**

Figure 7.3: (a) Channel charge injection and (b) clock feedthrough.

where W and L are the width and length of the transistor, C_{ox} is the gate capacitance per unit area, V_{DD} is the gate voltage, and V_{th} is the threshold voltage of the transistor.

Once transistor M_1 is turned off, the transistor channel's charge is transferred to capacitor C. This causes an additional variation of the voltage across the capacitor equal to:

$$\Delta V_C = \frac{W \, L \, C_{ox}(V_{DD} - V_{in} - V_{th})}{C} \quad [5] \tag{7.16}$$

Moreover, because the threshold voltage is given by the expression:

$$V_{th} = V_{TH0} + \gamma\left(\sqrt{V_{SB} + |2\,\phi_F|} - \sqrt{|2\,\phi_F|}\right) \tag{7.17}$$

the output voltage variation, due to channel charge injection, can also be expressed as

$$\Delta V_C = -\frac{W \, L \, C_{ox}}{C}\left[V_{in} - V_{DD} + V_{TH0} - \gamma\sqrt{2|\phi_F|} + \gamma\sqrt{V_{in} + |2\phi_F|}\right] \tag{7.18}$$

Note that the impact of channel charge injection on the voltage V_c is reduced by using transistors with small width W. However, this decreases the value of the constant A, and increases the time needed for the voltage to rise to the value $V_{DD} - V_{th}$. Hence, there is a tradeoff between the accuracy of SC-based processing and the speed of the processing [5].

C. Clock Feedthrough

Clock feedthrough represents the capacitive coupling of the clock signal applied to the MOSFET gate and the MOSFET gate-to-source and gate-to-drain capacitances [4]. Figure 7.3(b) illustrates the clock feedthrough phenomenon. Capacitance C_{gs} is the gate-to-source capacitance of a MOSFET, and capacitance C_{gd} is the gate-to-drain capacitance.

If clock ϕ_2 switches, then because of the capacitive coupling, the output voltage also changes as shown by Equation [4]:

$$\Delta V_{out} = \frac{-C_{gd,2} \, \Delta V_{\phi_2}}{C_{gd} + C} \tag{7.19}$$

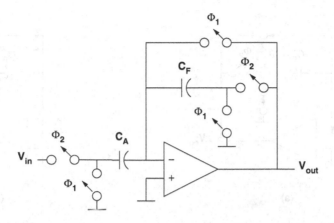

Figure 7.4: Switched capacitor fixed gain amplifier.

The effect of clock feedthrough is reduced for a small capacitance C_{gd}, thus for transistors with small width W. However, small width also lowers the speed of the circuits.

7.2 Active Switched Capacitor Circuits

This section presents the structure and operation of popular switched capacitor circuits, such as amplifiers, comparators, integrators, and differentiators. The utilization of these circuits in different applications is also illustrated.

7.2.1 Fixed Gain Amplifier

Figure 7.4 shows a simple architecture for a fixed gain amplifier. It consists of an OpAmp, an input capacitor C_A, a feedback capacitor C_F, and five switches. This circuit has two distinct phases of operation, namely (1) ϕ_1 *phase*, the acquisition of signals, and (2) ϕ_2 *phase*, the transfer of charge:

- ϕ_1 phase: Figure 7.5(a) shows that three switches are closed during the signal acquisition (ϕ_1) phase. This results in the OpAmp being configured as a follower. Negative feedback causes the voltage at the inverting input to be close to ground potential. This deviation from true ground potential (i.e., zero) is referred to as the OpAmp's input offset error (V_{offset}). The input side of capacitor C_A is also at ground potential, as is the output side of capacitor C_F. Measurement of this offset error, and storage, on both capacitors removes the effect this offset error may have on the output during the charge transfer phase. Because this measurement is done automatically during the acquisition phase, it is known as an "autozero" adjustment. During the transition period between the phases, all the switches are open for a brief period and the charge stored on the capacitors does not change.

- ϕ_2 phase. Figure 7.5(b) shows that two switches are closed during the charge transfer (ϕ_2) phase. Equation (7.20) defines the amount of charge needed to charge the input capacitor now that its input has been connected to input voltage V_{in}:

$$\Delta Q = V_{in} C_A \tag{7.20}$$

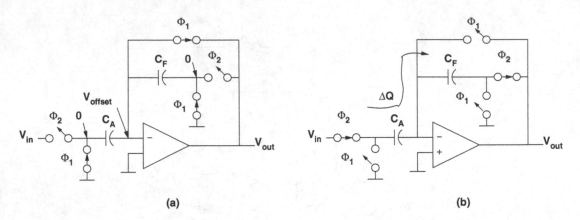

Figure 7.5: Set for (a) signal acquisition (ϕ_1) phase and (b) charge transfer (ϕ_2) phase.

This charge has no other path to take except by the feedback capacitor. Therefore, the feedback capacitor must receive the same amount of charge. Equation (7.21) describes how much the output voltage will change due to this transfer of charge:

$$V_{out} = -\frac{\Delta Q}{C_F} \tag{7.21}$$

Combining equations (7.20) and (7.21) to produce the amplifier's transfer function, yields:

$$Gain = \frac{V_{out}}{V_{in}} = -\frac{C_A}{C_F} \tag{7.22}$$

The result is an inverting amplifier with its gain determined by the ratio of its two capacitors. Note that the output voltage is available after the transfer of charge and near ground (V_{offset}) during signal acquisition.

The following paragraph explains the "auto-zeroing" behavior of the amplifier in more detail. The charge stored at the end of the acquisition phase, on the capacitors C_A and C_F, is described by the following equations.

$$Q_A^i = C_A V_{offset} \tag{7.23}$$
$$Q_F^i = C_F V_{offset} \tag{7.24}$$

The output voltage is equal to V_{offset}.

The charge stored at the end of the transfer phase on the two capacitors is equal to

$$Q_A^f = -C_A (V_{in} - V_{offset}) \tag{7.25}$$
$$Q_F^f = C_F (V_{offset} - V_{out}^f) \tag{7.26}$$

Due to the charge conservation law, the charge at the end of the acquisition phase and transfer phase is equal:

$$Q_A^i + Q_F^i = Q_A^f + Q_F^f \tag{7.27}$$

Hence, the amplifier output voltage at the end of the transfer phase is expressed as

$$V_{out}^f = V_{offset} - \frac{Q_F^f}{C_F} = V_{offset} - \frac{Q_A^i + Q_F^i - Q_A^f}{C_F} \tag{7.28}$$

$$V_{out}^f = V_{offset} - \frac{(C_A + C_F)V_{offset} - C_A(V_{offset} - V_{in})}{C_F} \tag{7.29}$$

$$V_{out}^f = -\frac{C_A}{C_F}V_{in} \tag{7.30}$$

Expression (7.30) shows that the amplifier output voltage, at the end of the transfer phase does not depend on the value of the offset voltage.

Example (Impact of the MOSFET nonzero resistance on the amplifier transfer function): This exercise analytically characterizes the modification of the amplifier transfer function due to the nonzero resistance of the MOSFET transistors used to implement the switches in a switched capacitor design. Equations (7.20)–(7.22) assumed a zero resistance of the MOSFETSs controlled by clock ϕ_2 in Figure 7.5(b). In reality, conducting transistors have a nonzero resistance. Let R_{on1} be the "on" resistance of the MOSFET, between input and capacitor C_A, and R_{on2} the "on" resistance of the MOSFET connected between capacitor C_F and output.

Assuming that the OpAmp has zero offset voltage, the following differential equation describes the voltage variation across the capacitor C_A:

$$\frac{d}{dt}[V_{C_A}(t)] = \frac{1}{R_{on1} C_A}(V_{in} - V_{C_A}(t)) \tag{7.31}$$

Thus,

$$V_{C_A}(t) = V_{in}\left(1 - \exp\left[-\frac{t}{R_{on1} C_A}\right]\right) \tag{7.32}$$

and the charge stored on a capacitor C_A, as a function of time, is given by:

$$Q(t) = C_A V_{in}\left(1 - \exp\left[-\frac{t}{R_{on1} C_A}\right]\right) \tag{7.33}$$

Finally, the output voltage as a function of time is given by

$$V_{out}(t) = -\frac{C_A}{C_F} V_{in}\left(1 - \exp\left[-\frac{t}{R_{on1} C_A}\right]\right) - \frac{R_{on2}}{R_{on1}} V_{in}\, exp\left[-\frac{t}{R_{on1} C_A}\right] \tag{7.34}$$

$$V_{out}(t) = -\frac{C_A}{C_F} V_{in} + V_{in}\, \exp\left[-\frac{t}{R_{on1} C_A}\right]\left(\frac{C_A}{C_F} - \frac{R_{on2}}{R_{on1}}\right) \tag{7.35}$$

The speed of the circuit increases with the value $\left[\frac{C_A}{C_F} - \frac{R_{on2}}{R_{on1}}\right]/[R_{on1} C_A]$. Hence, for a fixed gain, the product $R_{on1} C_A$ must be large for a fast circuit.

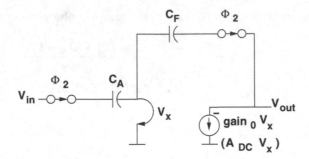

Figure 7.6: Impact of finite OpAmp gain of the SC amplifier performance.

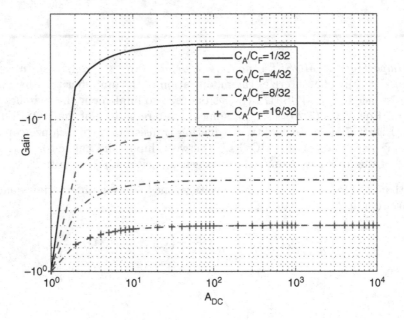

Figure 7.7: Dependency of the amplifier gain on the OpAmp's finite gain.

Example (Impact of the finite OpAmp gain on the amplifier transfer function). This example describes the effect of the finite OpAmp gain on the transfer function of the amplifier. Figure 7.6 shows the circuit used in the analysis, in which the OpAmp was replaced by its small-signal model. The model assumes that the input impedance of the OpAmp is infinite. The charge conservation law applied after the end of the transfer phase results in the following equation:

$$(V_{in} - \frac{V_o}{gain_0}) \, C_A \quad = \quad +(\frac{V_o}{gain_0} - V_o) \, C_F \tag{7.36}$$

$gain0$ is the finite OpAmp DC gain. The transfer function of the circuit is expressed as

$$\frac{V_o}{V_{in}} \quad = \quad -\frac{C_A}{C_F} \left[\frac{1}{1 - \frac{1}{gain_0} + \frac{C_A}{C_F \, gain_0}} \right] \tag{7.37}$$

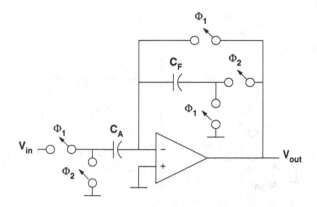

Figure 7.8: Amplifier with input switches swapped.

Thus,

$$\frac{V_o}{V_{in}} = Gain \left[\frac{1}{1 - \frac{1}{gain_0} - \frac{Gain}{gain_0}} \right] \tag{7.38}$$

where $Gain$ is the amplifier gain for an ideal OpAmp (with infinite OpAmp gain). Hence, the gain precision of amplifiers decreases with the value of the amplifier gain and the OpAmp gain. Figure 7.7 is a graph of the dependency of the amplifier's gain on the OpAmp's finite DC gain for different values of the ration C_A/C_F.

Selectable gain polarity amplifier. The circuit in Figure 7.8 is similar to that shown in Figure 7.4 except that the input switches have swapped phase. Now the input capacitor acquires V_{in} during the acquisition phase ϕ_1. The charge needed to pull the input back to ground during the transfer phase ϕ_2 moves in the opposite direction to that of the first example. This causes the output voltage to also move in the opposite direction. The result is a positive gain amplifier with a gain defined in Equation (7.39):

$$Gain = \frac{V_{in}}{V_{out}} = \frac{C_A}{C_F} \tag{7.39}$$

Figure 7.9 shows the modification that allows the correct phasing for both positive and negative gain operation. For the rest of these examples the control signal $Sign$ is assumed to be positive. That is, the C_A acquires V_{in} during ϕ_1 and moves back to zero during ϕ_2.

7.2.2   Comparators

Figure 7.10 shows the structure of a switched capacitor comparator. As compared to the amplifier, the comparator does not have the structure including the feedback capacitor C_F andthe

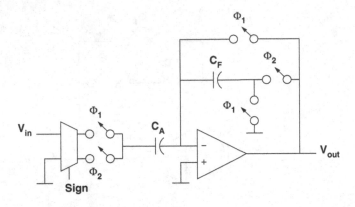

Figure 7.9: Selectable gain polarity amplifier.

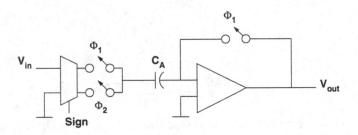

Figure 7.10: Switched capacitor comparator.

two switches. Intuitively, this is equivalent to setting the capacitance value, C_F, to zero in equation (7.39) (the amplifier's gain as the ratio of the values of the two capacitors). If the feedback capacitance is zero, then the gain is infinite, and the circuit acts as a comparator.

The functionality of the comparator can be explained based on the charge transfer during operation. During the acquisition phase (switches ϕ_1 are closed), the input voltage V_{in} is connected to the capacitor C_A, and the charge $Q = C_A\,V_{in}$ is stored on the capacitor. The OpAmp is configured as a voltage follower, and hence the output voltage V_o is zero. During the transfer phase (switch ϕ_2 is closed), the capacitor C_A, charged with charge Q, applies the voltage V_{in} to the OpAmp input. Because of the very large OpAmp gain, the circuit output is saturated to the corresponding positive or negative supply voltage (depending on whether the input voltage, V_{in}, is positive or negative).

7.2.3 Switched Capacitor Integrator

Figure 7.11 shows a switched capacitor integrator. Unlike a fixed gain amplifier, the feedback path of the circuit does not have the switch ϕ_1 to ground. This keeps the charge on the feedback capacitor C_F from being removed during the acquisition phase ϕ_1 while still allowing the transfer of the input charge during phase ϕ_2.

During the acquisition phase, the circuit is configured as shown in Figure 7.12(a). The charge stored on capacitor C_A is given by

$$Q \;=\; C_A\,V_{in} \tag{7.40}$$

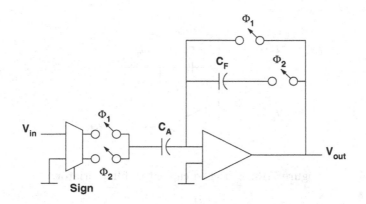

Figure 7.11: Switched capacitor integrator.

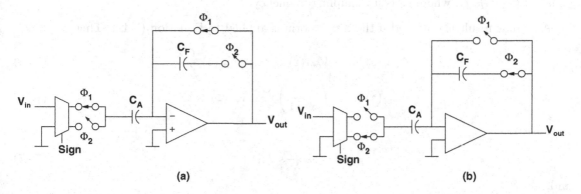

Figure 7.12: Integrator operation.

Figure 7.12(b) shows the circuit configuration during the transfer phase ϕ_2. The charge stored on capacitor C_A is transferred onto capacitor C_F, because both terminals of capacitor C_A are connected to ground. The overall charge stored on capacitor C_F (after the transfer) is expressed by the following:

$$Q_{tot} = Q' + C_A V_{in} \qquad (7.41)$$

where Q' is the charge on the capacitor C_F before the transfer phase. The output voltage, after the transfer phase, Φ_2, is given by

$$V_{out}(t) = V_{out}(t - T_s) + \frac{C_A}{C_F} V_{in} \qquad (7.42)$$

$V_{out}(t)$ is the output voltage at the end of the current transfer phase, and $V_{out}(t - T_s)$ is the voltage at the end of the previous transfer phase (thus, the output voltage corresponding to charge Q' is stored on capacitor C_F). T_s is the sampling period of the circuit. Given that

$$\frac{d}{dt}[V_{out}(t)] \approx \frac{V_{out}(t) - V_{out}(t - T_s)}{t - (t - T_s)} = \frac{C_A}{C_F} \frac{V_{in}}{T_s} \qquad (7.43)$$

thus,

$$\frac{d}{dt}[V_{out}(t)] \approx \frac{C_A}{C_F} f_s V_{in} \qquad (7.44)$$

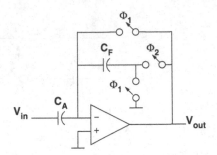

Figure 7.13: Switched capacitor differentiator.

Equation (7.44) shows that the switched capacitor circuit behaves as an integrator with a gain equal to $C_A/C_F\, f_s$, where f_s is its sampling frequency.

A similar result is obtained if the z transform is applied to Equation (7.42). Then

$$V_{out}(z) \;=\; V_{out}(z)\, z^{-1} \;+\; V_{in}(z)\, \frac{C_A}{C_F} \tag{7.45}$$

$$V_{out}(z) \;=\; \left(\frac{z}{z-1}\right) \frac{C_A}{C_F}\, V_{in}(z) \tag{7.46}$$

Because

$$z \;=\; \exp[-s\, T_s] \tag{7.47}$$

and,

$$\left(\frac{z}{z-1}\right) \approx \frac{1}{s\, T_s} \tag{7.48}$$

the transfer function of the circuit is given by:

$$\frac{V_{out}}{V_{in}} \;=\; \frac{1}{s}\, \frac{C_A}{C_F}\, f_s \tag{7.49}$$

Equation (7.49) is the transfer function of an integrator with a gain of $C_A/C_F\, f_s$, as shown by the previous analysis. Equations (7.44) and (7.49) state that the circuit is an adjustable gain integrator for which the gain can be varied by changing the capacitor ratio C_A/C_F.

7.2.4   Switched Capacitor Differentiator

Figure 7.13 shows that the input is permanently connected to the input capacitor. This topology makes the feedback capacitor function as a resistor and the input capacitor as a capacitor. It does allow for the operation shown in equation (7.50):

$$(V_{in} \;-\; V_{in}\, z^{-1})\, C_A \;=\; -V_{out}\, C_F \tag{7.50}$$

Manipulating equation (7.50) results in the transfer equation (7.51):

$$Gain \;=\; \frac{V_{out}}{V_{in}} \;=\; -(1 \;-\; z^{-1}) \frac{C_A}{C_F} \;=\; -s\, \frac{C_A}{C_F}\, \frac{1}{f_s} \tag{7.51}$$

Equation (7.51) shows that this is an adjustable differentiator.

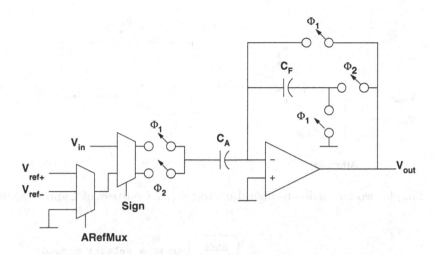

Figure 7.14: Improved reference selection for noninverting amplifier.

7.2.5   Reference Selection

So far, all of the examples discussed have had their input voltage referenced to ground. Although ground is often a convenient reference point, other reference points are possible. Figure 7.14 shows the noninverting amplifier with an improved reference selection. Signal *ARefMux* allows for the selection of two other references besides ground. Equations (7.52)-(7.54) define the output voltage for an amplifier. The equations assume that the reference is ground, V_{ref+}, or V_{ref-}, respectively:

$$V_{out} = \frac{C_A}{C_F} V_{in} \tag{7.52}$$

$$V_{out} = \frac{C_A}{C_F}(V_{in} - V_{ref+}) \tag{7.53}$$

$$V_{out} = \frac{C_A}{C_F}(V_{in} - V_{ref-}) \tag{7.54}$$

The first expression was already justified for the traditional noninverting amplifier. The second expression is explained next. During the acquisition phase (switches ϕ_1 are closed), the capacitor C_A is charged with the charge $Q = C_A\, V_{in}$. Let's assume that the selection signal *ARefMux* selects the input voltage V_{ref+}. During the transfer phase, as the capacitor C_A is charged with the charge $Q = C_A\, V_{ref+}$, the charge $C_A\,(V_{in}$ - $V_{ref+})$ is transferred to the capacitor C_F. The resulting output voltage is the following.

$$V_o = \frac{C_A}{C_F}\,(V_{in} - V_{ref+}) \tag{7.55}$$

The third equation can be explained by similar reasoning.

Example (Two-bit ADC [7]. Courtesy of Cypress Semiconductor Corporation): Combining the comparator in Figure 7.10 with the reference selection in Figure 7.14 results in a comparator with multiple compare points shown in Figure 7.15.

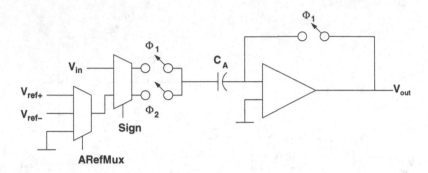

Figure 7.15: Simple two bit analog-to-digital converter [7]. Courtesy of Cypress Semiconductor Corporation.

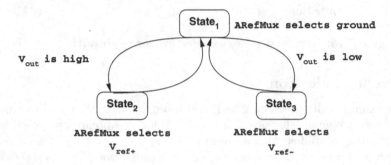

Figure 7.16: Simple two bit analog-to-digital converter [7]. Courtesy of Cypress Semiconductor Corporation.

With proper control of signal *ARefMux*, it is possible to determine which of the following four conditions is satisfied by voltage V_{in}: (a) greater than voltage V_{ref+}: (b) less than voltage V_{ref+}, but greater than ground: (c) less than ground, but greater than voltage V_{ref-} and (d) less than voltage V_{ref-}. These four situations make this circuit a two-bit analog-to-digital converter (ADC).

The functioning of the two-bit ADC corresponds to the three states of the finite state machine (FSM) illustrated in Figure 7.16. In state one, signal *ARefMux* selects ground as the reference signal. The output of the comparator is high if the input voltage is greater than ground, and the comparator output is low otherwise. Then, if the input signal is larger than ground the FSM moves to state two, otherwise the FSM goes to state three. In state two, the signal *ARefMux* selects the signal V_{ref+} as the reference signal. If the input voltage is larger than the reference V_{ref+} then the comparator output is high, and the comparator output is low otherwise. Similarly, in state three, the control signal *ARefMux* selects the signal V_{ref-} as reference. The comparator output is high if the input voltage is larger than the reference V_{ref-}, and the output is low if the input is smaller than signal V_{ref-}. The FSM returns to state one after being in states two and three.

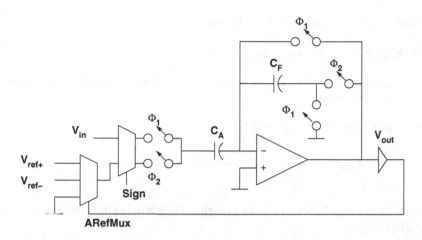

Figure 7.17: Analog to digital modulator [7]. Courtesy of Cypress Semiconductor Corporation.

In conclusion, the values of the two output bits encode the following four situations:

- Value "00": the input voltage is smaller than the reference V_{ref-}.

- Value "01": the input voltage is smaller than ground and larger than the reference V_{ref-}.

- Value "10": the input voltage is larger than ground, and smaller than the reference V_{ref+}.

- Value "11": the input voltage is larger than the reference V_{ref+}.

Section 7.3 discusses the PSoC-based implementation of the two-bit ADC.

7.2.6   Analog-to-Digital Conversion

A slight modification of the reference selection allows this circuit to function as an analog-to-digital modulator. Figure 7.17 shows that a comparator has been added to the output and connected to the reference selection multiplexer.

Selecting the proper signal *ARefMux* value allows the output of the comparator to determine the reference voltage. The relationship is:

- The reference is set to voltage V_{ref+} when the comparator is high (positive output value).

- The reference is set to voltage V_{ref-} when comparator is low (negative output value).

The switches associated with the feedback capacitor are configured to make an integrator. The comparator control causes a reference, with the same polarity as the voltage V_{out}, to be subtracted from the input. This negative feedback attempts to move the voltage V_{out} back towards zero.

Assume that:

- The initial condition of voltage V_{out} is zero.

- The switch cycle is preformed a number of n times.

- Voltage V_{out} is greater than zero (comparator is high) a number a of those times. The frequency, f_s, is constant.

Then, the output voltage is expressed as

$$V_{out} = \frac{C_A}{C_F} \left[n\, V_{in}\ -\, a\, V_{ref+}\ -\ (n-a)\, V_{ref-} \right] \tag{7.56}$$

If the references are of equal but opposite polarity, then solving for voltage V_{in} results in the following equation

$$V_{in} = V_{ref} \left(\frac{2a - n}{n} \right) + V_{out}\, \frac{1}{n}\, \frac{C_F}{C_A} \tag{7.57}$$

Because,

$$V_{ref} = V_{ref+} = -V_{ref-} \tag{7.58}$$

Voltage V_{in} is a function of voltages V_{ref} and V_{out}.

As stated earlier, the negative feedback causes voltage V_{out} to move back towards ground every cycle. This makes voltage V_{out} less than the value $C_A/C_F V_{ref}$. As the value n becomes larger, the contribution of voltage V_{out} to Equation (7.57) becomes negligible. This allows for a more simplified Equation (7.59):

$$V_{in} \approx V_{ref} \left(\frac{2a}{n} - 1 \right) \tag{7.59}$$

assuming that,

$$V_{out} < \frac{C_A}{C_F}\, V_{ref} \approx 0 \tag{7.60}$$

Voltage V_{in} is not dependent on the ratio of the two capacitors. It is only a function of voltage V_{ref}, and the ratio of value a and value n. Measuring voltage V_{in} is just a function of counting the number of times the comparator is high (value a) during a sequence of n switch cycles. The range is -V_{ref} (for $a = 0$) to +V_{ref} (for $a = n$), and the resolution is $2\,V_{ref}/n$. The longer the period is (larger value n) the better the resolution of the voltage measurement.

7.3 Switched Capacitor PSoC Blocks

PSoC's switched capacitor architecture is actually quite versatile, and can provide many different functions simply by altering the circuit's switch closure configurations. This architecture is used as the basis for the switched capacitor blocks in the PSoC microcontroller, and is implemented with Type C and Type D switched capacitor blocks [6]. However, there are small but significant differences between the various block types.

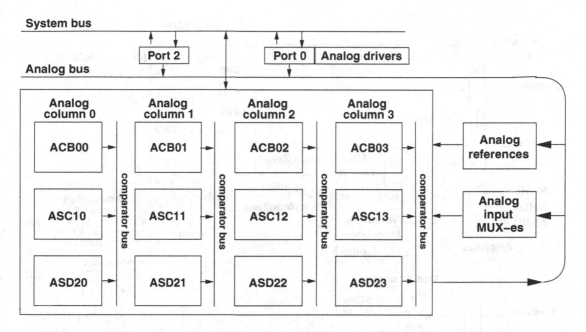

Figure 7.18: The analog section of the PSoC architecture [6].

Table 7.1: Control register addresses [6].

	ASC10	ASC12	ASC21	ASC23
register **ASCxxCR0**	80H	88H	94H	9CH
register **ASCxxCR1**	81H	89H	95H	9DH
register **ASCxxCR2**	82H	8AH	96H	9EH
register **ASCxxCR3**	83H	8BH	97H	9FH

7.3.1   Type C Switched Capacitor Blocks

Figure 7.18 shows the array of the programmable analog PSoC blocks. There are four columns of blocks with each column having its own analog bus, comparator bus, and clock to generate the ϕ_1 and ϕ_2 clocks. Chapter 3 detailed the analog interconnect structure, and also the available system clocks. (Additional discussion of the interconnections between continuous-time and switched capacitor blocks is provided in Chapters 3 and 6.) Each column contains one type C switched capacitor block and one type D switched capacitor block for a total of eight switched capacitor blocks.

The type C switched capacitor block, shown in Figure 7.19, consists of the following elements:

- *Control registers*: Four control registers are available for each type C block . These registers program the (i) functionality, (ii–iii) input and output configuration, (iv) power mode, and (v) sampling procedure of each switched capacitor type C block. The registers corresponding to block *xx* are denoted as *ASCxxCR0, ASCxxCR1, ASCxxCR2,* and *ASCxxCR3.* Table 7.1 presents the addresses of the control registers for each of the four type C blocks.

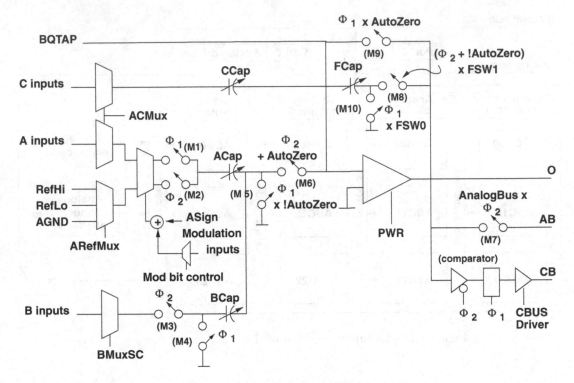

Figure 7.19: Type C Switched capacitor block [6].

- *OpAmp*: The OpAmp is the central component of a type C switched capacitor block. The power mode of the OpAmp can be selected as one of four possible power states, i.e., off, low, medium, and high. Two control bits, *PWR*, are used for this purpose, such as, bits 1–0 of the related register *ASCxxCR3*. If the bits are "00" then the OpAmp is off. The OpAmp is in low power mode if the control bits are set to "01", in medium power mode, if set to "10", and high power mode, if set to "11".

The functionality of the switched capacitor block is programmed by reconfiguring the topology of the network surrounding the OpAmp. Bits *FSW1* and *FSW0* aid in determining the block's functionality.

- *Bit FSW1*: This is a one-bit field that determines whether capacitor array *FCap* is connected. If the bit is zero then *FCap* capacitor is not connected, and the OpAmp functions as a comparator. If the bit is set, capacitor *FCap* becomes the feedback path so that the circuit can function either as a gain stage, or an integrator. Bit *FSW1* corresponds to bit 5 of register *ASCxxCR3*.

- *Bit FSW0*: This is a one-bit field that determines whether the capacitor array *FCap* is discharged during clock phase ϕ_1. If the bit is set to one, capacitor *FCap* is discharged, and the OpAmp functions as a gain stage. If the bit is set to zero, the capacitor is not discharged, and the circuit functions as an integrator. Bit *FSW0* corresponds to bit 4 in register *ASCxxCR3*.

- *Programmable capacitor matrices*: Each PSoC block has four programmable capacitor arrays, called *ACap* array, *BCap* array, *CCap* array, and *FCap* array [6]. The first three capacitor arrays appear in the input of the OpAmp, and the $FCap$ array is in the feedback path of the OpAmp.

The programmable features of the $ACap$ capacitor array consist of the following items.

- $ACap$ *capacitor value*: The value of this capacitor is programmed by bits 4-*0* of the corresponding register $ASCxxCR0$. These five bits determine a capacitance value ranging from 0 to 31 units.

- $ASign$ *bit*: A one-bit field sets the gain of the block to either positive or negative. The control bit corresponds to bit 5 of register $ASCxxCR0$. The block gain is positive if the bit is zero, and negative if the bit is one, similar to the scheme shown in Figure 7.9. For a positive gain, the input signal (if selected) is sampled by clock $\phi1$, or if the reference signal is selected then it is sampled by clock $\phi2$. The reference selection is implemented in a similar manner to that of the selectable gain polarity amplifier shown in Figure 7.14.

- $AutoZero$ bit: This is a one-bit field that, when set, forces an autozero during the ϕ_1 signal acquisition phase. The $AutoZero$ bit corresponds to bit 5 of the control register $ASCxxCR2$. If this bit is zero then there is no shorting of the feedback capacitor C_F, because switch M_9 is permanently open. If the bit is one, then switch M_6 is always closed, thereby connecting the input capacitor $C_A(ACAP)$ to the OpAmp inverting input. Finally, switch M_5 is always open. Switch M_9 is controlled by clock ϕ_1 to short circuit capacitor C_F. Switch M_5, connecting capacitor C_{ACap} to ground, is controlled by clock ϕ_1. Switch M_6, controlled by clock ϕ_2, connects capacitor C_A to the OpAmp input.

The programmable $BCap$, a switched capacitor array, connects to the summing node of the OpAmp. The $BCap$ array acts in many ways the same as the $ACap$ array except that it only samples its input on clock ϕ_2, and is always referenced to ground. The $BCap$ array can be programmed as follows.

- $BCap$ capacitor value: A five-bit field to set the value of the $BCap$ capacitor array from 0 to *31* units. The control field corresponds to bits 4-0 in register $ASCxxCR1$.

The programmable $CCap$ capacitor array connects to the summing node of the OpAmp. It is primarily used for multiple input amplifiers, for example in filter design. $CCap$ array can be programmed as follows:

- $CCap$ capacitor value: A five-bit field to set the value of $CCap$ capacitor array from 0 to 31 units. The five bits are implemented as bits 4–0 in register $ASCxxCR2$.

The $FCap$ array can be programmed as follows:

- $FCap$ capacitor value: A one-bit field sets the value of the $FCap$ array to either 16 or 32 units. Bit 7 of register $ASCxxCR0$ selects 16 units if the bit is set to zero, and 32 units if the bit is set to 1. Each unit of capacitance is approximately *50fF*.

- *Programmable inputs*: The nature of the inputs to the three capacitor arrays $ACap$, $BCap$, and $CCap$ is determined by the settings of the associated control registers. In addition, the reference voltage to the capacitor array $ACap$ is also selectable by the control registers.

The inputs to the $ACap$ capacitor array are programmable as follows:

Table 7.2: Programmable inputs to *AMuxSC* array [6].

	ACMux bits (3 bits)	**ASC10**	**ASC21**	**ASCB12**	**ASC23**
1	000	ACB00	ASD11	ACB02	ASD13
2	001	ASD11	ASD20	ASD13	ASD22
3	010	RefHi	RefHi	RefHi	RefHi
4	011	ASD20	Vtemp	ASD22	ABUS3
5	100	ACB01	ASC10	ACB03	ASC12
6	101	ACB00	ASD20	ACB02	ASD22
7	110	ASD11	ABUS1	ASD13	ABUS3
8	111	P2[1]	ASD22	ASD11	P2[2]

- *Bits ACMux*: A three-bit field (bits 7--5 of register *ASCxxCR1*) is used to select the inputs to the *ACap* and the *CCap* arrays. Table 7.2 shows the inputs selected by different control bits values.

- Bits *ARefMux*: A two-bit field (bits 7–6 of register *ASCxxCR3*) selects the voltage potential to which the *A* input is referenced. It can be analog ground (*AGND*), the positive reference voltage V_{ref+}, the negative reference voltage V_{ref-}, or a voltage reference determined by the state of the output comparator. Bits "00" select the analog ground, bits "01" the reference voltage high, and bits "10" the reference voltage low. For bits "11", the reference voltage is *RefHi* if the comparator output is high, and *RefLo* for a low comparator output.

Table 7.3: Programmable inputs to *BMuxSC* array [6].

	BMuxSC bits (2 bits)	**ASC10**	**ASC21**	**ASCB12**	**ASC23**
1	00	ACB00	ASD11	ACB02	ASD13
2	01	ASD11	ASD20	ASD13	ASD22
3	10	P2[3]	ASD22	ASD11	P2[0]
4	11	ASD20	TrefGND	ASD22	ABUS3

Table 7.4: Programmable inputs to *CMuxSC* array [6].

	ACMux bits (3 bits)	**ASC10**	**ASC21**	**ASCB12**	**ASC23**
1	000	ACB00	ASD11	ACB02	ASD13
2	001	ACB00	ASD11	ACB02	ASD13
3	010	ACB00	ASD11	ACB02	ASD13
4	011	ACB00	ASD11	ACB02	ASD13
5	100	ASD20	ASD11	ASD22	ASD13
6	101	ASD20	ASD11	ASD22	ASD13
7	110	ASD20	ASD11	ASD22	ASD13
8	111	ASD20	ASD11	ASD22	ASD13

The inputs to the *BCap* capacitor array are programmed as follows.

  – Bits *BMuxSC*: A two-bit field selects the inputs to the capacitor *BCap*. The bits correspond to bits 3–2 in register *ASCxxCR3*. Table 7.3 summarizes the input selection for different control bit values.

The inputs to the capacitor array *CCap* are also selected by the control bits *ACMux* used to select the inputs to the capacitor *ACap*. However, there is much less flexibility than for capacitor array *ACap*. Table 7.4 presents the input selection by the three control bits *ACMux*.

- *Programmable outputs*: The output of a Type C switched capacitor block can be connected to either analog blocks (continuous-time or switched capacitor block) or digital blocks.

  – Bit *AnalogBus*: A one-bit field that, when set, connects the output to an analog buffer. The control bit corresponds to bit 7 of the register *ASCxxCR2* corresponding to the block *xx*. If the bit is set then the connection is enabled, and disabled, if the bit is zero.

  – Bit *CompBus*: A one-bit field that, when set, connects the comparator to the data inputs of the digital blocks. The bit is mapped to bit 6 of register *ASCxxCR2*. If the bit is zero then the connection to the comparator bus is disabled, and the connection is enabled if bit *CompBus* is one.

- *Clocking scheme*: The clocking scheme of a type C switched capacitor block is selectable by bit *ClockPhase*, bit 6 of register *ASCxxCR0*. If the *ClockPhase* bit is set then the external clock ϕ_2 becomes the internal clock ϕ_1. If the bit is zero then the external clock ϕ_1 is also used for the internal clock ϕ_1. It is primarily used to match the input signal sampling to the output of a switched capacitor block.

7.3.2   Type D Switched Capacitor Blocks

Figure 7.20 shows the type D switched capacitor block. Type D blocks are quite similar to the type C switched capacitor blocks shown in Figure 7.19 with the following differences.

- Instead of a multiplexed input to the *CCap* capacitor array, there is a connection to the output of the block. The other side of the capacitor *CCap* connects to the summing node of the type C switched capacitor block next to it. This configuration is used to build biquad filters.

- The control field *BSW* allows the capacitor *BCap* to function as either a switched capacitor or fixed capacitor.

- A programmable *BCap* switched capacitor connects to the summing node of the OpAmp.

- An *AnalogBus* switch connects the OpAmp's output to an analog buffer.

- A *CompBus* switch connects the comparator to the digital blocks.

Capacitor *BCap* acts, in many ways, the same as the capacitor *ACap* except that it only samples its input on the clock phase ϕ_2 and is always referenced to ground. It is primarily used for multiple input amplifiers. Capacitor *CCap* is of primary benefit in designing filters.

There are 16 different parameters for a type D switched capacitor block. Thirteen are the same as the type C switched capacitor block, and have been discussed in the previous subsection. The remaining three are shown below along with a description of operation:

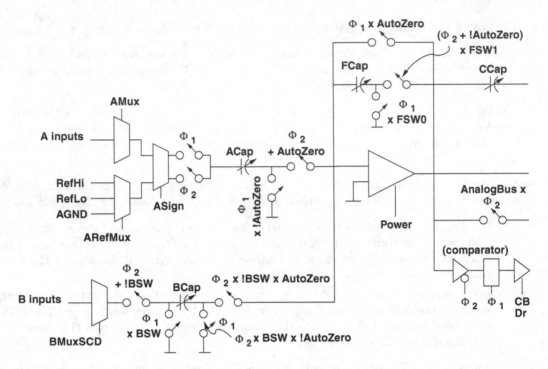

Figure 7.20: Type D switched capacitor block [6].

- *Field AMux*: A three-bit field to set the inputs to the *ACap*.

- *Field BSW*: A one-bit field that when selected causes *BCap* to function as a switched capacitor input. If not set then it functions as a capacitor.

- *Field BMux(SCB)*: A one-bit field to set the inputs to the *BCap*.

Example (Differential amplifier with common mode output [7]. Courtesy of Cypress Semiconductor Corporation). Figure 7.21 shows the architecture for a differential amplifier with a common mode output. A common mode output signal is useful to many signal processing applications, and so on when common mode feedback is used to drive a shield or a signal guard. The PSoC implementation of the circuit is shown in Figure 7.22.

The two input buffers are PGA (programmable gain amplifier) user modules with matched gains. The differential output is defined by Equation (7.61):

$$V_{differential} = PosInput - NegInput \tag{7.61}$$

The amplifier is an $A - B$ amplifier with a gain of one. Its parameters are shown in Table 7.5. For a difference amplifier with a gain of one, the values of capacitors *FCap*, *ACap*, and *BCap* must be the same. Capacitor *CCap* is not used so its value is set to zero. The continuous-time blocks source both input signals, so that there is no phasing problem with sampling their outputs. Field *ClockPhase* remains set to value *Norm*. The field *ACMux* is set to connect its input to block

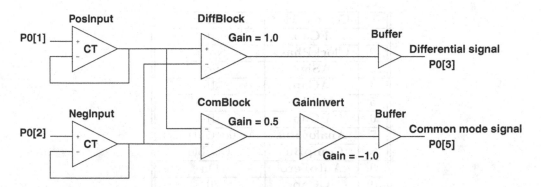

Figure 7.21: Differential amplifier with common mode output [7]. Courtesy of Cypress Semiconductor Corporation.

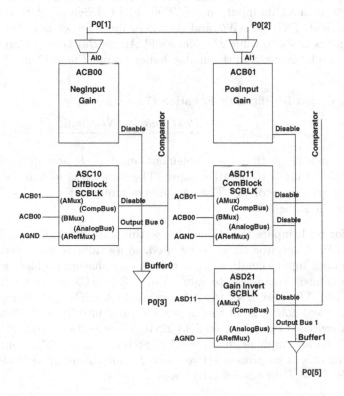

Figure 7.22: PSoC block placement for the differential amplifier with common mode output [7]. Courtesy of Cypress Semiconductor Corporation.

Table 7.5: Parameters for *DiffBlock SC* block

	Control fields	Value
1	**FCap**	16
2	**ClockPhase**	Norm
3	**ASign**	Pos
4	**ACap**	16
5	**ACMux**	ACB01
6	**BCap**	16
7	**AnalogBus**	AnalogOutBus0
8	**CompBus**	Disable
9	**AutoZero**	On
10	**CCap**	0
11	**ARefMux**	AGND
12	**FSW1**	On
13	**FSW0**	On
14	**BMux**	ACB00
15	**Power**	High

ACB01. Field *ARefMux* is set to analog ground, so both input references have the same potential. Field *BMux* is set to connect its input into *ACB00*. Field *ASign* is set to value *Pos* (positive). To be an amplifier, fields *FSW1*, *FSW0*, and *AutoZero* must be set *On*. The comparator is not used, so field *CompBus* is set to value *Disable*. Field *AnalogBus* is set to value *AnalogOutBus0*, so that the output can be brought to the analog buffer on pin *P0[3]*. Control bits *Power* are set to value *High*.

The common mode output is defined by Equation (7.62):

$$V_{common} \;=\; \frac{PosInput \,+\, NegInput}{2} \tag{7.62}$$

What is desired is a stage that can implement an $A + B$ amplifier. Unfortunately, the architecture limits the B input to negative gains. The solution is to build a $-A - B$ amplifier, and follow it with a gain stage of -1. The parameters required for a $-A-B$ amplifier, with a gain of 0.5, are show in Table 7.6.

A gain of 0.5 for both inputs is obtained by setting capacitor *FCap* to 32, and capacitors *ACap* and *BCap* to 16. Capacitor *CCap* is not used, so its value is set to zero. The continuous-time blocks source both input signals, so that there is no phasing problem with sampling their outputs. Field *ClockPhase* remains set to value *Norm*. Bits *ACMux* are set to connect its input to block *ACB01*. Field *ARefMux* is set to analog ground ($AGND$), so both input references have the same potential. Field *BMux* is set to connect its input into block *ACB00*. Capacitor *BCap* is required to be a switched capacitor, so field *BSW* is set to value *On*. Field *ASign* is set to *Neg* (negative). To be an amplifier, fields *FSW1*, *FSW0*, and *AutoZero* must be set on. The comparator is not used, so *CompBus* is set to value *Disable*. *AnalogBus* is also not used so it is set to value *Disable*. Field *Power* is set value *High*.

This block is to be followed by a gain inversion stage. The parameters for a -1 gain stage are shown in Table 7.7. For a gain of -1, the values of capacitors *Fcap* and *ACap* must equal. Capacitors *BCap* and *CCap* are not used, so their values are set to zero. Bit *ASign* is set to *Neg* (negative). Its input is sampled on the same phase as a valid *ComBlock* output signal, so field *ClockPhase* can remain set to *Norm*. Control bits *ACMux* are set to connect its input to block

Table 7.6: Parameters for the *ComBlock SC* block.

	Control fields	Value
1	**FCap**	32
2	**ClockPhase**	Norm
3	**ASign**	Neg
4	**ACap**	16
5	**ACMux**	ACB01
6	**BCap**	16
7	**AnalogBus**	Disable
8	**CompBus**	Disable
9	**AutoZero**	On
10	**CCap**	0
11	**ARefMux**	AGND
12	**FSW1**	On
13	**FSW0**	On
14	**BSW**	On
15	**BMux**	ACB00
16	**Power**	High

Table 7.7: Parameters for the *GainInvert SC* block

	Control fields	Value
1	**FCap**	16
2	**ClockPhase**	Norm
3	**ASign**	Neg
4	**ACap**	16
5	**ACMux**	ASD11
6	**BCap**	0
7	**AnalogBus**	AnalogOutBus1
8	**CompBus**	Disable
9	**AutoZero**	On
10	**CCap**	0
11	**ARefMux**	AGND
12	**FSW1**	On
13	**FSW0**	On
14	**BMux**	Not Set
15	**Power**	High

ASD11. Bit *ARefMux* is set to *AGND* (analog ground). Control field *BMux* is not needed so it is not set. Again *FSW1*, *FSW0*, and *AutoZero* must be set on. The comparator is not used, so bit *CompBus* is disabled. Field *AnalogBus* is set to *AnalogOutBus1* so that the output can be

connected to the analog buffer on pin *P0[5]*. Power for this example is set to high. The actual setting for other applications is determined by signal bandwidth.

These five blocks implement the circuit shown in Figure 7.21. The column clocks are set to 1 MHz. This sets the sample rate to 250 Ksps. When used as amplifiers, the SC blocks should not be sampled faster than 350 Ksps.

Example (Two-bit analog-to-digital converter [7]. Courtesy of Cypress Semiconductor Corporation). In Figure 7.17, a four-state ADC was shown that used a switched capacitor block as a comparator and changed the references to determine four different conditions. Figure 7.23(a) shows that architecture. The PSoC-based implementation is shown in Figure 7.23(b).

The input is brought in from pin *P2[2]* and connected to a SC Block (*TwoBit*). It is configured as a comparator. Software manipulates the field *ARefMux* in control register *TwoBit_SCBLOCKcr3* to select the reference to be analog ground *AGND*, voltage V_{ref+}, or voltage V_{ref-}. Bit 7 of the Analog Comparator control register (register *CMP_CR*) allows the software to determine the state of the column 3 comparator. Software determines whether the input is (i) greater than voltage V_{ref+}, (ii) less than voltage V_{ref+} but greater than ground, (iii) less than ground but greater than voltage V_{ref-}, and less than voltage V_{ref-}.

The parameters for a comparator with a configurable reference are shown in Table 7.8. For a comparator, capacitor *FCap* is not connected, so its value is unimportant. Capacitor *ACap* needs to be some nonzero value. Capacitors *BCap* and *CCap* are not used, so their values are set to zero. The input is from a continuous input signal, so there is no phase sampling problem, and bit *ClockPhase* can be set to *Norm*. Bit *ACMux* is set to connect its input to port *Port_2_2*. Bit *ARefMux* is to be controlled by software, but is set to a default value of *AGND* (analog ground). Bit *BMux* is not needed so it is not set. Bit *ASign* is set to *Pos* (positive). To implement a comparator, bits *FSW1* and *FSW0* must be set to *Off*, but bit *AutoZero* must be set *On*. This disconnects the feedback capacitor. Setting bit *CompBus* to *ComparatorBus_3* allows the CPU access to the state of the comparator. Bit *AnalogBus* is set to *Disable*. Bit *BSW* is not used, and is set to *Off*. Bit *Power* is set *High*. The column clock is set to 8 MHz. This sets the sample rate to 2 Msps (megasamples per second). When used as a comparator, a switched capacitor should not be sampled faster than 2 Msps.

The code in Figure 7.24 shows the software used to control this application. The program runs in a loop where the input is continuously sampled, and compared with the selectable references to determine one of four different levels. *Dout1* and *Dout0* are set accordingly.

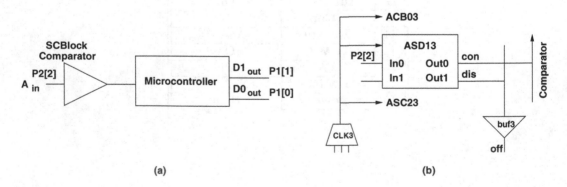

Figure 7.23: PSoC implementation of a four state ADC [7]. Courtesy of Cypress Semiconductor Corporation.

```
        ;SCBlocks already on
        loop:
          //set ARefMux to Agnd
          and reg[TwoBit_SCBLOCKcr3],3fh
          nop
          nop
          tst reg[CMP_CR], CMP_CR_COMP3
          if1: jz else1;(signal AGND)
          //set to REHI
            or reg[TwoBit_SCBLOCKcr3],40h
            nop
            nop
            tst reg[CMP_CR], CMP_CR_COMP3
            if2: jz else2;(signal REFHI)
            // REFHI input
              or reg[PRT1DR],03h ;Dout = 11
              jmp endif2
            else2:;(signal REFHI)
            // REFI input AGND
              mov A,reg[PRT1DR] ;Dout = 10
              or A,02h
              and A,~01h
              mov reg[PRT1DR],A
            endif2:
              jmp endif1
          else1:;(signal AGND)
          //set to RefLow
            or reg[TwoBit_cr3],80h
            nop
            nop
            tst reg[CMP_CR], CMP_CR_COMP3
            if3: jz else3;(signal REFLOW)
            // RELOW input AGND
              mov A,reg[PRT1DR] ;Dout = 01
              or A,01h
              and A,~02h
              mov reg[PRT1DR],A
              jmp endif3
            else3:;(signal REFREFLOW)
            // input REFLOW
              and reg[PRT1DR], ~03h ;Dout = 00
            endif3:
          endif1:
        endloop:
          jmp loop
        ret
```

Figure 7.24: Assembly code for two bit ADC [7]. Courtesy of Cypress Semiconductor Corporation.

Table 7.8: Parameters for the *TwoBit SC* block.

	Control fields	Value
1	**FCap**	0
2	**ClockPhase**	Norm
3	**ASign**	Pos
4	**ACap**	31
5	**ACMux**	Port_2_2
6	**BCap**	0
7	**AnalogBus**	Disable
8	**CompBus**	ComparatorBus_3
9	**AutoZero**	On
10	**CCap**	0
11	**ARefMux**	AGND
12	**FSW1**	Off
13	**FSW0**	Off
14	**BSW**	Off
15	**BMux**	Not Set
16	**Power**	High

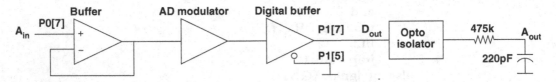

Figure 7.25: Isolated analog driver [7]. Courtesy of Cypress Semiconductor Corporation.

Example (Isolated analog driver [7]. *Courtesy of Cypress Semiconductor Corporation).* There are times when it is necessary to get an analog signal across an isolated barrier. For a higher frequency signal with no DC component, this can easily be done with capacitor or transformer coupling. For lower frequencies, the transformers and capacitors become increasing larger. The expense and size of these components for lower frequency and DC coupled signals makes this solution prohibitively expensive.

A cheaper solution requires using an *SC* block as an analog-to-digital (AD) modulator. Figure 7.17 shows an architecture for an AD modulator that converts an input signal to a series of pulses where a one represents voltage V_{ref+}, a zero represents voltage V_{ref-}, and the average is equal to the input signal.

Figure 7.25 shows how it can be assembled to pass the signal across an isolation barrier. The input signal (voltage A_{in}) is buffered and passed to the AD modulator, where the signal is converted to a series of digital pulses. These pulses are brought out of the chip via the digital buffer. The pulses (D_{out}) pass through an opto-isolator. Now isolated, these pulses are averaged to reconstruct an analog signal (voltage A_{out}). The lowpass filter is set to 1.5 kHz to eliminate the harmonics generated by the pulses.

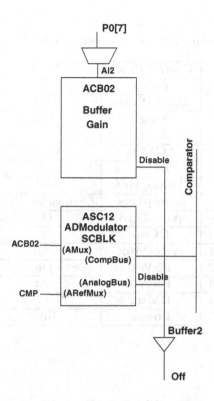

Figure 7.26: Analog isolator driver block placement [7]. Courtesy of Cypress Semiconductor Corporation.

The PSoC implementation is shown in Figure 7.26. The input buffer is a PGA user module. Its parameter selection is left as an exercise for the reader. The *ADmodulator* is an integrator with a comparator controlled input reference. Its parameters are shown in Table 7.9.

In theory, the ratio of *ACap/FCap* is unimportant. Practical considerations of loop gain and comparator offset dictate that capacitor *FCap* should be set to 32 and capacitor *ACap* set to 16. Capacitors *BCap* and *CCap* are not used, so their values are set to zero. The continuous-time blocks source the input signal, so that there is no phasing problem with sampling their outputs. Signal *ClockPhase* can remain set to value *Norm*. The signal *ACMux* is set to connect its input to block *ACA02*. Control bit *ARefMux* is set to value *CMP*, allowing the output comparator to control the reference selection. Bit *BMux* is not needed, so it does not need to be set. Bit *ASign* must be set to *Pos* (positive). To be an integrator, bit *FSW1* and bit *AutoZero* must be set *On* and bit *FSW0* must be set *Off*. The comparator needs to connect to the digital blocks, so the *CompBus* field is set to *ComparatorBus2*. Field *AnalogBus* is set to value *Disable*. Bits *Power* are set *High*.

The DigitalBuffer is an *SPI* slave user module that allows the comparator bus to be output on pin *P1[7]*. This output is connected to an optoisolator followed by a filter.

Table 7.9: Parameters for *ADModulator SC* block parameters

	Control Fields	Value
1	**FCap**	32
2	**ClockPhase**	Norm
3	**ASign**	Pos
4	**ACap**	16
5	**ACMux**	ACA02
6	**BCap**	0
7	**AnalogBus**	Disable
8	**CompBus**	ComparatorBus2
9	**AutoZero**	On
10	**CCap**	0
11	**ARefMux**	CMP
12	**FSW1**	On
13	**FSW0**	Off
14	**BMux**	Not Set
15	**Power**	High

7.4　Conclusions

This chapter has presented an introduction to switched capacitor (SC) circuits, and the programmable SC blocks that are provided by PSoC.

The basic principles of operation for SC circuits has been discussed in some detail. This principle is based on the controlled movement of charge, which is then stored on capacitors. The charge movement approximates an average current through a resistor. The value of the equivalent resistance depends on the capacitance value and the switching frequency of the capacitor. Switches are implemented by transistors controlled by clock signals. The behavior of SC circuits is influenced by a number of circuit nonidealities, and the like the nonzero resistance of "on" switches, channel charge injection, and clock feedthrough.

Basic SC circuits have also been presented (e.g., fixed gain amplifier circuit selectable gain amplifier, comparator, integrator, and differentiator) and the concept of auto-zeroing was explained for SC amplifiers. The effect of the OpAmp finite gain and nonzero switch resistance on the gain value was also presented. Two examples illustrated the use of basic SC circuits in building a two-bit quantizer and an analog-to-digital converter.

PSoC provides an array of programmable SC circuits. There are two types of circuits: type C and type D SC circuits. Chapter 3 detailed the interconnect structure for SC blocks, which can be connected to the neighboring continuous-time and SC blocks to form more complex networks, such as filters and ADCs. This chapter detailed the structure and control registers of the programmable SC blocks. It was shown that the functionality of each SC block can be programmed as inverting and noninverting amplifiers, integrators, and comparators by programming the registers *ASCxxCR3*. Four programmable capacitor arrays (*ACap*, *BCap*, *CCap*, and *FCap*) can be programmed to set the gain value of the block. Registers *ASCxxCR2*, *ASCxxCR1*, and

ASCxxCR0 control the values of the four capacitor arrays. The inputs, outputs, and clocks of a SC block can also be configured by the four control registers.

The chapter ended by presenting three design examples based on PSoC's programmable SC blocks: a differential amplifier with common mode output, a two-bit analog-to-digital converter, and an isolated analog driver.

7.5   Recommended Exercises

1. Assuming a switching frequency of 2.5 KHz and an equivalent resistance of 1000 ohms, calculate the required capacitance for a capacitor that could be used to replace the resistor.

2. For the circuit in Figure 7.1(b), explain what would happen if switches ϕ_1 and ϕ_2 were closed at the same time.

3. For the circuit in Figure 7.1(c) with $W/L = 20$ and $C = 1pF$, compute the expressions for the voltage V_c across the capacitor, if switch ϕ_2 is open and switch ϕ_1 is closed. Assume that the transistor is in saturation, and then repeat the computations if the transistor is in the linear region. Consider a $0.6\,\mu$ CMOS process.

4. Find the voltages computed in Exercise 3 by simulating the circuit (using a circuit simulator). Compare the simulation results with the values computed in Exercise 3. Comment on the accuracy of the analytical computations.

5. Repeat Exercise 3 for both decreasing and increasing values of the ratio W/L and the capacitance C. Explain the effect of the changing ratio and capacitance values on the behavior of the voltage V_c as a function of time.

6. For the circuit in Exercise 3, estimate the output error that is due to charge injection.

7. For a fixed-gain SC amplifier, size the circuit such that the maximum output error due to the nonzero resistance of the switches is less than 1%. Circuit sizing means finding all constraints for the capacitor values, switch resistances, and OpAmp gain, such that the accuracy requirement is met. Repeat the exercise if the analyzed error is due to charge injection.

8. For the ideal SC integrator, compute and plot the output voltage of the integrator for different values of the capacitor ratio, C_A/C_F, and for a fixed sampling frequency. Repeat the exercise if the capacitance ratio is constant but the sampling frequency is changing. Discuss your results.

9. Propose a scheme to simulate the differential amplifier in Section 7.4, and then use the scheme to study the effect of nonidealities on the circuit's performance. The considered nonidealities are nonzero switch resistance, and OpAmp finite gain. Propose a procedure also to analyze the effect of channel charge injection and clock feedthrough on the circuit performance.

10. Using the simulation method in Exercise 9, size the circuit such that the total output error due to nonidealities is less than 1%.

11. Repeat Exercise 9 for the common mode part of the differential amplifier with common mode in Section 7.4.

12. Repeat Exercise 10 for the common mode part of the differential amplifier with common mode in Section 7.4.

13. Using programmable SC blocks, implement the successive approximation register (SAR) method for analog-to-digital conversion. The method matches the input signal to be converted with the bitstring of the result by applying a binary search method. The method starts with the bitstring for the mid-scale voltage. This value splits the range into two equal halves. The bitstring is then converted into an analog voltage by the means of adigital-to-analog converter.

If the DAC output is lower than the input voltage then the method selects the mid-value of the upper half, and the process is repeated. A result bit for that position is set to high. If the DAC output is higher then the input then it selects the mid-point of the lower half before continuing the process. The result bit is reset. The process continues until all bits of the representation are produced.

14. Develop a simulation model for the two-bit comparator in Section 7.4. Find a simple way of co-simulating the SC and digital circuits of the design.

15. Build the simulation model for the isolated analog driver in Section 7.4. Find a simple solution for co-simulating the SC circuits and digital part of the design.

Bibliography

[1] R. J. Baker, *CMOS Circuit Design, Layout, and Simulation*, second edition, Hoboken, NJ: J. Wiley Interscience, 2005.

[2] S. Franco, *Design with Operational Amplifiers and Analog Integrated Circuits*, New York: McGraw Hill, 2002.

[3] P. Gray, P. Hurst, S. Lewis, R. Meyer, *Analysis and Design of Analog Integrated Circuits*, fourth edition, Hoboken, NJ: J. Wiley & Sons, 2001.

[4] R. Gregorian, G. Temes, *Analog MOS Integrated Circuits for Signal Processing*, Hoboken, NJ: J. Wiley & Sons, 1986.

[5] B. Razavi, *Design of Analog CMOS Integrated Circuits*, New York: McGraw Hill, 2001.

[6] PSoC Mixed Signal Array, Technical Reference Manual, *Document No. PSoC TRM 1.21*, Cypress Semiconductor Corporation, 2005.

[7] D. Van Ess, Understanding Switch Capacitor Analog Block, Application Note AN 2041, Cypress, March 30, 2004.

Chapter 8

Analog and Digital Filters

This chapter provides an overview of the subject of analog and digital filters describing the basic types, their primary characteristics, some key design criteria, practical considerations in component selection, and modern design tools. Although the subject of filters is quite broad and widely discussed in the literature, fortunately for designers much of the work required to design filters for modern applications has been greatly facilitated by advances in hardware and software.

There has, perhaps, been more written on the subject of filters than any other single topic in electrical engineering and yet it remains terra incognita for many. However, there exists a wealth of literature that illuminates and summarizes the basic concepts of filters and attempts to simplify filter design.[1] This chapter discusses filters and filtering in general terms and uses PSoC's inherent filter support to provide illustrative examples.

However, the material covered in this chapter is otherwise completely general and provides the basic filter concepts and principles that are of broad applicability in mixed-signal embedded systems. A brief discussion of passive filters is followed by a detailed analysis of some of the more popular active filter types, their characteristics, and implementations. The treatment in this chapter is based on each filter's transfer function, because it is the transfer function that ultimately determines parameters such as phase shift, overshoot, ringing, etc. That is, the characteristics of the transfer function for a given filter type uniquely determine the filters behavior in terms of magnitude response, phase response, etc., as long as the physical circuit remains linear.

The chapter has the following structure:

- Section 1 introduces filter fundamentals.

- Section 2 discusses filter design.

- Section 3 presents a discussion of analog filters.

- Section 4 discusses digital filters.

- Section 5 describes popular filter design tools.

- Section 6 presents concluding remarks.

---

[1]References [4], [10], [22], [26], [30], [31], [36], [39], and [45]–[47] are but a small sampling of such literature.

A. Doboli, E.H. Currie, *Introduction to Mixed-Signal, Embedded Design*, DOI 10.1007/978-1-4419-7446-4_8, © Springer Science+Business Media, LLC 2011

8.1   Filter Fundamentals

One might well ask why so much has been written about filters. And more to the point, why include a chapter on filters in a text about mixed-signal and embedded systems. The answer is simply that embedded systems typically involve data acquisition (via sensors and other input devices), data processing, some decision making, and subsequently output of data and control signals that drive actuators and other devices. And it is in the nature of things that applications involved with signal handling and/or signal processing are inherently susceptible to the introduction of unwanted signals.

Such signals may arise as the result of poor shielding, thermal noise sources, transients coupling into the signal path, etc. As a result, filters are found in virtually all modern electronic devices, e.g., cell phones, televisions, stereo systems, amateur and professional communication systems, etc.

There are a number of reasons for filters being so important some of the most common are: 1) the need to separate signals that have been combined with noise (thermal, atmospheric, interference (EMI sources), etc.) 2) to restore a signal that has otherwise been distorted or altered, and 3) to separate a given signal from one or more other signals, e.g., as a result of simple "mixing" of two or more signals, as in the case of telemetry systems and transmitters/receivers. Because embedded systems typically employ sensors and actuators, both of which are capable of serving either as noise sources or pathways for the introduction of noise into an embedded system, filtering of both the latter and former arises naturally. Unfortunately, as a general rule, "Where there is a signal, there is noise...".

In an embedded system, sensors are often used to provide information on linear/angular position and displacement, linear and/or rotational velocity and acceleration, temperature, pressure, vibration, etc., in the form of voltages, currents and frequencies.

In such applications, specific parameters to be measured include amplitude, waveshape, phase, timing, frequency, pattern recognition, etc. In order to facilitate computation and decision making by the embedded system, analog signals are converted into their digital equivalents, processed and, based on predefined criteria, output signals are produced in analog and/or digital formats. At each stage of this process spurious signals (noise) may be introduced and/or signals may be degraded, or otherwise corrupted.

Filters are often used in conjunction with analog-to-digital converters to remove electromagnetic interference (EMI), radio frequency interference (RFI) and electrical noise in general from a wide other sources. Filters are typically used in conjunction with a wide variety of sensors for the same reason. One of the most common sources of EMI noise has proven to be 50–60 Hz signals from local AC power sources. Filters can be very effective in removing much of this type of interference.

There is a wide variety of technologies to consider when designing a filter. However, broadly speaking there are two basic classifications of filters: active and passive. Although active filters require power, they assure, inter alia, that each stage of the filter has a sufficiently high input impedance to minimize loading of the previous stage. Loading does occur with passive filters and when employing multiple passive stages can result in a substantial loss of signal. Filters of the active type can be further classified as either digital or analog. Digital filters can have excellent filter characteristics, as shown and their use has been greatly facilitated by the advent of the digital signal processor (DSP). However, digital filters are not currently capable of providing as much dynamic range as active filters.

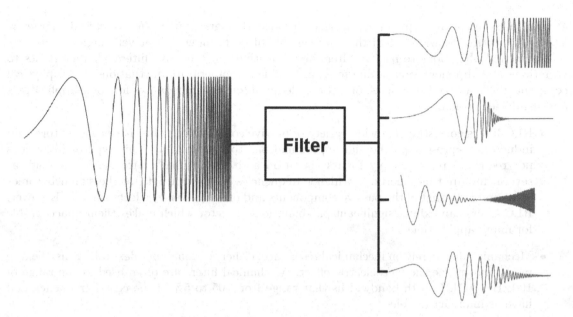

Figure 8.1: Classical filter functionality.

However, in some situations the designer must be content with reducing "noise" to an "acceptable" level. Eliminating noise completely is indeed a " a consummation devoutly to be wished ..."[2], but it is seldom achieved.

Each type of filter has certain advantages and disadvantages [35]. Therefore the designer must carefully weigh the pros and cons of each, and determine the best filter technology for a given application. It is also the case that in some embedded designs more than one filter technology will be required.

Historically, filters have been used to provided the types of functionality shown in Figure 8.1. As shown, an input signal of constant amplitude is applied to a generic filter while being "swept" over an arbitrary frequency range. Depending on the type of filter used, the output may be one of the four shown, or some permutation thereof. Filters may also have much more complex characteristics than those shown. In real world applications, this type of arrangement is often used to test filters. The generic filter is referred as the DUT (Device Under Test) and the DUT's output is viewed on an oscilloscope, or spectrum analyzer, to determine its characteristics. A combination of tracking generator, to provide the swept frequency of a signal in the range of interest, and spectrum analyzer, sometimes referred to as a scalar network analyzer, provide a graphical representation of the performance of a filter with respect to magnitude vs. frequency and serves as an excellent method of determining actual performance of a given filter. A vector network analyzer provides both magnitude and phase performance graphically.

8.1.1   Passive Filters

Passive filters of the types shown in Figure 8.2 were, for many years, the most prevalent form of filter, but in recent years, as more and more integrated circuit technology has emerged, and with the introduction of the operational amplifier, active filters have become commonplace in a wide variety of applications and they have largely replaced passive filters. However, passive filters

---

[2]William Shakespear, *The Tragedy of Hamlet, Prince of Denmark*, Act 3, Scene 1.

are still used in certain types of applications because they are simple, their electrical characteristics are easy to determine, and they are capable of performing well at very high frequencies. The principal objections to passive filters are: sensitivity of a passive filter's characteristics to variations in component values (due to aging, ambient temperature fluctuations, etc.), physical component size at low frequencies, objectionable input/output impedances for some applications and insufficient rolloff.

- RLC filters consisting of a wide variety of passive combinations of resistors, capacitors, and inductors, represent polynomial approximations to ideal filters. This type of filter does not require a power supply for its operation and can be used at frequencies as high as several hundred megahertz. At higher frequencies, the wavelength of the signal becomes comparable to that of the passive components and other (adverse) effects emerge. However, RLC filters can exhibit significant passband loss, a factor which makes them unacceptable for many applications.

- Mechanical filters rely on mechanical resonance, either torsional, bar flexural, or disk flexure mode and utilize the piezoelectric effect. Mechanical filters are often used in the range of 100 kHz–700 kHz with bandwidths that range from .05 to 5% of the center frequency and have as many as 12 poles.

- Crystal filters are based on the mechanical characteristics of quartz crystals and the piezoelectric effect and are most commonly employed where very high Q[3] steep bandpass characteristics, and a specific center frequency are required. This type of filter is particularly useful in communications devices. Crystal filters generally provide much better characteristics than lumped or discrete component counterparts. It is possible to obtain Qs well above 10,000 with crystal (quartz) resonators. Crystal filters can have temperature dependencies that require that in some applications they be housed in a temperature controlled oven.

- SAW (surface acoustic wave) filters convert electrical signals into a mechanical waves in a piezoelectric crystal. A wave propagates across the crystal and is then converted back into an electrical signal. The characteristics of the wave and the resulting delay introduced by the speed of propagation of the surface acoustic wave can be altered as it propagates through the SAW filter. Its characteristics allow it to be used in much the same way that digital filters are implemented.

However, RLC, mechanical, crystal, and SAW filters are most commonly found in high frequency applications such as cell phones and communications equipment as a result of several factors: cost, frequency ranges supported by each of these technologies, and size.

8.1.2   Linear Active Filters

Active RC filters, for example of the type shown in Figure 8.8, are characterized by their relatively small size, weight and external power requirements and, generally speaking, are without attenuation concerns in the passband. However, a power supply is required and, depending on the specific application, active filters can prove significantly more expensive than their passive counterparts. Additionally, the active filters can, in some applications, be the source of a substantial amount of noise that otherwise would be absent with passive filters. One of the most common disadvantages of active filters is that they can give rise to serious distortion of the output signal if overdriven by the input signal. In addition, the introduction of an operational amplifier introduces additional factors that must be taken into account: amplifier bandwidth limitations,

---

[3] Q is a figure of merit for filters and defined in a later section of this chapter.

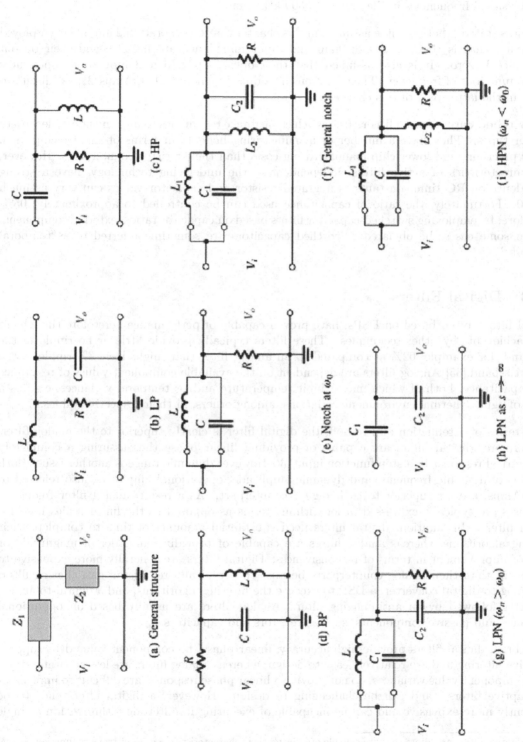

Figure 8.2: Examples of lumped, LRC, passive filters. (Reproduced with permission) [27].

amplifier noise, noise introduced by the OpAmp's and power supply. Active filters are typically used for signal frequencies in the 0.1 Hz to 500 kHz band.

Unless stated otherwise, it is assumed in this chapter that the operational amplifiers employed are "ideal", that is their open loop gain and input impedance are infinite, and their output impedance is zero. It is also assumed that their open loop gain and input/out impedances are not functions of frequency. The deviations introduced by nonideal OpAmps [21] are discussed briefly in the latter part of this chapter.

Switched capacitance filters use switch capacitance techniques to implement a wide variety of filter types. They have a number of manufacturing benefits in terms of smaller chip area, higher precision, and lower chip manufacturing costs than their resistor counterparts. However, the characteristics of active filters, independent of the underlying technology, have a strong dependence on RC time constants. Integrated resistor and capacitor values can vary as much as ±30. Fortunately, the ratio of capacitance used can be controlled to approximately .05%. Therefore, by employing switched-capacitor filters based on capacitor ratios, external components can, in some cases, be obviated. Switched capacitors are sometime referred to as "temporal resistors."

8.1.3   Digital Filters

Digital filters, often based on DSPs, have proven capable of performance heretofore thought to be unachievable by other techniques. These filters typically provide little or no ripple in the passband, for example .02% as compared to an analog filter that might have 6% ripple in the same passband.[35] Analog filters are dependent on the available component values of resistance and capacitance, both of which may exhibit temperature and/or temporal variances, e.g., as a result of aging, thermal environment variations, among others, in their respective values.

In terms of attenuation and roll-off, the digital filter is clearly superior to the analog filter. Furthermore, digital filters are capable of providing linear phase thus assuring a completely symmetrical response to a step function input [3]. However dynamic range is another issue. Both in terms of dynamic frequency and dynamic amplitude, operational amplifiers, also referred to as OpAmps, are far superior to their digital counterpart. As a result, analog filter frequency responses are typically expressed in logarithmic terms as opposed to the linear scales used for digital filters. In addition, digital filters need substantial amounts of time to complete their filtering algorithms whereas analog filters are capable of providing an order of magnitude or more of improvement in terms of responsiveness. Digital filters are generally more expensive to implement than their analog counterparts because they typically require an anti-aliasing filter, an analog-to-digital converter, a DSP to execute the filtering algorithms, and a digital-to-analog converter followed by an anti-imaging filter. Analog filters are usually based on operational amplifiers and passive components, such as resistors and capacitors.

In brief, digital filters provide high accuracy, linear phase, no component value drift support adaptive filtering and are relatively easy to design, whereas analog filters are less accurate, suffer from component value variances, do not provide a linear phase response, are difficult to implement as adaptive filters,[4] and can be challenging to design. However, a digital filter tends to be inherently more expensive and can be incapable of executing its filter algorithm within a single

---

[4] Adaptive filters are filters that are capable of altering their characteristics, i.e. their transfer function, in real time thus enabling them to "adapt" to changing noise situations. This is particularly useful when the type of noise to be encountered is not known a priori. Such filters are typically digital filters.

sampling clock period. Analog filters are capable of operating at extremely short sampling periods and are often cheaper than their digital counterparts.

Thus analog filters offer greater dynamic range in terms of frequency and amplitude, are fast, cheaper and arguably are easier to design than digital filters. However, digital filters represent a programmable solution, offer superior performance in terms of roll-off in the transition band, have relatively little or no ripple (particularly when compared to their analog counterpart), and are capable of providing phase linearity and excellent stopband attenuation. Also, digital filters are relatively unaffected by component value variations.

There are a number of technologies available for removing unwanted "noise" and/or the recovery of signals from noise. Fortunately, many mixed-signal, embedded system applications do not require the sophisticated filtering capabilities of digital filters and often lowpass analog filters are a quite acceptable solution to an embedded system's filtering requirements.

It has been suggested, in the vernacular, that the distinction between noise and signal is simply that noise is the "stuff you don't want" and signal, is the "stuff you do want." Fortunately, there are actually much more substantive ways for describing noise and signal. The real concern about noise is simply how to remove it.

Noise can often obscure the signal to the point that it is virtually impossible to recover the "true signal" without some form of filtering. Digitizing analog signals was thought, many years ago, to be a way of completely removing or at least minimizing noise in a system, but the process of converting an analog signal to a digital equivalent can introduce "noise" that, in some cases is even more of a problem than would have been encountered by merely processing the signal in a purely analog environment. Furthermore, in some systems it is not intuitively obvious what is noise and what is signal. For this reason it is generally a good practice to design systems that have a "naturally" high noise immunity because it is often easier to keep noise out than remove it, once it is present. However, it is not possible, in general, to completely avoid noise problems simply by employing good engineering techniques. It is this simple fact that occupies our attention for the balance of this chapter.

8.1.4 Filter Components

Nonideal operational amplifiers have been dealt with in some detail in Chapter 6. Real-world resistors and capacitors can bear little resemblance to their ideal counterparts [17] as shown in Figure 8.3.[5] Because of their physical design resistors, capacitors, and inductors can exhibit resistive, inductive, and capacitive characteristics. When designing analog filters, where possible, resistor values should be chosen that have values ranging from $1\,K\Omega$ to $100\,K\Omega$. Resistors are generally of three types: carbon composition, wire-wound, and metal-film. Carbon composition is the most widely available, and are typically the cheapest of these three types.

Noncritical use in room temperature environments can often warrant using carbon-based resistors. For filter design, when using carbon composition resistors, tolerances should be 5% or less for best results. In critical applications, for which the ambient temperatures are higher than normal room environments and best performance is required, metal-film or wire wound resistors are recommended. Fifth- and sixth-order filters should be implemented with 2% or better resistors and seventh- and eighth-order filters should use 1% tolerance resistors. Staying

---

[5]Note that ESR in this figure refers to "Effective Series Resistance" which arises as a result of resistance introduced by various constituents of a real world capacitor, e.g., dielectric resistance, plate resistance, resistance of terminal leads, etc. It is of particular concern for electrolytic capacitors, which are often used when large capacitance values are required (e.g., tens of microfarads, or greater).

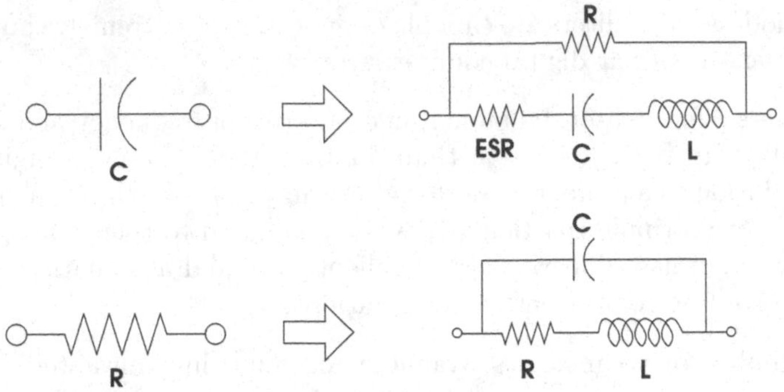

Figure 8.3: Ideal versus nonideal resistors and capacitors.

within these ranges, whenever possible, will help to limit current requirements for low-value resistors and the inherent noise associated with high-value resistors.

The choice of capacitance values is less restrictive as long as the values chosen are well above stray capacitances present. Typical values range from 1 nF to $1-10\,\mu F$. Ceramic disk capacitors can be used in noncritical applications, but they have a nonlinear temperature dependence that can degrade filter performance by 10% or more in some environments. Mylar, polystyrene, teflon, among others capacitors are excellent choices for filter implementations for critical applications. Resistors and capacitors can both undergo substantial changes in parameter value when subjected to high temperatures, widely varying temperatures, vibration, shock, and aging. This, in turn, can have a significant impact on a filter's characteristics and may actually degrade an embedded system's performance. Thus the choice of component types must be considered carefully when designing a filter for a particular environment and especially if the environmental conditions are unspecified.

8.2 Filter Design

Perhaps the simplest of passive analog filters are shown in Figures 8.12 (a) and (b). The reader may note that neither circuit has an inductive component. This is in part due to the fact that inductors tend to be large and more expensive than capacitors and resistors. In today's world, of ever-decreasing component size, inductors are often expensive, difficult to manufacture, and introduce other concerns, all of which can generally be avoided by utilizing resistors (or their switched capacitor equivalent), capacitors, and active components, that is OpAmps. As shown, life without inductors is not as difficult as one might think. Combining resistors, capacitors, and OpAmps has proven to be very effective alternative to inductor-based filters. And there are additional benefits to replacing inductors with a combination of resistors, capacitors and an OpAmp (e.g., gain can be added, inductive coupling is largely eliminated the active filter's higher input impedance and lower output impedance make it easy to cascade filters for increased performance, and perhaps the greatest benefit is the ability to tune the active filter by varying passive, noninductive components). Therefore, inductor-based filters have proven to be much less common and are not explicitly treated except in passing in the discussion that follows. However, that is not to suggest that the functionality provided by inductors can be ignored. It is incumbent on the OpAmp to provide any of the benefits that an inductor might provide which would otherwise be lost in relying solely on capacitors and resistors.

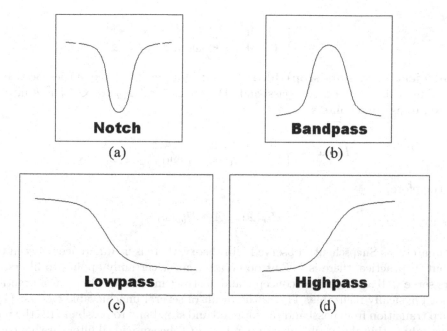

Figure 8.4: A graphical representation of basic filter types.

Before continuing it is helpful to define some terms to be used in this chapter. Filters can be broadly characterized as one of four types: bandpass, lowpass, highpass, and bandreject. The term "passband" refers to the band of frequencies that a given filter will pass.

There are four fundamental types of ideal analog and digital filters in common use.

- **Lowpass filter:** Rejects all frequencies ω above a given frequency,[6] typically from 0 Hz, that is DC, to the corner frequency, ω_c.

$$H_{lp}(\omega) = \begin{cases} A & \text{if } |\omega| \leq \omega_c \\ 0 & \text{if } |\omega| < 0 \end{cases}$$

- **Highpass filter:** Rejects all frequencies below a given frequency, and its passband ranges from ω_c to ∞. (It should be noted that such a highpass filter is not possible in real-world designs based on OpAmps due to the inherent limitations of operational amplifier bandwidth.)

$$H_{hp}(\omega) = \begin{cases} 0 & \text{if } |\omega| \leq \omega_c \\ A & \text{if } |\omega| > 0 \end{cases}$$

Thus,

$$H_{hp}(\omega) = A - H_{lp}(\omega) \tag{8.1}$$

- **Bandpass filter:** Rejects all frequencies below a given frequency ω_{c1} and all frequencies above a given frequency ω_{c2}.

---

[6]In this discussion it is assumed that $\omega \geq 0$ and the term "frequency" in this context refers to angular frequency unless otherwise noted.

$$H_{bp}(\omega) = \begin{cases} A & \text{if } \omega_{c1} \leq |\omega| \leq \omega_{c2} \\ 0 & \text{if } \omega_{c1} < |\omega| \text{ and } \omega_{c2} < |\omega| < \infty \end{cases}$$

- **Bandreject (aka Bandstop) filter** blocks: "Rejects" all frequencies between ω_{c1} and ω_{c2}. Thus this filter has two passbands DC to ω_{c1} and ω_{c2} to ∞. These filters are also referred to as "notch filters."

$$H_{br}(\omega) = \begin{cases} 0 & \text{if } \omega_{c1} \leq |\omega| \leq \omega_{c2} \\ A & \text{if } \omega_{c1} < |\omega| \text{ and } \omega_{c2} < |\omega| < \infty \end{cases}$$

and therefore,

$$H_{br}(\omega) = A - H_{bp}(\omega) \tag{8.2}$$

Unfortunately, as Snepscheut [7] observed, "In theory, there is no difference between theory and practice, but in practice, there is." This observation most certainly applies to filters. Although ideal filters serve well as theoretical concepts and abstract models for filters, as a practical matter they are not physically realizable. This is the result of several unachievable factors: (1) the ideal filter's sharp transition from passband to stopband and stop band to passband (2) the requirement for constant gain within the passband/stopband, and (3) because ideal filters are noncausal. Thus the designer must determine the necessary tradeoffs that are required to design an acceptable filter that can actually be built. Figure 8.6 shows diagrammatically how one might approximate an ideal bandpass filter with a physical filter.

Any filter can be uniquely characterized by its impulse response, step response, or frequency response, and having done so the other two responses are contemporaneously determined. Typically, filter characteristics are represented as graphs of the magnitude of the gain and phase versus frequency, called Bode plots and named for Hendrik Wade Bode, a recognized pioneer in the fields of control theory and electronic telecommunications. An example of a Bode graph, or plot, is shown in Figure 8.5.

Figures of merit for filters include passband amplitude characteristics, stopband amplitude characteristics and stopband attenuation. In the case of an ideal filter, the passband amplitude should be constant (i.e., "flat') the stop band amplitude should also be constant (i.e., zero), and the transition from passband to stopband, and vice versa should be a step function. The ideal filter should have amplitude and delay characteristics which are independent of frequency.

8.2.1   Specific Filter Types

There is a broad range of filter types and implementations, whoever, the following are the most commonly encountered whether implemented individually as redundant stages in some cases with varying parameters, or as a permutation of the basic types described below.

- **Bessel:** This filter's response is smooth and flat in the passband. However, near the passband attenuation rises slowly and far from the passband it rises at the rate of $nX6$ dB/octave where n is the filter's order. The transition region change is slower than that of Butterworth [5], Chebychev, inverse Chebychev, and so on. Bessel filters do not exhibit ringing or significant overshoot in the passband, as is the case for Butterworth, Chebychev, and inverse Chebychev.

---

[7]Jan L.A. van de Snepscheut, 1958–1994, Professor of Computing Science. California Institute of Technology.

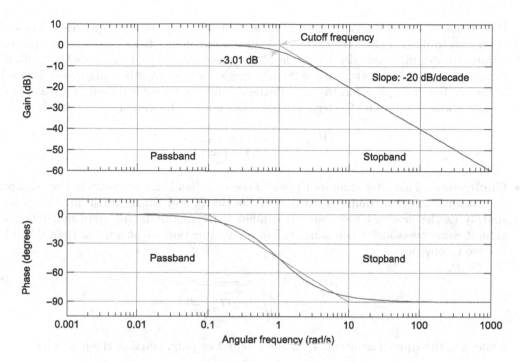

Figure 8.5: Example of a typical Bode plot.

The transfer function for a Bessel filter is given by,

$$H(s) = \frac{\theta_n(0)}{\theta_n\left(\frac{s}{\omega_0}\right)} \tag{8.3}$$

where

$$\theta_n(x) = \sum_{k=0}^{n} \frac{(2n-k)!}{(n-k)!} \frac{x^k}{2^{n-k}} \tag{8.4}$$

Table 8.1: Reverse Bessel polynomials.

$$\theta_0(x) = 1$$
$$\theta_1(x) = x^1 + 1$$
$$\theta_2(x) = x^2 + 3x^1 + 3$$
$$\theta_3(x) = x^3 + 6x^2 + 15x^1 + 15$$
$$\theta_4(x) = x^4 + 10x^3 + 45x^2 + 105x^1 + 105$$
$$\theta_5(x) = x^5 + 15x^4 + 105x^3 + 420x^2 + 945x^1 + 945$$
$$\theta_6(x) = x^6 + 21x^5 + 210x^4 + 1260x^3 + 4725x^2 + 10395x^1 + 10395$$

- **Butterworth:** This filter has a gain versus frequency characteristic that is also completely flat in the passband but its attenuation in the stopband rises much faster than that of the

Bessel filter, but not as steeply as that of the Chebychev and inverse Chebychev filters. The Butterworth filter is often referred to as having a maximally flat response in the passband, because it exhibits virtually no ripple in the passband. The Butterworth's "roll-off" is monotonic and smooth at a rate of 20 dB/decade, or equivalently 6 dB/octave for each of its poles. For example a fourth-order Butterworth would have a roll-off of 80 dB per each decade above cutoff. The Butterworth also does not exhibit ringing in the stopband.

$$|H(j\omega)| = \frac{1}{\sqrt{\left[1 + \left(\frac{\omega}{\omega_c}\right)^{2n}\right]}} \tag{8.5}$$

- **Chebychev:** This filter exhibits ripples in the pass band. Its attenuation rises sharply in the transition region compared to that of the Bessel and Butterworth filters, but not as sharply as the inverse Chebyshev. It exhibits little ringing in the stopband, but it does exhibit some overshoot when subjected to a step function, although less than that of the inverse Chebychev.

$$|H(j\omega)| = \frac{1}{\sqrt{1 + \epsilon^2 T_n^2\left(\frac{\omega}{\omega_0}\right)}} \tag{8.6}$$

where ϵ is the ripple factor and T_n^2 is the Chebyshev polynomial of the nth-order.

Table 8.2: Chebyshev Polynomials.

$$
\begin{aligned}
T_0(x) &= 1 \\
T_1(x) &= x \\
T_2(x) &= 2x^2 - 1 \\
T_3(x) &= 4x^3 - 3x \\
T_4(x) &= 8x^4 - 8x^2 + 1 \\
T_5(x) &= 16x^5 - 20x^3 + 5x \\
T_6(x) &= 32x^6 - 48x^4 + 18x^2 - 1 \\
T_7(x) &= 64x^7 - 112x^5 + 56x^3 - 7x \\
T_8(x) &= 128x^8 - 256x^6 + 160x^4 - 32x^2 + 1 \\
T_9(x) &= 256x^9 - 576x^7 + 432x^5 - 120x^3 + 9x
\end{aligned}
$$

- **Inverse Chebychev:** This filter provides a very flat response in the passband. It has a much steeper response in the transition region than Butterworth, Bessel, or Chebychev but does tend to exhibit more ringing, overshoot, and ringing in response to a step function.

The basic form of this filter is given by

$$|H_n(j\omega)| = \frac{1}{\sqrt{\left[1 + \epsilon^2 T_n^2\left(\frac{\omega}{\omega_0}\right)\right]}} \tag{8.7}$$

where T_n is the Chebyshev polynomial of order n as shown in Table 8.1 and ϵ is the ripple factor[8].

---

[8]The ripple factor is the parameter that controls the amount of ripple in the passband.

- **Elliptic filter:** (aka Cauer) This a filter can have the same amount of ripple in the passband, as in the stopband or the ripple in both bands can be varied independently. This type of filter also has the fastest transition from passband to stopband of any of the filters discussed in this chapter.

The basic form of its transfer function is given by

$$|H(jw)| = \frac{1}{\sqrt{1 + \epsilon^2 R_n^2(\xi, \frac{\omega}{\omega_0})}} \tag{8.8}$$

where R_n are the Chebyshev rational functions given by

$$R_n(\xi, x) = cd\left(n \frac{K(\frac{1}{L_n})}{K(\frac{1}{\xi})} cd^{-1}\left(x, \frac{1}{\xi}\right) \frac{1}{L_n} \right) \tag{8.9}$$

(cf. Table 8.2).[9][28]

There are many types of filters of which the five most common are Bessel, Butterworth, Chebychev 1, Chebychev II, and elliptic filters. It is useful to compare each to an ideal filter that has the following characteristics: unit gain in the passband(s), zero gain in the stopband(s), and zero phase shift in the passband. Such a filter is of course not obtainable as a practical matter, but it is possible to achieve progressively sharper passband–stopband transitions by increasing the order of the filter. However, all five exhibit nonzero phase shifts no matter what the order of the respective filter. Nonetheless, there is a type of filter referred to as an ideal linear phase filter which has as its primary characteristic that its phase shift is proportional to frequency. For an arbitrary transfer function:

$$H(s) = |H(j\omega)|e^{j\theta(\omega)} = G(\omega)e^{j\omega(\omega)} \tag{8.10}$$

and the phase components are defined by:

$$PD = Phase\ Delay = -\frac{\theta(\omega)}{\omega} \tag{8.11}$$

$$\tau(\omega) = Group\ Delay = -\frac{\partial\theta(\omega)}{\partial\omega} \tag{8.12}$$

Therefore if the variation in phase is a linear function of the frequency, then the group delay is constant, and there is no distortion. If this is not the case, the energy corresponding to each frequency component may be unchanged, but the phase relationship between each component will be changed resulting in distortion of the original signal.

As an illustrative example, consider the second-order lowpass filter whose transfer function is given by

$$H(s) = \frac{\omega_0^2}{s^2 + \frac{1}{Q\omega_0}s + \omega_0^2} \tag{8.13}$$

with

$$\tau(\omega) = \frac{1}{\omega_0 Q}\left[1 + \left(\frac{\omega^2}{\omega_0^2}\right)\right]^{-1}\left[1 - \left(1 - \frac{1}{2Q^2}\right)\right]\left(\frac{2\omega}{\omega_0}\right)^2 + \left(\frac{\omega}{\omega_0}\right)^4\right]^{-1} \tag{8.14}$$

---

[9]Note that cd is the Jacobi elliptic cosine function, K a "complete" elliptic integral of the first kind and $L_n(\xi). = R_n(\xi, \xi)$.

and therefore for $\omega << \omega_0$,

$$\tau(\omega) \approx \frac{1}{\omega_0 Q} = \frac{2\xi}{\omega_0} \qquad (8.15)$$

Thus as shown in Figure 8.25, the group delay is relatively constant with respect to a normalized angular frequency in the range $0 < \omega \leq 0.2$, where $0.05 \leq \xi \leq 1.0$.

8.2.2 Filter Parameters

Filter design is based on consideration of a number of factors which, when taken together, completely characterize the filter required. The key design parameters are allowable passband and stopband ripple, the filters transient response, its monotonicity, rate of attenuation at/near cutoff, the roll-off rate beyond cutoff and the filter's order.

- *Allowable passband and stop band ripple:* A filter that fails to exhibit monotonicity in the passband will exhibit what is referred to as ripple.

- *Transient response:* This can be a much more serious issue in filter design than other filter-related issues, because filters can "ring". Such oscillations can introduce noise into the embedded system, give rise to inaccurate input data and/or unduly influence the systems response to sensor input, and so on. High Q and extremely sharp cutoff may be highly desirable, but unfortunately the higher the Q and the sharper the cutoff, the greater the ringing is. One way of determining the degree of ringing that a filter can produce is to subject the filter to a step function. Ringing can be difficult to control and the best approach is often to impose no sharper cutoff or higher Q than is really needed in a given design.

- It may seem that the order of a filter is not a particularly critical design criterion. However, the higher the order is, the more components are required, and therefore the higher the cost and the greater the physical volume. Perhaps even more important is that higher order filters involve correspondingly greater complexity in terms of the design effort. The order of the filter should be based primarily on the required roll-off characteristics, and wherever possible Occam's razor[10] should be employed.

In thinking about, and designing filters, there are four rather simple parameters that can be relied upon to characterize a reasonably good filter for a particular application:

1. f_c: The cutoff frequency

2. f_s: The frequency at which the stopband begins

3. A_{min}: The maximum allowable change, or equivalently the minimum allowable attenuation, within the passband

4. A_{max}: The maximum allowable variation in gain within the passband.

Note that the gain axis represents the "normalized gain" and that the corner or "cutoff" frequency is shown as the $-3\,\text{dB}$ point on the magnitude plot in Figure 8.5. The cutoff frequency

---

[10]Occam's razor is attributed to a 14th century monk and can be simplistically stated as "The simplest solution is typically the best".

is defined as the point at which the magnitude of the gain has dropped by a factor of $1/\sqrt{2}$ or .707.

Also the unit of measure for gain is given in dB defined as

$$dB = 20 \log \left[\frac{V_{out}}{V_{in}} \right] \tag{8.16}$$

Furthermore, the slope shown as -20 dB/decade represents the rate of attenuation. This can also be expressed as -6 dB/octave. One of the principal figures of merit of a filter is its Q which is defined as

$$Q = \frac{f_c}{\Delta f} \tag{8.17}$$

where

$$\Delta f = f_H - f_L \tag{8.18}$$

is defined as the width in Hz of the half-power points. Recall that the -3 dB point also represents the half-power point because

$$20 \log \left[\frac{V_{out}}{V_{in}} \right] = -3 \text{ dB} \tag{8.19}$$

which implies that,

$$\frac{V_{out}}{V_{in}} = \frac{\sqrt{2}}{2} = .707 \tag{8.20}$$

However, power is proportional to the square of the maximum amplitude of the voltage and therefore because $.707^2 \approx .5$, the -3 dB point corresponds to the "half-power point." The -3 dB point also defines the cutoff frequency f_C. In the case of digital filters, the cutoff frequency is specified as the point at which amplitude is $50 - 99\%$ of the maximum value. For biquadratic filters with low Q factors (e.g., $Q \leq 2$) there is relatively little sensitivity to component tolerances. For $2 < Q < 20$ component sensitivity issues begin to arise and for $20 < Q$, depending on the technology involved, component sensitivity can become a serious issue.

8.2.3   Scaling and Normalization

One technique that can be used to reduce the amount of work required to design a particular filter is to employ normalization[26, 29]. This allows a designer to focus on the desired filter characteristics and then use scaling to "scale" the filter for a particular frequency, or range of frequencies, without altering the filters characteristics. It is also possible to scale the filter's impedance by using a similar technique. Note that for a capacitor, doubling the frequency halves the impedance and for an inductor doubles the impedance which represents a 6 dB change.

Typically, a cutoff frequency of 1 radian per second, or equivalently $1/2\pi$ Hz, and an impedance of 1 ohm are chosen to "normalize" a particular filter configuration being analyzed. This choice is easily understood when one considers that for a combination of a one farad capacitor and one ohm resistor, the cutoff frequency becomes one radian per second. However, designers work with filters in two contexts: (a) the analysis of a particular filter and (b) the "synthesis" of a filter to meet a specific set of requirements [40, 44, 48]. In the latter case, the design is often facilitated by using more realistic values, for example $R = 10K\Omega$ and a cutoff frequency of 1KHz.

8.2.4   Cascading Analog Filters

The characteristics of a nonideal filter can often be improved by passing the signal though a series of filter stages. This is a result of the fact that a second-order filter has a maximum phase shift of 180 degrees and is therefore unconditionally stable, meaning no chance for oscillation to occur. Thus second-order filters[11] are attractive as basic building blocks for higher-order filters.

Fortunately, any linear time invariant (LTI) system can be represented in terms of combinations of first and second-order stages, sometimes referred to as sections (cf. Table 8.3). One of the significant benefits of analog filters is the fact that they can be cascaded. Care must be exercised in doing so to avoid loading successive stages in such an arrangement, but the high input impedance of many OpAmps reduces the loading to a level that for most applications, is insignificant. Furthermore, adding additional filtering stages comes at the cost of additional components (i.e., resistors, capacitors, and OpAmps), increased power requirements, increased printed circuit board real estate, and makes the design more susceptible to component tolerance variations, noise, and so on. Nonetheless, the benefits often outweigh the costs.

The transfer function for N cascaded filter stages is given by[12, 13]:

$$H(s) = \prod_{j=1}^{N} H_j(s) = H(s)_1 H(s)_2 H(s)_3 \cdots H(s)_N \tag{8.21}$$

where $H_j(s)$ is the transfer function for the jth filter section and $s = \sigma + j\omega$. Because multiplication of the transfer functions is commutative, it may appear that the order of application of each stage is irrelevant. However, in order to avoid loss of dynamic range and signal clipping each transfer function is typically applied to the input signal in order of increasing Q for each stage. To minimize noise, the gain of each successive stage should be less than that of the preceding stages.

Any physically realizable circuit can be represented by a transfer function that is a rational polynomial which is a function of s.

The jth stage of order n can be expressed as

$$H_j(s) = \frac{N(s)_j}{D(s)_j} = K \frac{b_0 + b_1 s^1 + b_2 s^2 + b_3 s^3 + b_4 s^4 + \ldots + b_m s^m}{a_0 + a_1 s^1 + a_2 s^2 + a_3 s^3 + a_4 s^4 + \ldots + a_n s^n} \tag{8.22}$$

$$H_j(s) = \frac{N(s)_j}{D(s)_j} = K \frac{(s - z_1)(s - z_2) \ldots (s - z_{m-1})(s - z_m)}{(s - p_1)(s - p_2) \ldots (s - p_{n-1})(s - p_n)} \tag{8.23}$$

where a_n and b_m are real, and the order of $N(s)$ is less than the order of $D(s)$.[12]

The "frequency response" of a filter, section or stage is defined as:

$$H_j(s = j\omega) = K \frac{(j\omega - z_1)(j\omega - z_2) \cdots (j\omega - z_{m-1})(j\omega - z_m)}{(j\omega - p_1)(j\omega - p_2) \cdots (j\omega - p_{n-1})(j\omega - p_n)} \tag{8.24}$$

However, terms such as $(j\omega - p_m)$ represent the distance in the s-plane from $j\omega$ to p_m and therefore:

$$|j\omega - p_i| = \sqrt{\sigma^2 + (\omega - \omega_i)} \quad \text{and} \quad |j\omega - p_i| = \sqrt{\sigma^2 + (\omega - \omega_i)} \tag{8.25}$$

---

[11] Equally important is the fact that second-order characteristic equations are readily solved by the quadratic "formula."

[12] If this is not the case then the numerator should be divided by the denominator until m< n.

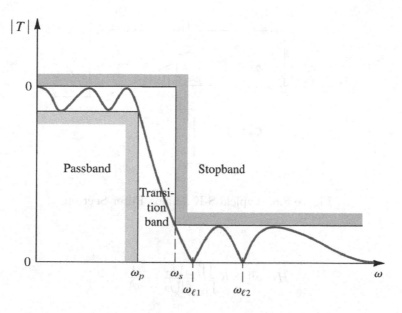

Figure 8.6: Sixth order lowpass filter characteristics [27].

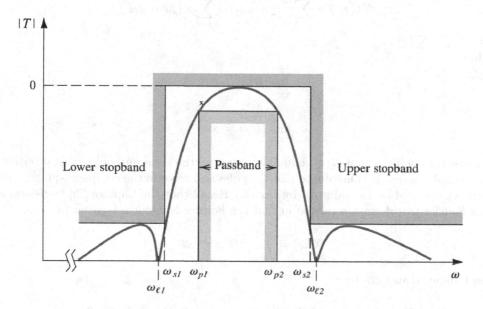

Figure 8.7: Fifth Order Bandpass filter characteristics [27].

and,

$$\angle(s - p_i) = tan^{-1}\left[-\left(\frac{\omega - \omega_i}{\sigma_i}\right)\right] \quad \text{and} \quad \angle(s - z_i) = tan^{-1}\left[-\left(\frac{\omega - \omega_i}{\sigma_i}\right)\right] \tag{8.26}$$

Thus it is the frequency response on the $j\omega$ axis defined by the distance from each of the poles and zeros to a point $s = j\omega$ in the s-plane, that determines the magnitude and phase of the transfer function for a particular value of s.

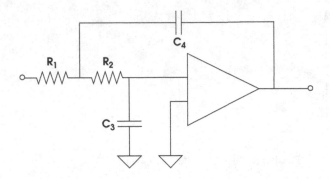

Figure 8.8: Typical S-K Analog Filter Section.

Therefore,

$$|H(j\omega)| = K\frac{\prod_{i=1}^{m}(j\omega - z_i)}{\prod_{i=1}^{n}|(j\omega - p_i)|} \tag{8.27}$$

and,

$$\angle H(j\omega) = \sum_{i=1}^{m}\angle(j\omega - z_i) - \sum_{i=1}^{n}\angle(j\omega - p_i) \tag{8.28}$$

Note that:

$$\lim_{s \to z_i} H(s) = 0 \tag{8.29}$$

$$\lim_{s \to p_i} H(s) = \infty \tag{8.30}$$

and because the coefficients of $N(s)_j$ and $D(s)_j$ are real, the zeros and poles must be either real or occur in complex pairs. It is helpful to plot the poles and zeros in the complex s-plane. (Typically poles are represented by xs and zeros by circles.) Recall that the Laplace [29] transform and the complex Fourier transform are related in that the Fourier transform is given by

$$X(\omega) = \int_{-\infty}^{\infty} f(t)e^{-j\omega t}dt \tag{8.31}$$

and the Laplace transform by

$$X(\sigma, \omega) = \int_{-\infty}^{\infty} [f(t)e^{-\sigma t}]e^{-j\omega t}dt = \int_{-\infty}^{\infty} f(t)e^{-(\sigma + j\omega)t}dt \tag{8.32}$$

Note also that the sign of σ determines the stability, or lack thereof, of a filter section and whether it is critically damped, underdamped, or overdamped. Therefore because $s = \sigma + \omega$, Equation 8.32 becomes:

$$\mathcal{L}(s) = \int_{-\infty}^{\infty} f(t)e^{-st}dt \tag{8.33}$$

The term $e^{-st} = e^{-(\sigma + j\omega)}$ represents both exponential and sinusoidal components. In the case of a real pole (i.e., $p_i = -\sigma$), an exponentially decaying component is introduced, where the rate of decay is determined by the location of the pole: the farther to left of the origin, the faster the rate of decay of a pole is. Poles to the right of the origin, introduce instability and therefore must be avoided when designing filters. Poles at the origin (i.e. $p_i = 0$) are defined as "marginally" stable and contribute a constant amplitude component. Similarly, a pair of poles on the imaginary axis introduce a constant amplitude, oscillatory component. A three dimensional graph of σ, ω, and d is shown in Figure 8.29 for a biquadratic filter.

The Bode plot can be constructed directly from the poles and zeros of a filter section to determine the shape of a section's response as a function of angular frequency.

The so-called "break frequencies" are defined by

$$\omega_b = \sqrt{(\sigma^2 + \omega^2)} \tag{8.34}$$

and can be derived from the transfer function for a given section or from the pole–zero diagram.

The distance to each pole is measured along the imaginary axis, for example by using a compass to draw an arc on the s-plane where the arc is centered at the origin and the radius is the distance to each pole and zero in the left-hand side of the plane. At frequencies above each pole and zero, the slope of their respective asymptotes changes by ± 20 dB per decade beginning at the break frequency of each pole or zero.

Complex conjugate poles, or zeros, add two breaks of ± 20 dB per decade resulting in an asymptote with a slope of 40 dB per decade. The slope of the asymptotic function representing the magnitude at any point ω is determined solely by on the number of breakpoints at values less than ω. For Z zero breakpoints to the left of ω and P pole breakpoints, the slope at ω is given by $slope = 20(Z - P)$ dB per decade. Although this method does not provide the actual gain of the section, it does provide insight into the behavior of the filter as a function of frequency.

Without any loss of generality, attention is restricted in the following discussion to values of n such that $0 < n \le 2$, that is

$$H(s) = \frac{a_0 + a_1 s^1 + a_2 s^2}{b_0 + b_1 s^1 + b_2 s^2} \iff Generic\ Biquad \tag{8.35}$$

and the following case for $a_2 = a_1 = 0$:

$$H(s) = \frac{a_0}{b_0 + b_1 s^1 + b_2 s^2} \iff Lowpass \tag{8.36}$$

Substituting $j\omega = s$ yields, after some rearranging,

$$H(jw) = \frac{a_0}{(j\omega)^2 + (b_1/b_2)\omega + (b_0/b_2)} \tag{8.37}$$

and because a_0 is just a scaling factor, let $a_0 = \omega_0^2$ so that

$$H(jw) = \frac{1}{(j\omega/\omega_0)^2 + (b_1 j\omega/b_2\omega_0^2) + (b_0/b_2\omega_0^2)} \tag{8.38}$$

$a_2 = a_0 = 0$,

$$H(s) = \frac{a_1 s^1}{b_0 + b_1 s^1 + b_2 s^2} \iff Bandpass \tag{8.39}$$

Table 8.3: Filter order versus first- and second-order factors.

n	a_n
1	$(s+1)$
2	$(s^2 + 1.4142s + 1w)$
3	$(s+1)(s^2 + s + 1)$
4	$(s^2 + 0.7654s + 1)(s^2 + 1.8748s + 1)$
5	$(s+1)(s^2 + 0.6180s + 1)(s^2 + 0.6180s + 1)$
6	$(s^2 + 0.5176s + 1)(s^2 + 1.4142s + 1)(s^2 + 1.9391s + 1)$
7	$(s+1)(s^2 + 0.4450s + 1)(s^2 + 1.2470s + 1)(s^2 + 1.8019s + 1)$
8	$(s^2 + 0.3902s + 1)(s^2 + 1.1111s + 1)(s^2 + 1.6629s + 1)(s^2 + 1.9619s + 1)$

$a_1 = a_0 = 0$,

$$H(s) = \frac{a_2 s^2}{b_0 + b_1 s^1 + b_2 s^2} \Longleftrightarrow Highpass(Biquad) \tag{8.40}$$

$a_1 = 0$,

$$H(s) = \frac{a_0 + a_2 s^2}{b_0 + b_1 s^1 + b_2 s^2} \tag{8.41}$$

8.3   Analog Filters

In the following sections, several types of common analog filters are discussed. Those considered are only a small sample of known designs, however, they represent some of the more popular architectures.

8.3.1   Time-Continuous Integrators as Filters

Perhaps the simplest of all active filters is the OpAmp-based integrator[49]. Figure 8.9 shows one such integrator and its symbolic representation. Because the transfer function for an integrator can be expressed simply as

$$H(s) = -\frac{1}{sRC} \tag{8.42}$$

and therefore

$$H(j\omega) = -\frac{1}{(j\omega)RC} = \left| \frac{1}{\omega RC} \right| \angle -90° \tag{8.43}$$

so that the phase shift is 90 degrees and the gain decreases with increasing frequency. Note that at $\omega = 0$, that is for DC input, this filter exhibits a single pole because the output voltage is limited only by the power supply. Although not a particularly interesting filter, it can be used

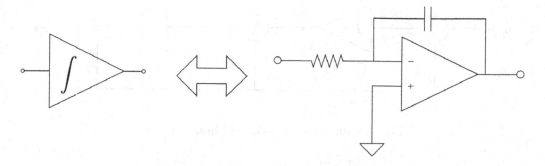

Figure 8.9: A simple integrator-based filter.

as the basis for creating much more sophisticated filters. Consider the transfer function for a second-order, lowpass filter, namely

$$H(s) = \frac{\omega_0^2}{s^2 + \frac{\omega_0}{Q} + \omega_0^2} \tag{8.44}$$

Because $H(s) = V_{out}/V_{in}$, Equation 8.44 can be expressed as,

$$\omega_0^2 V_{in} = s^2\, V_{out} + \frac{\omega_0}{Q} s\, V_{out} + \omega_0^2 V_{out} \tag{8.45}$$

and,

$$\frac{\omega_0^2}{s^2} V_{in} = V_{out} + \frac{\omega_0^2}{Qs} V_{out} + \frac{\omega_0^2}{s} V_{out} \tag{8.46}$$

so that

$$V_{out} = \omega_0^2 \left[\frac{1}{s^2}\right] V_{in} - \omega_0^2 \left[\frac{1}{s^2}\right] V_{out} - \frac{\omega_0^2}{Q}\left[\frac{1}{s}\right] V_{out} \tag{8.47}$$

$$= \frac{\omega_0^2}{s^2}[V_{in} - V_{out}] - \frac{\omega_0^2}{sQ} V_{out} \tag{8.48}$$

V_{in} must undergo successive integrations to produce V_{out} , as shown in Figure 8.10. It should be noted that by convention the summands are assumed to be noninverting and the integrators inverting. Because the integral of a sum is equal to the sum of the integrals, that is,

$$\int (f + g) = \int f + \int g \tag{8.49}$$

and

$$\left[\frac{1}{s}\right] \Rightarrow \int \quad \text{and} \quad \left[\frac{1}{s^2}\right] \Rightarrow \int\!\!\int$$

The filter can be based on the circuit shown in Figure 8.11 provided that values of R_1C_1 and R_2C_2 are determined by

$$\frac{\omega_0}{Q} = \frac{1}{R_2C_2} \quad \Rightarrow \quad R_2C_2 = \frac{Q}{\omega_0} \tag{8.50}$$

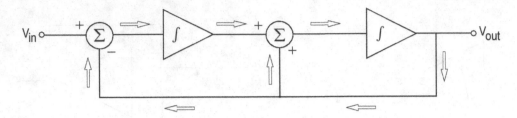

Figure 8.10: Integrator-Based Filter Network.

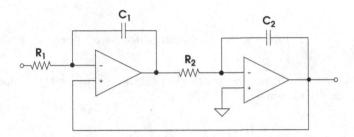

Figure 8.11: Integrator-based filter.

and,

$$\omega_0^2 = \frac{1}{R_1C_1R_2C_2} \quad \Rightarrow \quad R_1C_1 = \frac{1}{\omega_0^2 R_2 C2} = \frac{1}{\omega_0 Q} \tag{8.51}$$

8.3.2 The Passive Lowpass Filter

It is helpful to consider the classical lowpass filter. This simple RC network and its converse shown in Figures 8.12 (a) and (b), respectively, are used in many applications where relatively simple filtering is required.

Obviously the transfer function for Figure 8.12 a) is given by

$$H_{lp}(s) = \frac{1}{1 + sRC} \tag{8.52}$$

It proves useful to introduce an angular frequency ω_c defined as

$$\omega_c = \frac{1}{RC} \tag{8.53}$$

Thus Equation 8.52 becomes:

$$H_{lp}(s) = \left[1 + \frac{s}{\omega_c}\right]^{-1} = \left[1 + \frac{j\omega}{\omega_c}\right]^{-1} = \left[\frac{1}{1 + j\Omega}\right] \tag{8.54}$$

where $\Omega = \omega/\omega_c$, a normalized angular frequency, so that the magnitude of $H(s)_{lp}$ is given by

$$|H_{lp}(s)| = \frac{1}{\sqrt{1 + \Omega^2}} \tag{8.55}$$

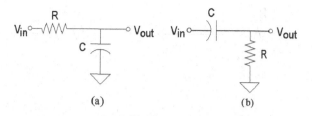

Figure 8.12: Simple Passive LP and HP Filters.

In order to understand in general terms how this simple filter will behave as a function of frequency, consider expressing the transfer function in terms of dB

$$G(\text{dB}) = 20 \ \log(|H(s)|) = 20 \ \log\left[\frac{1}{\sqrt{1+\Omega^2}}\right] = 20 \ \log(1) - 20 \ \log\left(1+\Omega^2\right)^{1/2} \tag{8.56}$$

Note that when $\omega = \omega_c$, $\Omega = 1$ and therefore Equation 8.56 becomes:

$$G(dB) = 20 \ \log(1) - 20 \ \log\left(1+1^2\right)^{1/2} = -20 \ \log(2) = 3 \ dB \tag{8.57}$$

As the frequency, f, of the input signal increases:

$$1 + \Omega^2 \approx \Omega^2 \tag{8.58}$$

and therefore,

$$G(w) \approx -20\log\Omega \tag{8.59}$$

Note that the transfer function for the RC filter shown in Figure 8.12 (b) is given by

$$H_{hp}(s) = \frac{sRC}{1+sRC} \tag{8.60}$$

which also has a single pole at $s = -1/RC$.

Therefore as in the previous case:

$$G(\text{dB}) = 20 \ \log(|H(s)|) = 20 \ \log\left[\frac{\Omega}{\sqrt{1+\Omega^2}}\right] = 20 \ \log(\Omega) - 20 \ \log\left(1+\Omega^2\right)^{1/2} \tag{8.61}$$

which for $\Omega \gg 1$ yields: $G(\text{dB}) \approx 0$. Note that for sufficiently large Ω the slope of the response curve becomes $-20 \, \text{dB}$ per decade, or equivalently $-6 \, \text{dB}$ per octave.

It is quite easy to provide isolation to this passive circuit to avoid any loading effects on the filter by the next stage by simply adding a voltage follower using an OpAmp [14] as shown in Figure 8.18. Although arguably now an active filter, in actuality the sole function of the OpAmp is to minimize loading. Figure 8.18 shows an example of a fourth-order filter in which each stage is buffered from the previous stage by a voltage follower. In later examples, the OpAmp will actually provide feedback to the filter and contribute some energy to minimize loss effects due to Ohmic dissipation.

As shown in Figure 8.13 it is a simple matter to add gain to these types of filters. The transfer function then becomes

$$H(s) = \frac{K}{(1+j\Omega)} \tag{8.62}$$

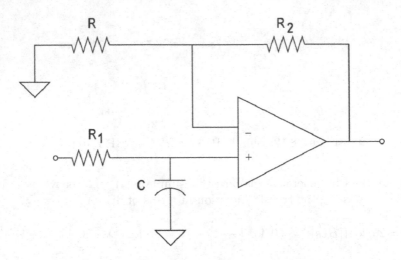

Figure 8.13: Simple lowpass filter with gain K.

Note that K–1 is used for the value of the feedback resistor for the noninverting OpAmp. The gain provided by the OpAmp is determined by

$$V_{out} = \left[\frac{R_1 + R_2}{R_1}\right] V_{in} \tag{8.63}$$

so that the gain contribution for the filter shown in Figure 8.13 is simply:

$$\frac{V_{out}}{V_{in}} = \left[\frac{1 + (K-1)}{1}\right] = K \tag{8.64}$$

If the relative positions of the capacitor C and resistor R are reversed as shown in Figure 8.12 then the passive network has a transfer function given by

$$H = \frac{K \ j\Omega}{(1 + j\Omega)} \tag{8.65}$$

and therefore,

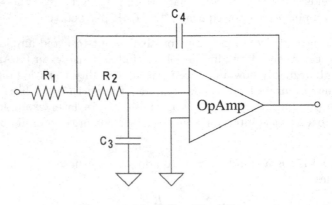

Figure 8.14: Sallen-Key Filter.

$$G(\text{dB}) = 20 \log \left[\frac{\Omega^2}{1 + \Omega^2} \right]^{1/2} = 20 \log[\Omega] - 20 \log[1 + \Omega^2] \tag{8.66}$$

This result illustrates an important feature or features sometimes referred to as "turning a filter inside out." This refers to the fact that replacing f with $1/f$ in a transfer function for a filter causes a lowpass filter to become a highpass filter and conversely a highpass filter to become a lowpass filter. This is equivalent to replacing the filter's capacitors with resistors, and resistors with capacitors. Note, however, that in both cases the OpAmp does not contribute any energy to the RC filter or feedback, merely circuit gain.

8.3.3   The Sallen–Key Lowpass Active Filter

Although there appear to be a virtually unlimited number of basic filter designs, one of the most common is the so-called Sallen-Key[26],[32], VCVS,[13] two pole filter which is shown in Figure 8.14. This filter has become popular for several reasons: (a) It is based on one OpAmp, two resistors, and two capacitors, and is therefore relatively easy to implement; (b) its gain characteristic is flat in the passband with "good" characteristics outside the passband; (c) multiple S–K filters can be cascaded to create arbitrarily complex, multipole filters; and (d) it is relatively inexpensive. Therefore this type of filter is suitable for a wide range of applications.

Such filters can be designed based on either continuous-time, or switched capacitance devices[25]. In the following, examples of each are presented together with illustrative examples based on PSoC technology. The gain equation for this particular configuration of the Sallen-Key filter is given by [26] as

$$\frac{V_{out}}{V_{in}} = \frac{K}{s^2(R_1 R_2 C_3 C_4) + s\left[R_1 C_3 + R_2 C_3 + R_1 C_4(1 - K) + 1\right]} \tag{8.67}$$

or,

$$\frac{V_{out}}{V_{in}} = \frac{\frac{K}{R_1 R_2 C_3 C_4}}{s^2 + s\left[\frac{1-K}{R_2 C_4} + \frac{R_1 + R_2}{R_1 R_2 C_4}\right] + \frac{1}{R_1 R_2 C_3 C_4}} \tag{8.68}$$

and defining:

$$\omega^2 = \frac{1}{R_1 R_2 C_3 C_4} = \omega_0 \omega_n \tag{8.69}$$

and

$$\frac{1}{d} = \frac{\sqrt{R_1 R_2 C_3 C_4}}{R_1 C_3 + R_2 C_3 + R_2 C_4(1 - K)} \tag{8.70}$$

which yields, after rearranging:

$$\frac{V_{out}}{V_{in}} = \frac{K}{s^2 + d\omega s + \omega^2} \tag{8.71}$$

---

[13]Voltage controlled voltage source.

Note that d, the damping factor given by Equation (8.70), is an explicit function of the gain K and frequency ω that are determined by the chosen values of resistance and capacitance. Obviously this degree of coupling is not very desirable in real-world applications. This dependance can be reduced somewhat by setting $R_1 = R_2 = R$ and $C_1 = C_2 = C$ so that:

$$\omega_0^2 = \left[\frac{1}{RC}\right]^2 \tag{8.72}$$

and,

$$Q = \frac{RC}{2RC + RC(1-K)} = \frac{1}{3-K} \tag{8.73}$$

Thus Q and frequency are no longer mutually dependent. Note also that $K > 3$ results in the filter becoming an oscillator.

As an illustrative example, consider the following

$$\frac{V_{out}}{V_{in}} = \frac{\frac{K}{(RC)^2}}{s^2 + s\left[\frac{1-K}{RC + \frac{2R}{R^2C}}\right] + \frac{1}{(RC)^2}} \tag{8.74}$$

and if

$$\omega_0^2 = \frac{1}{RC} \tag{8.75}$$

then Equation (8.74) becomes:

$$\frac{V_{out}}{V_{in}} = \frac{K\omega_0^2}{s^2 + s\left[\frac{1-K}{\omega_0 + \frac{2R}{R^2C}}\right] + \omega_0^2} \tag{8.76}$$

and

$$d = \frac{1-K}{RC + \frac{2R}{R_2C}} \tag{8.77}$$

Assuming that d, c_3, C_4, K, ω_0, ω_n are constants, Equations (8.69) and (8.70) can be solved for R_1 and R_2 [34]. Note that the values for R_1 and R_2 are, in the case of PSoC, switched capacitors whose equivalent resistance value is determined by the switching frequency f_s.

Thus, solving Equation (8.70) for R_2 and substituting it into Equation (8.69) yields, after rearranging,

$$R_1 = \frac{d + \sqrt{d^2 - 4([C_3/C_4] + 1 - K)}}{2\omega_n\omega_0[C_3 + (1-k)C_4]} \tag{8.78}$$

and similarly,

$$R_2 = \frac{1}{R_1C_3C_4\omega_0^2\omega_n^2} \tag{8.79}$$

However, $R_1 \in \mathbb{R}$ and therefore, $\sqrt{d^2 - 4\left(\frac{C_3}{C_4} + 1 - K\right)}$ must also be $\in \mathbb{R}$, and therefore it follows that,

$$d^2 > 4\left(\frac{C_3}{C_4} + 1 - K\right) \tag{8.80}$$

There are several important things to note about the results:

1. Damping is not a function of frequency.
2. If the gain $K = 1$, then $C_4 \geq 4C_3/d^2$
3. C_3 can be chosen arbitrarily to provide the required circuit impedance.

These filters are trivial to implement using PSoC's continuous time (CT) programmable gain amplifier with a PSoC buffer and external components for R_1, R_2, C_3, and C_4. However, caution should be used in such cases as rolloff frequencies should be restricted to one tenth of the gain bandwidth of the amplifier.

8.3.4 The Switched-Capacitance Filter

The basic RC-Biquad filter is shown in Figure 8.15 and its transfer function is given by

$$\frac{V_{out}}{V_{in}} = \frac{R_2}{R_1} \frac{1}{R_2 R_3 C_A C_B} \left[s^2 + \frac{C_4}{R_3 C_A C_B} s + \frac{1}{R_2 R_3 C_A C_B} \right]^{-1} \tag{8.81}$$

Because the middle OpAmp is functioning solely as a unity gain inverter stage, it has been replaced in the PSoC implementation of this filter. The PSoC equivalent circuit is obtained by replacing all of the resistors used in this figure with their switched capacitor equivalent [43]. The switched capacitance version of this same circuit is shown in Figure 8.16. Its transfer function is given by

$$\frac{V_{out}}{V_{in}} = \frac{-\dfrac{C_1 C_B}{C_2 C_3}\left[\dfrac{1-\frac{s}{2f_s}}{\frac{C_A C_B}{C_2 C_3}-\frac{1}{4}-\frac{1}{2}\frac{C_4}{C_2}}\right] f_s^{\,2}}{s^2 + \dfrac{C_4}{C_2}\left[\dfrac{s f_s}{\frac{C_A C_B}{C_2 C_3}-\frac{1}{4}-\frac{1}{2}\frac{C_4}{C_2}}\right] + \left[\dfrac{f_s^2}{\frac{C_A C_B}{C_2 C_3}-\frac{1}{4}-\frac{1}{2}\frac{C_4}{C_2}}\right]} \tag{8.82}$$

Note that

$$Q = \frac{C_2}{C_4}\left[\frac{C_A C_B}{C_2 C_3} - \frac{1}{4} - \frac{1}{2}\frac{C_4}{C_2}\right]^{-1/2} \tag{8.83}$$

$$G = -\frac{C_1}{C_2}\frac{C_B}{C_3} \tag{8.84}$$

and

$$f_c = \frac{1}{2\pi}\frac{f_s}{\left[\frac{C_A C_B}{C_2 C_3} - \frac{1}{4} - \frac{1}{2}\frac{C_4}{C_2}\right]^{1/2}} \tag{8.85}$$

8.3.5 Biquad Switched Capacitor Filter

It is quite easy to construct a biquad switched capacitor filter [43] by employing two of PSoC's programmable switched capacitor blocks which results in a filter that consists of two OpAmps, eight capacitors, and has a programmable sample frequency. Referring to Figure 8.16, and assuming both switched capacitor blocks are clocked at the same rate

$$V_{out1} = V_{out1}z^{-1} - V_{in}\frac{C_1}{C_A} - V_{out2}\frac{C_2}{C_A} - \left[V_{out2} - V_{out2}z^{-1}\right]\frac{C_4}{C_A} - \left[V_{in} - V_{in}z^{-1}\right]\frac{C_p}{C_A} \quad (8.86)$$

$$V_{out2} = V_{out2}z^{-1} - V_{out1}z^{-1}\frac{C_3}{C_B} - \left[V_{in} - V_{in}z^{-1}\right]\frac{C_{pp}}{C_B} \quad (8.87)$$

Therefore,

$$\left[\frac{V_{out1}}{V_{in}}\right] = \left[\frac{V_{out1}}{V_{in}}\right]z^{-1} - \frac{C_1}{C_A} - \frac{C_2}{C_A}\left[\frac{V_{out2}}{V_{in}}\right] - \left[1 - z^{-1}\right]\frac{C_4}{C_A}\left[\frac{V_{out2}}{V_{in}}\right] - \frac{C_p}{C_A}\left[1 - z^{-1}\right] \quad (8.88)$$

$$\left[\frac{V_{out2}}{V_{in}}\right] = \left[\frac{V_{out2}}{V_{in}}\right]z^{-1} - \frac{C_3}{C_B}\left[\frac{V_{out1}}{V_{in}}\right]z^{-1} - \frac{C_{pp}}{C_B}\left[1 - z^{-1}\right] \quad (8.89)$$

and,

$$\left[\frac{V_{out1}}{V_{in}}\right][1 - z^{-1}] = -\frac{C_1}{C_A} - \frac{C_2}{C_A}\left[\frac{V_{out2}}{V_{in}}\right] - \left[1 - z^{-1}\right]\frac{C_4}{C_A}\left[\frac{V_{out2}}{V_{in}}\right] - \frac{C_p}{C_A}\left[1 - z^{-1}\right] \quad (8.90)$$

$$\left[\frac{V_{out2}}{V_{in}}\right][1 - z^{-1}] = -\frac{C_3}{C_B}\left[\frac{V_{out1}}{V_{in}}\right]z^{-1} - \frac{C_{pp}}{C_B}\left[1 - z^{-1}\right] \quad (8.91)$$

so that,

$$\left[\frac{V_{out1}}{V_{in}}\right] = -\frac{C_1}{C_A}\frac{1}{[1 - z^{-1}]} - \frac{C_2}{C_A}\left[\frac{V_{out2}}{V_{in}}\right]\frac{1}{[1 - z^{-1}]} - \frac{C_4}{C_A}\left[\frac{V_{out2}}{V_{in}}\right] - \frac{C_p}{C_A} \quad (8.92)$$

$$\left[\frac{V_{out2}}{V_{in}}\right] = -\frac{C_3}{C_B}\left[\frac{V_{out1}}{V_{in}}\right]\frac{z^{-1}}{[1 - z^{-1}]} - \frac{C_{pp}}{C_B} \quad (8.93)$$

which can be treated as a system consisting of two equations in the two unknowns, $\left[V_{out1}/V_{in}\right]$ and $\left[V_{out2}/V_{in}\right]$, and therefore leads to:

$$\left[\frac{V_{out1}}{V_{in}}\right] = \frac{-C_BC_pz^2 + 2C_BC_pz - C_BC_1z^2 - C_BC_p + C_BC_1z}{C_BC_Az^2 - 2C_BC_Az + C_2C_3z + C_4C_3z - C_4C_3 + C_BC_A} \quad (8.94)$$
$$+ \frac{C_{pp}C_2z^2 + C_{pp}C_4z^2 - 2C_{pp}C_4z - C_{pp}C_2z + C_{pp}C_4}{C_BC_Az^2 - 2C_BC_Az + C_2C_3z + C_4C_3z - C_4C_3 + C_BC_A}$$

and,

$$\left[\frac{V_{out2}}{V_{in}}\right] = \frac{C_{pp}C_Az^2 - 2C_{pp}C_Az + C_{pp}C_A + C_pC_3z - C_pC_3 + C_1C_3z}{C_BC_Az^2 - 2C_BC_Az + C_2C_3z + C_4C_3z - C_4C_3 + C_BC_A} \quad (8.95)$$

The bilinear transform that maps the imaginary axis $i\omega$ into the unit circle $|z| = 1$ (to retain the frequency characteristics), and the left half of the s-plane onto the interior of the unit circle in the z-plane (to retain stability), is given by

$$z = \frac{1 + \left(\frac{T}{2}\right)s}{1 - \left(\frac{T}{2}\right)s} = \frac{1 + \left(\frac{1}{2f_s}\right)s}{1 - \left(\frac{1}{2f_s}\right)s} \tag{8.96}$$

where f_s is the sampling frequency.

Substituting Equation (8.96) in Equations (8.95) and (8.95) yields:

$$H(s)_{out1} = \frac{-\frac{C_{pp}}{C_3}\left[1 + \left(\frac{s}{f_s}\right)\left(1 + \frac{C_4}{C_2}\right)\right] + \frac{C_1 C_B}{C_2 C_3}\left(\frac{s}{f_s}\right)^2 + \frac{C_1 C_B}{C_2 C_3}\left(\frac{s}{f_s}\right)\left[1 + \frac{1}{2}\left(\frac{s}{f_s}\right)\right]}{\left[\frac{C_B C_A}{C_2 C_3} - \frac{C_4}{2C_2} - \frac{1}{4}\right]\left(\frac{s}{f_s}\right)^2 + \frac{C_4}{C_2}\left(\frac{s}{f_s}\right) + 1} \tag{8.97}$$

$$H(s)_{out2} = \frac{-\frac{C_{pp} C_A}{C_2 C_3}\left(\frac{s}{f_s}\right)^2 + \frac{C_p}{C_2}\left(1 - \frac{s}{f_s}\right)\left(\frac{s}{f_s}\right) + \frac{C_1}{C_2}\left[1 - \frac{1}{4}\left(\frac{s}{f_s}\right)^2\right]}{\left[\frac{C_B C_A}{C_2 C_3} - \frac{C_4}{2C_2} - \frac{1}{4}\right]\left(\frac{s}{f_s}\right)^2 + \frac{C_4}{C_2}\left(\frac{s}{f_s}\right) + 1} \tag{8.98}$$

The second-order transfer function for a filter can be expressed in a generalized form as

$$H(s) = \frac{h_{hp}\left(\frac{s}{2\pi f_0}\right)^2 + h_{bp}\left(\frac{s}{2\pi f_0}\right) + h_{lp}}{\left(\frac{s}{2\pi f_0}\right)^2 + d\left(\frac{s}{2\pi f_0}\right) + 1} \tag{8.99}$$

where f_0 is defined as the rolloff frequency and d is the damping factor with typical values in the range $0 \le d \le 2$. The parameters h_{hp}, h_{bp} and h_{lp} are the highpass, bandpass, and lowpass coefficients, respectively. Note that the transfer function for a "section" is uniquely determined by five parameters, namely, h_{hp}, h_{bp}, h_{lp}, d (or alternatively Q) and f_0.

Comparing Equation 8.99 with Equations 8.97 and 8.98, shows that

$$h_{hp1} = -C_p C_B\left[C_A C_B - \frac{1}{2}\frac{C_3}{C_4} - \frac{1}{4}C_2 C_3\right]^{-1} \tag{8.100}$$

$$\frac{h_{bp1}}{d} \approx -\frac{C_1 C_B}{C_4 C_3} \tag{8.101}$$

$$h_{hp2} = -C_{pp} C_A\left[C_B C_A - \frac{1}{2}C_3 C_4 - \frac{1}{4}C_2 C_3\right]^{-1} \tag{8.102}$$

$$\frac{h_{bp2}}{d} \approx -\frac{C_p}{C_4} \tag{8.103}$$

$$h_{lp2} \approx -\frac{C_1}{C_2} \tag{8.104}$$

In order for the denominators of Equations (8.99), (8.97), and (8.98) to be the equal the following equality must be true

$$\left(\frac{s}{2\pi f_0}\right)^2 + d\left(\frac{s}{2\pi f_0}\right) + 1 = \left[\frac{C_B C_A}{C_2 C_3} - \frac{C_4}{2C_2} - \frac{1}{4}\right]\left(\frac{s}{f_s}\right)^2 + \frac{C_4}{C_2}\left(\frac{s}{f_s}\right) + 1 \tag{8.105}$$

where f_0 is defined as

$$f_0 = \frac{f_s}{2\pi} \frac{\sqrt{C_2 C_3}}{\sqrt{C_a C_B - \frac{1}{2}\frac{C_4}{C_3} - \frac{1}{4}\frac{C_2}{C_3}}} = 2\pi \sqrt{\frac{C_A C_B}{C_2 C_3} - \frac{C_4}{2C_2} - \frac{1}{4}} \tag{8.106}$$

and d is defined as

$$d = \frac{C_4}{\sqrt{C_A C_B - \frac{1}{2}\frac{C_4}{C_2} - \frac{C_2 C_3}{4}}} \sqrt{\frac{C_3}{C_2}} = \frac{C_4}{C_2} \frac{1}{\sqrt{\frac{C_A C_B}{C_2 C_3} - \frac{C_4}{2C_2} - \frac{1}{4}}} \tag{8.107}$$

Note that typical values for d fall in the range $0 \leq d \leq 2$. depending on whether the response is oscillatory, underdamped, critically damped, or overdamped. Obviously oscillating filters are to be avoided.

The oversample ratio *OSR* is given by

$$OSR = \frac{f_0}{f_s} = \frac{1}{2\pi} \frac{\sqrt{C_2 C_3}}{\sqrt{C_A C_B - \frac{1}{2}\frac{C_4}{C_3} - \frac{1}{4}\frac{C_2}{C_3}}} \tag{8.108}$$

Note that the rollover frequency is directly proportional to the sampling frequency, a feature of switched capacitor filters, and the oversample ratio and damping factor are not a function of the parasitic capacitance terms (i.e. C_{pp} and C_p). Once the rolloff frequency, damping value, and pass coefficients for a particular filter section have been selected, Equations (8.106) and (8.107) can be used to determine the values of $C_A, C_B, C_2, C_3,$ and C_4. Equations (8.101)–(8.104) can then be used to determine C_1.

Values for the parasitic capacitances C_p, C_{pp}, if not explicitly known, are often assumed to be zero. Given that three parameters fully characterize the filter's behavior but two equations are used to determine the values for C_A, C_B, C_2, C_3, and C_4 the system is not uniquely determinant. This can be viewed as both an advantage and a disadvantage in that an iterative process may be required to determine these values and an advantage in that it provides the designer some latitude in choosing component values. Fortunately, PSoC is provided with software wizards for a lowpass and bandpass section design that programmatically select values for C_A, C_B, C_2, C_3, and C_4.

Traditionally, filter design has been an iterative process often involving tradeoffs. In what follows it is shown that much of the work has now been automated, allowing a designer to quickly converge on an acceptable design. The PSoC architecture provides both bandpass and lowpass modules together with tools that greatly facilitate the design process. However, a highpass module is not provided because such a filter would of necessity require a very high OSR which would in turn require substantially more silicon real estate at the chip level and substantially increase costs.

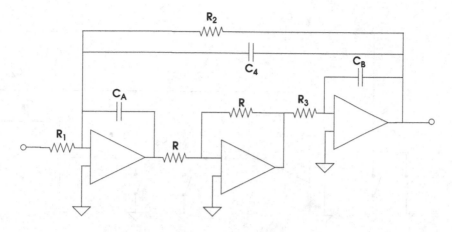

Figure 8.15: Biquad LP Filter.

PSoC's spreadsheet-based design tools for bandpass and lowpass filters allow the designer to select a filter type (e.g., Butterworth, .1 dB Chebyshev, 1 dB Chebyshev, Bessel or Gaussian), the desired center frequency, bandwidth, and gain. Once initial values have been chosen for C_2 and C_4, values are automatically calculated for C_A, C_B C_3 and C_1. It should be noted that available capacitance values are "quantized", for example C_1, C_2, C_3,C_4, can only have integer values in the range from 0 to 32 and C_A and C_B can have the values 16 or 32.

8.3.6 An Allpass Filter

One of the more curious forms of filters is the allpass filter. One might well ask why have a filter that passes all frequencies? The answer is simply that there are situations for which it is useful to be able to pass all frequencies, but with a constant phase shift for the frequencies of interest. The phrase "allpass filter" is, however, at the very least ambiguous and at most misleading. Such filters are better described as delay equalizers, phase-shift filters, or time-delay filters. One such filter is shown in Figure 8.23. Assuming that $R_1 = R_3$ and that $R_2 = R_1/2$, the transfer function for the allpass filter shown in Figure 8.23 can be expressed as

$$H(s) = \left[1 + \frac{R_F}{R_A}\right]\left[\frac{1}{R_4 + R_6}\right]\left[\frac{sR_4RC + R_4 - R_6}{1 + sRC}\right] \tag{8.109}$$

Setting $R_4 = R_6/2$ yields

$$H(s) = 0.33\left[1 + \frac{R_F}{R_A}\right]\left[\frac{sRC - 1}{1 + sRC}\right] \tag{8.110}$$

and if $R_f = R_A$, then

$$H(s) = -\left[\frac{1 - sRC}{1 + sRC}\right] \tag{8.111}$$

and if $\omega_c = 1/RC$, the transfer function in the frequency domain becomes

$$H(j\omega) = \left[\frac{1 - j\left(\frac{\omega}{\omega_c}\right)}{1 + j\left(\frac{\omega}{\omega_c}\right)}\right] \tag{8.112}$$

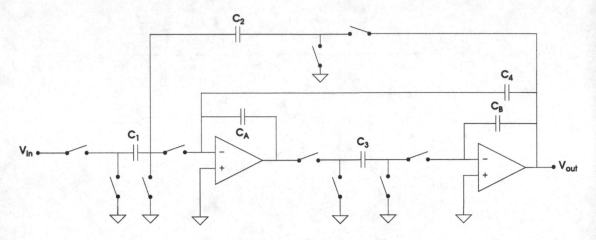

Figure 8.16: Switched Capacitance Lowpass Filter.

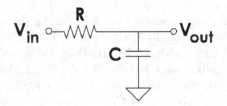

Figure 8.17: First-order passive lowpass, RC filter.

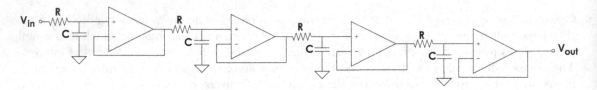

Figure 8.18: Fourth-order lowpass RC filter.

and therefore, the phase if given by

$$\theta = 180° - 2\tan^{-1}\left(\frac{\omega}{\omega_c}\right) \tag{8.113}$$

and is shown graphically in Figure 8.25.

A much simpler allpass filter is shown in Figure 8.24. In this case the transfer function is given by

$$H(s) = -\frac{[s - [1/RC)]}{[s + (1/RC)]} \tag{8.114}$$

and the phase as a function of frequency can be shown to vary from 0° to 180°. Reversing R and C changes the transfer function to

$$H(s) = +\frac{[s - (1/RC)]}{[s + [1/RC)]} \tag{8.115}$$

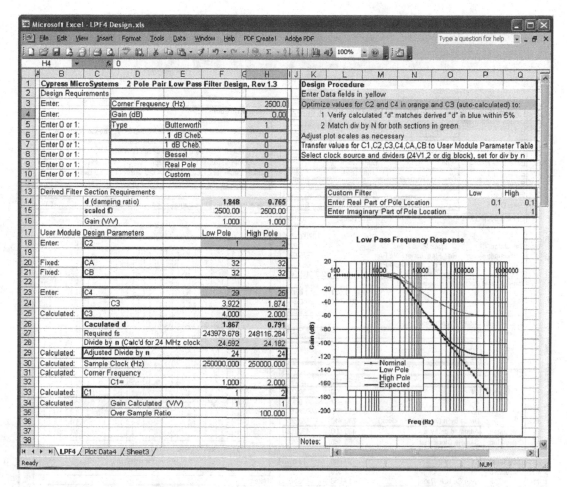

Figure 8.19: PSoC's two-pole lowpass design tool (Excel-based) [50].

and the phase in this case is given by

$$\theta = tan^{-1}\left(\frac{[2\omega/RC]}{[\omega^2 - (1/RC)]}\right) \tag{8.116}$$

which results in a circuit capable of varying phase from -180° to 0°.

8.4 Digital Filters

In typical embedded system applications it is often necessary to filter data that are in digital form. Digital filters[16][41] have becoming increasingly important in critical applications because among other things they are not affected by the temperature, passive component drift, OpAmp nonidealities, and so on, that are characteristic of active filters. The fact that the filtering is implemented in the digital domain means that digital filters can be employed as "adaptive filters" for speech recognition, noise cancelation and echo suppression.

Digital and analog filters are dramatically different in terms of implementation, but the design methodology for each is rather similar, in that the filter's desired characteristics must be specified

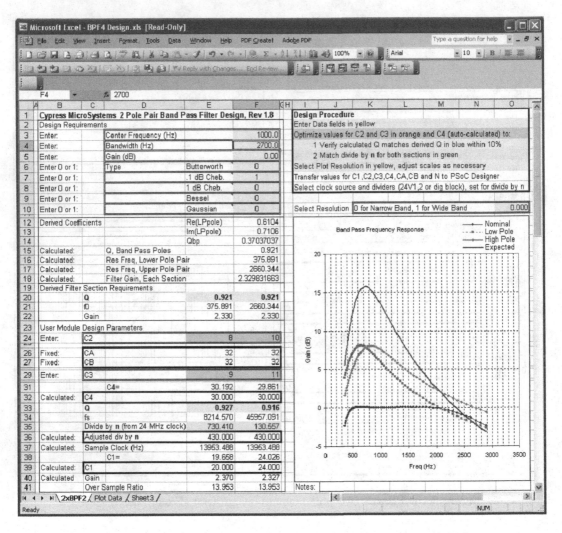

Figure 8.20: PSoC bandpass filter design tool (Excel-based) [50].

and the values of the associated parameters be determined. In the analog domain this requires determining the appropriate OpAmp characteristics, and resistance/capacitance values. In the digital domain, one must determine the necessary coefficient values and deal with the vagaries of developing the associated software algorithm(s).

Although implementing filters in the digital domain may seem, superficially at least, to be perhaps more aesthetically pleasing, it has some potentially serious drawbacks. Not the least of these is the fact that for real-time operation, the digital filter must be capable of completing execution of the necessary algorithm(s) between sample periods. Buffering of input data can be employed in some systems to provide sufficient time to complete the necessary computations, but as a practical matter this is not an acceptable alternative in the majority of real-time applications.

Given that the real world is largely analog, tradeoffs must be considered when designing an embedded system that take into account cost, performance, and stability that often lead to an embedded system implementation consisting of both analog and digital techniques. A comparison of analog and digital filters is provided in Table 8.5.

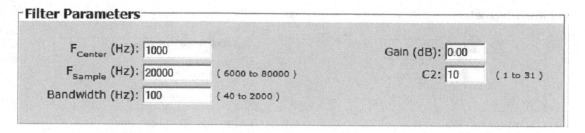

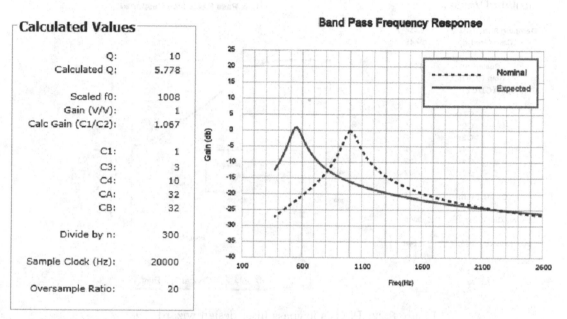

Figure 8.21: PSoC's filter design wizard [50].

In addition, digital filters can be based on convolving the input signal with the filter's impulse response which is perhaps the most general method of filtering, or by employing recursion. In the first case samples of the input signal are "weighted" and the resulting samples are summed. In the second case, previous output values of the filter input are fed back. These two digital filters are referred to as finite impulse response (FIR) and infinite impulse response (IIR), respectively. FIR filters do not strictly speaking have an analog, however, IIR filters can be considered to have as their analog counterparts, Bessel, Butterworth, Chebyshev, and Elliptic filters.

The IIR filter is a recursive filter in that it employs feedback and therefore an impulse applied to an IIR filter is, in theory at least, present for an infinite time, hence the name.

8.4.1 Digital FIR Filter

The defining form for a FIR filter, assuming LTI applies, is given by:

$$y[n] = b_0 x[n] + b_1 x[n-1] + \cdots + b_{M-1} x[n-N+1] \tag{8.117}$$

where the b_i are "weight" coefficients and N is the number of zeros.

Alternatively, Equation 8.117 can be expressed as:

$$y[n] = \frac{1}{N} \sum_{m=0}^{N-1} x[n-m] \tag{8.118}$$

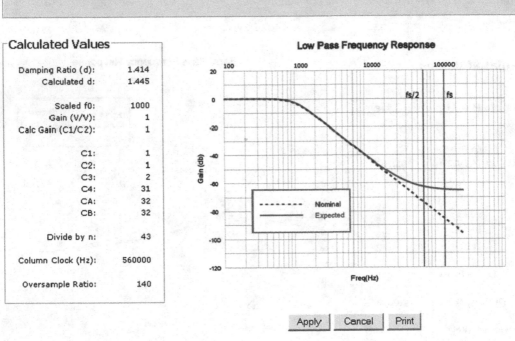

Figure 8.22: PSoC's lowpass filter design wizard [50].

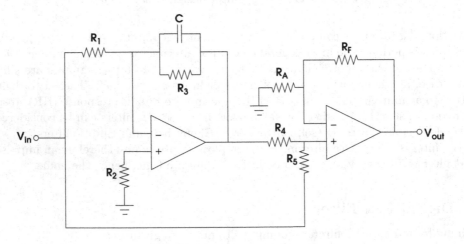

Figure 8.23: Allpass Filter.

$$H(e^{j\omega}) = \frac{1}{N}\sum_{m=0}^{N-1} e^{-jm\omega} = \frac{1}{N}\frac{[1 - e^{-jmN}]}{[1 - e^{j\omega}]} = \frac{1}{N}\left[e^{-j(N-1)\omega/2}\right]\frac{[e^{jm\omega/2} - e^{-jm\omega/2}]}{[e^{j\omega/2} - e^{-j\omega/2}]} \quad (8.119)$$

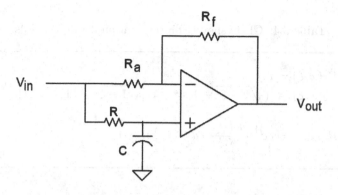

Figure 8.24: A very simple allpass filter [51].

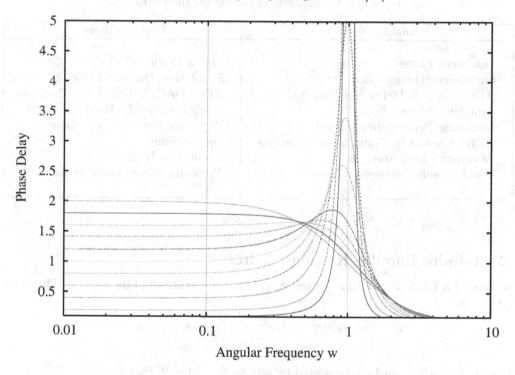

Figure 8.25: Linear phase second-order lowpass filter.

and because,

$$\sin(\theta) = \frac{e^{j\theta} - e^{-j\theta}}{2j} \tag{8.120}$$

$$H(e^{j\omega}) = \frac{1}{N}\left[e^{-j(1/2)(N-1)\omega}\right]\left[\frac{\sin\left(\omega\frac{N}{2}\right)}{\sin\left(\frac{\omega}{2}\right)}\right] \tag{8.121}$$

Note the transfer functions $\frac{1}{N}$ dependence. As shown in Figure 8.28, recursion can be used to perform the functions of both lowpass and highpass filtering.

Table 8.4: Chebyshev (Elliptic) Rational Polynomials.

$$R_1(x, \xi) = x$$
$$R_2(x, \xi) = \left[(\sqrt{1 - \xi^2} - 1)x^2 + 1 \right] \left[\sqrt{1 - \xi^2} + 1)x^2 - 1 \right]^{-1}$$
$$\cdots$$
$$R_n(x, \xi) = cd \left[n \frac{K\left(\frac{1}{L_n}\right)}{K(\frac{1}{\xi})} cd^{-1}(x, \frac{1}{\xi}), \frac{1}{L_n} \right]$$

Table 8.5: Comparison of analog vs. digital filters.

Analog Filters	Digital Filters
Nonlinear Phase	Linear Phase (FIR)
Simulation/Design - Easy	Simulation/Design - Difficult
ADC, DAC, & DSP - Not Required	ADC, DAC, & DSP - Not Required
Adaptive Filters - Easy	Adaptive Filters - Hard
Real-time Performance - Good	Real-time Performance - Depends
Drift - Caused by Component Variations	Drift - None
Accuracy - Less Accurate	Accuracy - High
Good Dynamic Range	Typically Not as Good as Analog

8.4.2   Infinite Impulse Response Filter

In the case of a LTIS (linear time invariant system) an infinite impulse response filter is represented [38] by

$$
\begin{aligned}
y[n] \quad &= \quad b_0 x[n] \ + \ b_1 x[n-1] \ + \ldots + \ b_m x[n - \mathrm{M}] \\
&\quad - \ a_1 y[n-1] \ - \ \cdots - \ a_N y[n - N]
\end{aligned}
\tag{8.122}
$$

A biquad filter[38] can be represented by setting $N = M = 2$, that is,

$$
y[n] = b_0 x[n] \ + \ b_1 x[n-1] \ + \ b_2[n-2] \ - \ a_1 y[n-1] - a_2 y[n-2]
\tag{8.123}
$$

and therefore by applying the z-transform to Equation (8.123), the corresponding transfer function is given by

$$
H(z) = \frac{b_0 + b_1 z^{-1} + b_2 z^{-2}}{1 + a_1 + a_2 z^{-2}}
\tag{8.124}
$$

This represents an infinite impulse response filter because the output is summed with the input. The topology for this transfer function is represented in Figure 8.27.

In implementing this filter, using PSoC, several factors must be taken in account. PSoC has an integral MAC as part of its architecture, therefore

$$
V_{hp} = V_{in} - V_{lp}
\tag{8.125}
$$

Table 8.6: Comparison of filter types.

Filter Type	Passband	Transition Region	Stopband
Bessel(B)	Flat	$B < W < C < I < E$	No Ringing
Butterworth(W)	Flat	$B < W < C < I < E$	No Ringing
Chebyshev (C)	Ripple	$B < W < C < I < E$	No Ringing
Inverse Chebyshev (I)	Flat	$B < W < C < I < E$	No Ringing
Elliptic (E)	Ripple	$B < W < C < I < E$	Ringing

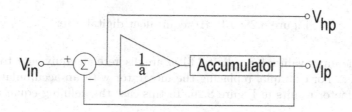

Figure 8.26: Sampled single pole passive filter.

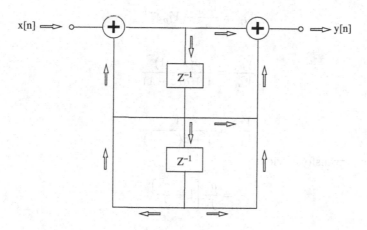

Figure 8.27: Second-order IIR filter topology.

$$V_{lp} = \frac{V_{hp}}{sRC} \tag{8.126}$$

so that,

$$\frac{V_{hp}}{V_{in}} = \frac{sRC}{1 + sRC} \tag{8.127}$$

and

$$\frac{V_{lp}}{V_{in}} = \frac{1}{1 + sRC} \tag{8.128}$$

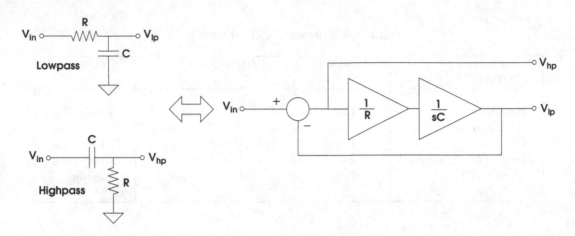

Figure 8.28: LP, HP equivalent digital filter.

But because the input values for a digital filter are discrete an equivalent topology is required for the digital filter. For example, replacing the integrator with an accumulator, and RC by an appropriate scale factor results in Figure 8.26. In this case the defining equations are given by

$$V_{hp} = V_{in} - V_{lp}z^{-1} \tag{8.129}$$

and

$$V_{lp} = \frac{V_{hp}}{a} - V_{lp}z^{-1} \tag{8.130}$$

Thus

$$\frac{V_{hp}}{V_{in}} = \frac{a(z-1)}{1 + a(z-1)} \tag{8.131}$$

and

$$\frac{V_{lp}}{V_{in}} = \frac{1}{1 + a(z-1)} \tag{8.132}$$

And applying the z transform yields

$$\frac{V_{hp}}{V_{in}} = \frac{as}{f_s}\left[1 + \frac{as}{f_s}\right]^{-1} \tag{8.133}$$

$$\frac{V_{lp}}{V_{in}} = \left[1 + \frac{as}{f_s}\right]^{-1} \tag{8.134}$$

where f_s is the sampling frequency and the rolloff frequency is:

$$f_0 = \frac{f_s}{2\pi a} \tag{8.135}$$

In this case the rolloff frequency is a function of both the sample frequency and the attenuation factor a. Alternatively the attenuation factor can be determined from the rolloff frequency. One way of simplifying the design is to restrict the values of a to integer powers of two, that is, $a = 2^n$ where n is an integer. Imposing this restriction yields the immediate benefit that division of the binary value of a reduces to a series of n right shifts. For example, if $a = 256$, then dividing the numerator merely requires a one-byte shift. In coding such a filter algorithm it will of course be necessary to initialize the filter to establish the starting value of V_{lp}.

8.5 Filter Design Software Tools

The personal computer [11] has done much to ease some of the burden of filter design by serving as an environment in which a variety of filter design applications can be employed. Everything from simple Java programs that graphically display the relationships between pole/zero placement in the complex plane to the resulting Bode plots, to sophisticated applications that allow the designer to specify a wide range of desired design parameters for an nth-order filter consisting of an arbitrary number of stages (sections). This is indeed a fortunate state of affairs in that much of filter design involves iteration and tradeoffs which the modern PC and available software are well equipped to handle.

Cypress provides several tools as part of its PSoC Designer development environment that greatly facilitate the design of lowpass and bandpass filters using integrated filter modules. Figures 8.19–8.21 show three such tools, the first two of which are Microsoft Excel-based and the third a windows-based application. The first two allow the designer to specify virtually any of the parameters involved and then provide a graphical representation of the resulting filter characteristics. The third allows the designer to specify the center frequency, sample frequency bandwidth and gain and provides calculated gain, Q, oversample ratio, and values for the associated capacitance values. The application shown in Figure 8.21 automatically populates the calculated values into the appropriate fields in PSoC Designer. However, it does not allow the designer the complete freedom to specify arbitrary values for the other parameters such as C_1 and C_3 among others. If this is required, the Excel-based tools must be employed.

Other tools include MatLab,[14] Mathcad,[15] Electronic Workbench,[16] and a number of Internet-based design tools are now available to aid the designer. However, each of these tools assumes some a priori knowledge of basic analog and/or digital filter concepts[28]. It has not been possible in the few pages of this chapter to provide much more than an overview of the subject of filters. Nonetheless, the authors have attempted to provide the necessary background that will allow the reader to profitably use any one of the tools discussed herein.

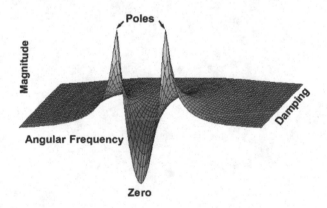

Figure 8.29: Biquadratic response as a function of gain, ω, and d.

---

[14] The MathWorks
[15] MathSoft
[16] National Instruments

8.6   Conclusion

The subject of filter design is far too complicated to be covered in a single chapter. However, the basic concepts have been presented in at least a cursory manner and the reader can rely on a number of PC-based tools to facilitate the design of filters for embedded applications. SoC technologies such as PSoC provide software support in the form of wizards and hardware modules for both bandpass and lowpass filters (e.g., Bessel, Butterworth, Chebychev, and custom filters with either real or complex poles). The designer has merely to specify a minimal set of parameters (e.g., f_c, f_{sample}, bandwidth, gain, etc.) to complete the filter's design. The resulting PSoC filter parameters are automatically entered into PSoC Designer. PSoC Designer also provides the designer with several alternatives in terms of filter module placement in order to obtain optimum utilization of PSoC's resources.

PSoC's bandpass modules allow the designer to set the midband gain, Q and center frequency programmatically. The filter's center frequency stability is determined by the clock accuracy and the sampling rates employed can be as high as 1 MHz. These modules also have an integral comparator referenced to ground and can be used as band-limited zero-crossing detectors. The center frequency is determined by PSoC's OpAmp characteristics, capacitor values, and clock frequency.

8.7   Recommended Exercises

1. Design a second-order, Sallen-Key lowpass analog filter, using $C = 0.33\,\mu f$ and $R_f = 47K$ ohms, that has a 3dB cutoff frequency of 750 Hz. Sketch the Bode plots for gain and phase as a function of angular frequency.

2. Q has been defined in this chapter as $Q = f_c/\Delta f$.

   a. Show that for a filter consisting of an inductor, capacitor and resistor connected in parallel, Q could also be defined by any of the following:

$$Q = \frac{\omega}{\Delta\omega_0} = \frac{[energy\ stored]}{[energy\ dissipated]} = \frac{R}{\omega_0 L} = \omega_0 RC \approx number\ of\ ringing\ cycles$$

   b. Show that at resonance $i_{in} = Q * i_R$

3. Derive the transfer function for the allpass filter given by Equation 8.109 and then design an allpass filter that has a phase shift of 53.13° at $\omega = 20,000$ radians/sec. Use commonly available values for resistance and capacitance to the extent practicable.

4. Using the circuit shown in Figure 8.9, design a integrator-based filter with $Q = .707$ and $\omega_0 = 31,415$ radians per second. Sketch the corresponding Bode graphs.

5. Beginning with the transfer function:

$$\frac{V_{out}}{V_{in}} = \frac{Ks^2}{s^2 + s(\frac{\omega_0}{Q}) + \omega^2}$$

Derive the equivalent OpAmp circuit and sketch the resulting Bode graphs for magnitude and phase. Show each step in your development.

6. Beginning with Equations (8.91) and (8.92), derive Equation (8.93).

7. A second-order filter (section) has complex poles at $s = -3 \pm j2$ and a single zero at the origin of the s-plane. Find this section's transfer function and evaluate it for $s = 0 + j6$. Sketch the Bode plot for this transfer function for $0.01 \leq \omega \leq 100$ radians per second.

8. The impulse responses for the ideal lowpass and highpass filters discussed in this chapter are given by

$$h_{lp}(t) = \left[\frac{\omega_{cA}}{\pi}\right] \text{sinc}\left(\frac{\omega_c t}{\pi}\right)$$

and

$$h_{hp(t)} = A - \left[\frac{\omega_{cA}}{\pi}\right] \text{sinc}\left(\frac{\omega_c t}{\pi}\right)$$

respectively. Sketch these functions. Are these filters causal or noncausal? Derive the impulse functions for the ideal bandpass and bandstop filters. Are either the bandpass or bandstop causal filters?

9. The normalized pole positions for a Butterworth filter are given by:

$$p_m = -\sin\left[\frac{(2m-1)\pi}{2n}\right] + j\cos\left[\frac{(2m-1)\pi}{2n}\right]$$

Find the poles for a seventh-order Butterworth filter.

10. Given that $d = \sqrt{2}$, $K = 1.0$, $C_3 = 0.0022\,\mu F$, $\omega_0 = 1.0$, and $\omega_n = 2\pi(1000)$, calculate the values for R_1 and R_2 for a lowpass Butterworth (Sallen–Key), two-pole filter. Sketch the Bode plots for magnitude and phase as functions of angular frequency.

11. Consider the multistage filter shown in Figure 8.18 and assume that it has n stages (sections) instead of four as shown. Assuming that all of the stages have the same cutoff frequency f_c, show (a) that the coefficients for this filter's transfer function are given by

$$a_1 = a_2 = \cdots = \alpha = \sqrt{2^{1/n} - 1} \tag{8.136}$$

and (b), that f_c of each stage is $1/\alpha$ higher than f_c for the n stage filter.[17]

---

[17]From an example suggested by Thomas Kugelstadt. "Active Filter Design Techniques," Texas Instruments SLOD006A (Op Amps For Everyone, Chapter 16).

Bibliography

[1] P. L. D. Abrie, *The Design of impedance Matching Networks*, Norwood, MA: Artech House, 1985.

[2] A. S. Sedra and P. Bracket: *Filter Theory and Design:Active Passive*, New York and London: Pitman, 1979.

[3] A. Antoniou. *Digital Filters: Analysis, Design and Applications*, New York: McGraw-Hill, 1993.

[4] H. Berlin. *Design of Active Filters with Experiments*, Indiana: Howard W. Sams & Co., 1979.

[5] S. Butterworth. On the Theory of Filter Amplifiers, *Wireless Engineer* (aka *Experimental Wireless and Radio Engineer*), 7, pp. 536–541: 1930.

[6] A. B. Carlson, *Communications Systems*, New York: McGraw-Hill, 1986.

[7] C. H. Chen. *Signal Processing Handbook*, New York: Marcel Deker, Inc., 1988.

[8] W. K. Chen. *Passive and Active Filters: Theory and Implementations*, John Wiley & Sons, New York: 1986.

[9] W.K. Chen. *Broadband Matching Theory and Implementations*, New Jersey and London: World Scientific, 1988.

[10] W.K Chen. *The Circuits and Filters Handbook*, Boca Raton, Florida: CRC Press/IEEE, 1995.

[11] T.R. Cuthbert. *Circuit Design Using Personal Computers*, John Wiley & Sons, New York: 1983.

[12] G. Daryanani. *Principles of Active Network Synthesis and Design*, John Wiley & Sons, New York: 1976.

[13] M.G. Ellis. *Electronic Filter Analysis and Synthesis*, Norwood MA: Artech House, 1994.

[14] S. Franko. *Design with Operational Amplifiers and Analog Integrated Circuits*, New York: McGraw-Hill, 1988.

[15] M. Van Falkenburg. *Analog Filter Design*, New York: Oxford University Press, 1982.

[16] R.W. Hamming. *Digital Filters*, Third Edition, Upper Saddle River, NJ: Prentice-Hall, 1989.

[17] P. Horowitz and W. Hill. *The Art of Electronics*, second edition, Cambridge: Cambridge Press, 1989.

[18] L.P. Huelsman, *Active and Passive Analog Filter Design*, New York: McGraw-Hill, 1993.

[19] L.P. Huelsman and P.E. Allen. *Introduction to the Theory and Design of Active Filters*, New York: McGraw-Hill, 1993.

[20] L.P. Huelsman, P. Lawrence. *Active Filters: Lumped, Distributed, Integrated, Digital and Parametric*, New York: McGraw-Hill, 1970.

[21] R.G. Irvine. *Operational Amplifier, Characteristics and Applications*, Upper Saddle River, NJ: Prentice-Hall, Inc., 1994.

[22] D. Johnson and J. Hilburn. *Rapid Practical Design of Active Filters*, New York: John Wiley & Sons, 1975.

[23] D.E. Johnson, J.R. Johnson and H.P. Moore. *A Handbook of Active Filters*, Upper Saddle River, NJ: Prentice-Hall, 1980.

[24] B.C. Kuo. *Automatic Control Systems*, Englewood Cliffs, New Jersey: Prentice-Hall, 1962.

[25] Kerry Lancaster. A Basic Introduction to Filters-Active Passive and Switched-Capacitor, *National Semiconductor*, App Note 779, 1991.

[26] D. Lancaster. *Active Filter Cookbook*, Second Edition, Boston: Newnes, 1996.

[27] K. Lee, Private communication. Korea Advanced Institute of Science and Technology (KAIST), Taejon, Korea.

[28] M.D. Lutovac, V. D. Tosic, B. L. Evans. *Filter Design for Signal Processing using MATLAB and Mathematica*, Englewood Cliffs, NJ: Prentice-Hall, 2000.

[29] D. Meador. *Analog Signal Processing with Laplace Transforms and Active Filter Design*, London: Delmar-Thomson Learning, 2002.

[30] W. Middlehurst. *Practical Filter Design*, Englewood Cliffs, NJ: Prentice-Hall, 1993.

[31] S. Niewiadomski. *Filter Handbook: A Practical Design Guide*, Boca Raton, Florida: CRC Press, 1989.

[32] R.P. Sallen. A Practical Method of Designing RC Active Filters, *IRE Transactions Circuit Theory*, CT-2, pp. 74–85, March 1955.

[33] M. Schaumann, S. Ghausi and K.R. Laker, K.R. *Design of Analog Filters*, Englewood Cliffs, NJ: Prentice-Hall, 1990.

[34] D. Sequine and C. McNeese. Adjustable Sallen and Key Low-Pass Filters, Application Note #AN2031, *Cypress Semiconductor*, 2004.

[35] S.W. Smith. *The Scientists and Engineer's Guide to Digital Signal Processing*, first edition, San Diego, California Technical Pub., 1997.

[36] R.C. Stephenson. *Active Filter Design Handbook*, New York: McMillan Press, 1985.

[37] K.L. Su. *Analog Filters*, London: Chapman and Hall, 1996.

[38] S. Sukittanon and S. G. Dame. nth Order IIR Filtering Graphical Design Tool for PSoC. Application Note #AN2312. *Cypress Semiconductor*, 2005.

[39] F.P. Tedeschi. *The Active Filter Handbooook*, Blueridge Summit, PA: Tab Books Inc., 1979.

[40] G.C. Temes and J.W. LaPatra. *Circuit Synthesis and Design*, New York: McGraw-Hill, 1977.

[41] T.J. Terrell. *Introduction to Digital Filters*, New York: Macmillan Press, 1980.

[42] G.H. Tomlinson. *Electrical Networks and Filters: Theory and Design*, New York: Prentice-Hall, 1991.

[43] D. Van Ess. Understanding Switched Capacitor Filters, Cypress Microsystems Application Note #*AN*2168, *Cypress Semiconductor*, 2004.

[44] L. Weinberg. *Introduction to Modern Network Synthesis*, New York: McGraw-Hill, 1962.

[45] H.P. Wettman. *Reference Data for Radio Engineers*, Indianapolis: Howard W. Sams & Co., Inc., 1999.

[46] A.B. Williams and F. Taylor. *Electronics Filter Designer's Handbook*, 2nd Ed., New York: McGraw-Hill, 1988.

[47] S. Winder, *Analog and Digital Filter Design*, second edition, Boston: Newnes Press, 1997.

[48] A.I. Zverev. *Handbook of Filter Synthesis*, New York: John Wiley & Sons, 1967.

[49] Design Using Integrator Blocks, Maxim Application Note 727, 2001.

[50] PSoC Designer 4.0 and 5.0. Cypress Semiconductor Corporation, 2009.

[51] D. Sequine. Private communication. Cypress Semiconductor Corporation, 2009.

Chapter 9

$\Delta\Sigma$ Analog-to-Digital Converters

The chapter discusses the basic concepts of Delta Sigma ($\Delta\Sigma$) analog-to-digital converters (ADCs), and details the design of first- and second-order $\Delta\Sigma$ ADCs using the PSoC architecture.

Analog-to-digital converters (ADCs) are important subsystems in many embedded mixed-signal systems. They convert continuous-valued, continuous-time signals into discrete-valued, discrete-time data by sampling and quantization. A $\Delta\Sigma$ ADC can be used to minimize the in-band quantization noise power by oversampling and noise-shaping to achieve the desired conversion accuracy.

- Oversampling is defined as the use of a sampling frequency that is much larger than the Nyquist frequency. High oversampling rates reduce the in-band, quantization noise power, and simplify the removal of the images by lowpass filtering.

- Noise-shaping results from selecting the $\Delta\Sigma$ modulator, transfer function such that the signal transfer function (STF) is an allpass filter, and the noise transfer function (NTF) is a highpass filter that eliminates the in-band quantization noise.

The PSoC-based implementation of an eight-bit, first-order, $\Delta\Sigma$ ADC implementation discussed in this chapter includes the modulator, decimator, and API routines. The modulator is based on PSoC's programmable SC blocks. The decimator for lowpass filtering and downconversion is a hardware/software design with the integration part of the decimator implemented using PSoC's Type 1 decimator blocks and the differentiation part in software. The API routines include subroutines for managing the ADC.

In addition, this chapter presents analytical expressions and simulation models for estimating the degradation in ADC performance due to nonidealities, for example clock jitter, switch thermal noise, integrator leakage, and OpAmp noise, finite gain, slew rate, and saturation.

This chapter has the following structure:

- Section 1 defines the concepts of Nyquist converts, including sampling and quantization.

- Section 2 presents the defining elements of $\Delta\Sigma$ ADCs, such as oversampling, noise-shaping, modulator performance, and first- and second-order $\Delta\Sigma$ modulators.

- Section 3 provides chapter conclusions.

A. Doboli, E.H. Currie, *Introduction to Mixed-Signal, Embedded Design*,
DOI 10.1007/978-1-4419-7446-4_9, © Springer Science+Business Media, LLC 2011

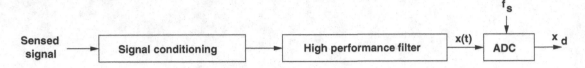

Figure 9.1: Analog signal sensing in embedded systems.

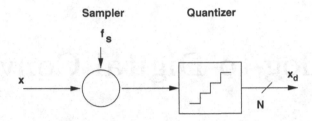

Figure 9.2: Sampling and quantization in analog-to-digital converters.

9.1 Nyquist ADCs-A Short Introduction

Analog-to-digital converters are systems that can be used to convert continuous-valued, continuous-time signals into discrete-valued, discrete-time data. Figure 9.1 shows a typical application of an ADC in a mixed-signal, embedded system. After amplification and filtering, the analog signals acquired from the sensing elements are converted into digital data. The latter are then processed by a microcontroller, DSP, or customized digital blocks, for example digital filters.

9.1.1 Sampling and Quantization

The ADC process involves two steps: *sampling* and *quantization*. Sampling is the representation of a continuous-time signal by a sequence of continuous-valued signals at discrete-time intervals. Quantization is the approximation of a continuous-valued signal as a discrete-valued signal. Figure 9.2 illustrates this two-step conversion process.

9.1.2 Sampling

The principal requirement for sampling is to collect sufficient data to accurately represent a continuous-time signal by the sampled discrete-time data. This requirement can be stated more

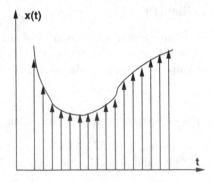

Figure 9.3: Signal sampling.

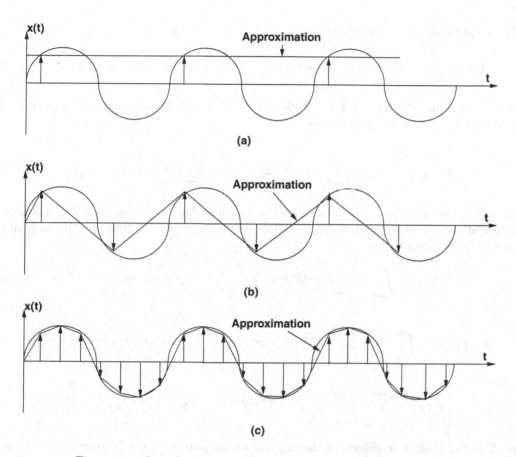

Figure 9.4: Signal sampling at increasing sampling frequencies.

definitively as the determination of the sampling frequency that allows a precise description of the continuous-time signals. This requirement can be illustrated as shown in Figures 9.3 and 9.4. A sufficient number of points must be sampled to allow a precise description of the signal. If only one point is collected for each time period of the signal, then the sampled points all have the same amplitude, which leads, erroneously, to a constant signal. If two points are sampled in each time period, then the approximation improves. The signal then resembles a saw tooth. If three points are collected in each time period, the approximation improves still more. The approximation continues to improve as a direct function of the number of points sampled in each time period. The following theorem defines the sufficient condition for an accurate representation of a continuous-time signal by sampled data.

Nyquist sampling theorem: A band-limited signal can be reconstructed exactly, if the sampling frequency is greater than the Nyquist frequency. (Note: The Nyquist frequency is defined as twice the signal bandwidth.)

For example, if the highest spectral component of a signal x, has the frequency f_H, then the related Nyquist frequency is $f_{Nyquist} = 2 \times f_H$, and the sampling frequency f_s should meet the requirement $f_s > f_{Nyquist} = 2 \times f_H$.

A simplified proof of Nyquist's sampling theorem: If the continuous-time signal is $x(\text{t})$, the sampling signal is $s(\text{t})$, and the sampled signal is $x_s(\text{t})$, then assuming a very short positive level for the sampling signal, the sampling signal can be expressed as [1]:

$$s(t) = \sum_{n=-\infty}^{\infty} \delta(t - nT_s) = \delta(t) + \delta(t \pm T_s) + \delta(t \pm 2T_s) + ... \delta(t \pm nT_s) + ... \tag{9.1}$$

where the function $\delta(t)$ is one for $t = 0$, zero for $t \neq 0$ and T_s is the sampling period. The sampled signal can then be expressed as

$$x_s(t) = x(t) \times s(t) = \sum_{n=-\infty}^{\infty} x(t)\, \delta(t - nT_s) = \sum_{n=-\infty}^{\infty} x(nT_s)\, \delta(t - nT_s) \tag{9.2}$$

Note that the sequence $x(nT_s)$ includes the sampled data at instances, nT_s. The Fourier transform of the sampled signal, $x_s(t)$, provides some insight into the frequency domain characteristics of the sampled signal:

$$X_s(f) = \int_{-\infty}^{\infty} x_s(t)\, e^{-j2\pi ft} dt = \int_{-\infty}^{\infty} \sum_{n=-\infty}^{\infty} x(t)\, \delta(t - nT_s)\, e^{-j2\pi ft} dt \tag{9.3}$$

$$X_s(f) = \int_{-\infty}^{\infty} \sum_{n=-\infty}^{\infty} x(t)\, e^{j2\pi nf_s t}\, e^{-j2\pi ft} dt = \int_{-\infty}^{\infty} \sum_{n=-\infty}^{\infty} x(t)\, e^{-j2\pi(f - nf_s)t} dt \tag{9.4}$$

$$X_s(f) = \sum_{n=-\infty}^{\infty} \int_{-\infty}^{\infty} x(t)\, e^{-j2\pi(f - nf_s)t} dt = \sum_{n=-\infty}^{\infty} X(f - nf_s) \tag{9.5}$$

where X is the Fourier transform of the original signal, and X_s is the Fourier transform of the sampled signal. Thus, the sampled signal is expressed in the frequency domain as the following series:

$$X_s(f) = X(f) + X(f \pm f_s) + X(f \pm 2f_s) + X(f \pm 3f_s) + ...X(f \pm nf_s) + ... \tag{9.6}$$

Equation (9.6) completes the proof of Nyquist's sampling theorem. The sampled signal includes the original signal $X(f)$ plus the images, that is copies, $X(f \pm n\ f_s)$. These images are located at distances equal to the positive and negative multiples of the sampling frequency f_s.

If the sampling frequency, f_s, is larger than twice the frequency, f_H, of the highest spectral component of the signal X, then the original signal and the images do not overlap, and a lowpass filter can be used to eliminate the images and retrieve the original signal X from the sampled signal X_s. As the sampling frequency increases, obviously the distance between the images increases, which helps filter out the images. If the sampling frequency is below the Nyquist frequency, the original signal and the images overlap, which makes recovering the original signal impossible. This phenomenon is called aliasing. Figure 9.5 presents the two situations: sampling without and with aliasing.

Reconstruction of the sampled signal: The original signal can be recovered by lowpass filtering of the sampled signal. Assuming that the transfer function of the ideal filter is given by [1, 8]:

$$H_{ideal}(f) = \text{rect}\left[\frac{f}{f_s}\right] \tag{9.7}$$

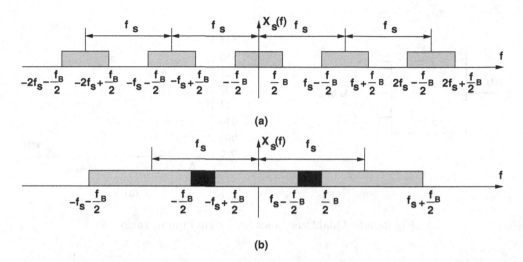

Figure 9.5: Sampling (a) without and (b) with aliasing.

where the function rect is zero for frequencies outside the range $[-f_s/2, \; f_s/2]$, and one within the range.

The signal $x(t)$ (the reconstructed, sampled signal) is the convolution of the lowpass filter transfer function $h(t)$ and the signal $X_{s(t)}$ [1, 8]:

$$x(t) = h(t) * x_s(t) \tag{9.8}$$

The time domain expression for the transfer function is the inverse Fourier transform of $H(f)$:

$$h(t) = \int_{-\infty}^{\infty} H(f)\, e^{j2\pi ft} df = \int_{-\infty}^{\infty} \mathrm{rect}\frac{f}{f_s}\, e^{j2\pi ft} df = \int_{-f_s/2}^{f_s/2} e^{j2\pi ft} df = \frac{\sin \pi f_s t}{\pi t} \tag{9.9}$$

$$h(t) = f_s\, \mathrm{sinc}(f_s t) \tag{9.10}$$

where the function *sinc* is defined as

$$\mathrm{sinc(x)} = \frac{\sin(\pi x)}{\pi x}. \tag{9.11}$$

Substituting Equation (9.10) into Equation (9.8), results in the following expression for the signal, *x(t)*:

$$h(t) * x_s(t) = f_s \mathrm{sinc}(f_s t) \sum_{n=-\infty}^{\infty} x(nT_s)\delta(t - nT_s) \tag{9.12}$$

$$h(t) * x_s(t) = f_s \sum_{n=-\infty}^{\infty} x(nT_s)\mathrm{sinc}(f_s t)\delta(t - nT_s) \tag{9.13}$$

$$h(t) * x_s(t) = f_s \sum_{n=-\infty}^{\infty} x(nT_s)\mathrm{sinc}\left[f_s(t - nT_s)\right] = \sum_{n=-\infty}^{\infty} x(nT_s)\mathrm{sinc}\left[\frac{t - n\,T_s}{T_s}\right] \tag{9.14}$$

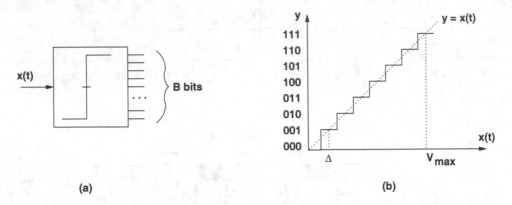

(a)

(b)

Figure 9.6: Quantizer block and signal quantization.

Equation (9.14) shows that according to the Whittaker–Shannon interpolation relationship, the convolution $h(t) * x_s(t)$ actually represents the signal $x(t)$. This mathematical proof explains that the original signal $x(t)$ can be retrieved by passing the signal $x_s(t)$ through a lowpass filter.

9.1.3   Quantization

Quantization is the process of converting a sampled continuous-valued signal into discrete-valued data. Figure 9.6(a) presents a B-bit quantizer block. If the quantizer block outputs B bits, and therefore the discrete data are represented using B bits, the continuous-valued range of the input signal $x(t)$ is divided into $2^B - 1$ subranges, each of the ranges being represented by a unique bitstring of length B. For a three-bit quantizer, Figure 9.6(b) describes the eight bitstring encodings and the corresponding subranges. Note that all the values within a subrange are represented as the same bitstring, which obviously introduces an approximation error. This error is called the *quantization error*. If all discretization ranges are equal, then the quantizer is called a *uniform quantizer*.

For a uniform quantizer, and input scaled to the range (−1, 1), the width Δ of a discretization range is:

$$\Delta = \frac{2}{2^B - 1} \tag{9.15}$$

Figure 9.7(a) illustrates the definition of Δ. The quantization error e_r is within the range $(-\Delta/2, \Delta/2)$, as shown in Figure 9.7(b):

$$e_r \in (-\frac{\Delta}{2}, \frac{\Delta}{2}) \tag{9.16}$$

The quantization noise exhibits very complex behavior that is obviously correlated to the input signal. R. Gray has provided a very detailed analysis of the quantization noise in $\Delta\Sigma$ ADCs[4]. Nevertheless, these models are very hard to use in analysis of the behavior and performance of ADCs. Instead, for the purpose of ADC analysis, the quantization noise is approximated as (i) white noise, and (ii) uncorrelated to the input signal [1, 4, 5]. Bennett's condition [4] states that the above two properties of the quantization noise hold if the following four conditions are true: (i) the input does not overload the quantizer, (ii) the number B of quantization bits is large, (iii) the width Δ is small, and (iv) the joint probability density function of the input at

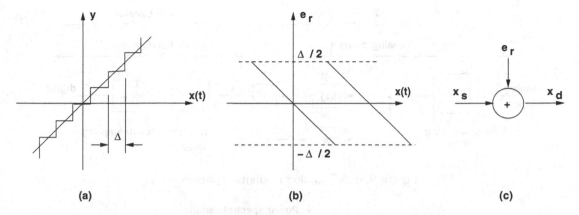

Figure 9.7: Signal quantization and quantizer modeling.

various sampling moments is smooth. In this case, the quantized data x_d can be modeled as the sum of the continuous-valued, sampled signal x_s, and the quantization noise e_r:

$$x_d = x_s + e_r \tag{9.17}$$

This modeling is shown in Figure 9.7(c).

The *power spectral density* of the noise e_r is flat, and is equally distributed over the range $(-\Delta/2, \Delta/2)$:

$$f_e(e) = \frac{1}{\Delta} \tag{9.18}$$

The *quantization noise power* is:

$$\sigma_e^2 = \int_{-\Delta/2}^{\Delta/2} e^2 \, f_e(e) de = \frac{1}{\Delta} \int_{-\Delta/2}^{\Delta/2} e^2 de = \frac{\Delta^2}{12} \tag{9.19}$$

Using Equation (9.15), the quantization noise power can be expressed as a function of the number of quantization bits as follows:

$$\sigma_e^2 = \frac{1}{3(2^B - 1)^2} \tag{9.20}$$

This result shows that increasing the number of quantization bits B reduces the quantization noise power σ_e^2, and thus improves the accuracy of quantization.

9.2   ΔΣ ADCs

Figure 9.8 illustrates the block structure of ΔΣ ADCs which consist of the ΔΣ modulator and the decimator block. The modulator provides the sampling part of the conversion process with the goal of providing the desired conversion accuracy by minimizing the in-band quantization noise power of the modulator. The ΔΣ modulator reduces the in-band quantization noise power by oversampling the input signal, and noise-shaping the quantizer noise by proper selection of the transfer function $H(z)$. The output of the modulator quantizer is the input to the decimator block,

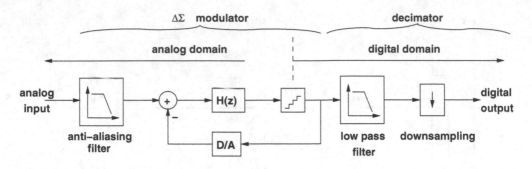

Figure 9.8: $\Delta\Sigma$ analog-to-digital converters.

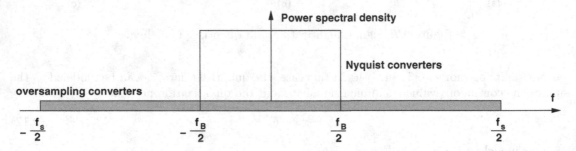

Figure 9.9: Quantization noise in oversampled converters.

which reconstructs the sampled signal as discussed in the previous subsection. The decimator circuit performs the down-sampling process to address the oversampling of the modulator and a lowpass digital filtering to eliminate the images that accompany the sampled signal. The modulator uses analog circuits, and the decimator uses digital circuits.

The following sections present the principle of $\Delta\Sigma$ ADC, the related performance metrics, and $\Delta\Sigma$ ADCs of different orders, for example first-order and second-order modulators.

9.2.1   Oversampling and Noise-Shaping

The operation of $\Delta\Sigma$ modulators relies on two basic concepts, oversampling and noise-shaping, for achieving accurate data conversion [1, 4]:

- *Oversampling*: As stated previously, oversampling refers to using a sampling frequency that is higher than the Nyquist frequency $f_{Nyquist}$. The oversampling ratio (OSR) is defined as the ratio of the sampling frequency to the Nyquist frequency:

$$OSR = \frac{f_s}{f_{Nyquist}} \tag{9.21}$$

OSR values are expressed as powers of two, for example 2, 4, 6, 8, 16, ... , 256, because these values simplify the implementation of the digital decimator.

Using high OSR values increases the distance between the sampled signal and its images (cf. Figure 9.5) and allows the images to be removed by simpler antialiasing filters.

The second advantage of oversampling is the reduction of the in-band noise power. Assuming the white noise model for quantization noise, the quantization noise power $\Delta^2/12$ is

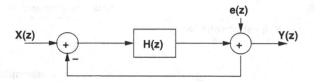

Figure 9.10: Linear model of a ΔΣ modulator.

uniformly distributed over the frequency range ($-OSR \times f_B/2$, $OSR \times f_B/2$), and therefore the in-band quantization noise power is reduced by the factor, OSR, that is,

$$P_{in-band} = \frac{\sigma_e^2}{OSR} \tag{9.22}$$

Figure 9.9 shows that the in-band noise power decreases with the value of OSR. The quantization noise outside the signal band is filtered out by digital filtering in the decimator block.

- *Noise-shaping*: The term noise-shaping refers to selecting the modulator transfer function $H(z)$ (cf. Figure 9.8), such that the signal transfer function (STF) of the modulator acts as an allpass filter, and the noise transfer function (NTF) is a highpass filter that removes the in-band quantization noise. Figure 9.11 illustrates the concepts of highpass NTF, and shows that the remaining in-band quantization noise power is low.

Using the quantizer model in Figure 9.7(c), a ΔΣ modulator can be analyzed using the model in Figure 9.10. The modulator output is given by:

$$Y(z) = \frac{H(z)}{1 + H(z)} \, X(z) + \frac{1}{1 + H(z}E(z) \tag{9.23}$$

Thus

$$STF(z) = \frac{H(z)}{1 + H(z)} \tag{9.24}$$

and

$$NTF(z) = \frac{1}{1 + H(z)} \tag{9.25}$$

ΔΣ modulators, for which the NTF is a highpass filter, are called lowpass modulators [5]. These modulators are used in converting signals with a narrow frequency band at low frequencies. In contrast, high-frequency signals have a narrow band at high center frequencies, and thus require modulators with NTFs that are bandstop filters instead of highpass filters. These ΔΣ modulators are called bandpass modulators. The discussion in this chapter is restricted to lowpass ΔΣ modulators because the targeted applications involve only the sampling of low-frequency signals.

9.2.2   ΔΣ ADC Performance

The performance attributes of ΔΣ ADCs determine the accuracy of the conversion process by relating the quantization noise power to the power of the input signal [1, 3, 5]. Signal-to-noise ratio (SNR) and dynamic range (DR) are the most popular performance figures of merit used in ΔΣ design.

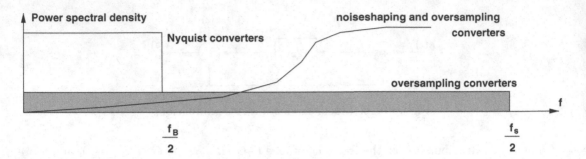

Figure 9.11: Noiseshaping in oversampled converters.

- *Signal-to-noise ratio*: Signal-to-noise ratio is defined as the ratio of the signal power to the in-band quantization noise power:

$$SNR\ (dB)\ =\ 10\ \log\left[\frac{signal\ power}{in-band\ quantization\ noise\ power}\right] \tag{9.26}$$

For a sinusoidal input signal of magnitude A, the corresponding SNR is computed by the following expression [3]:

$$SNR\ (dB) = 10\ \log\left[\frac{\frac{A^2}{2}}{\frac{\Delta^2}{12}}\right] = 10\ \log\left[\frac{(2^B-1)^2\frac{\Delta^2}{8}}{\frac{\Delta^2}{12}}\right] \approx 6.02\ B + 1.76\ (dB) \tag{9.27}$$

This shows that increasing SNR by 3 dB improves the accuracy by 0.5 bits. Also, a SNR of $\approx$ 50 dB corresponds to 8 bits accuracy, 62 dB to 10 bits, 74 dB to 12 bits, 86 dB to 14 bits , and 98 dB to 16 bits.

Given that the in-band quantization noise power decreases with OSR, SNR improves with higher OSR :

$$SNR\ (dB) = 10\ \log\left[\frac{\frac{(2^B-1)^2\Delta^2}{8}}{\frac{\Delta^2}{12\ OSR}}\right] \approx 6.02\ B + 10\ log\ OSR \tag{9.28}$$

This shows that doubling the OSR frequency increases the accuracy by 0.5 bits, or, equivalently increases the SNR by 3 dB.

- *Dynamic range*: Dynamic range is defined as the ratio of the output power for a sinusoidal input with full-range amplitude to the output power of the smallest input signal that it can distinguish and quantize [1, 3]. The full-range amplitude is defined by the quantizer used [3], and is $\Delta/2$ for a single-bit quantizer. The smallest signal that can be processed is of the same magnitude as the quantization error, and therefore the corresponding SNR is zero. DR is defined as:

$$DR\ (dB) = 10\ \log\left[\frac{\frac{1}{2}\frac{\Delta^2}{4}}{in-band\ quantization\ noise\ power}\right] \tag{9.29}$$

The dynamic range is linked to the ADC resolution by the following expression

$$B\ (bits) = \frac{DR\ (dB) - 1.76}{6.02} \tag{9.30}$$

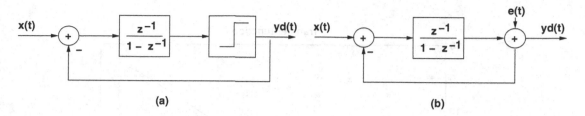

Figure 9.12: First-order $\Delta\Sigma$ modulator.

9.2.3 First-Order $\Delta\Sigma$ Modulator

Figure 9.12(a) shows the topology of a first-order $\Delta\Sigma$ modulator [5] where:

$$y_d(t) = (x(t) - y_d(t))\frac{z^{-1}}{\left[1 - z^{-1}\right]} + e(t) \tag{9.31}$$

$$y_d(t) = x(t)\frac{z^{-1}}{\left[1 - z^{-1}\right]} + e(t) \tag{9.32}$$

$$y_d(t) = z^{-1} x(t) + (1 - z^{-1}) e(t) \tag{9.33}$$

Hence, the signal transfer function (STF) of a first-order $\Delta\Sigma$ modulator is:

$$STF(z) = z^{-1} \tag{9.34}$$

and its noise transfer function (NTF) is:

$$NTF(z) = 1 - z^{-1} \tag{9.35}$$

Equation (9.34) shows that the STF delays the input signal, but otherwise leaves it unchanged. The NTF in expression (9.35) acts as a highpass filter for the quantization noise. Figure 9.13 illustrates the STF and NTF of the first-order $\Delta\Sigma$ modulator. Note the noise-shaping characteristics of the NTF.

The total in-band quantization noise power of a first-order $\Delta\Sigma$ modulator can be estimated by expressing the noise power as the cumulative value over the input signal frequency band of the oversampled and noise-shaped quantization noise power. The effect of oversampling is illustrated by equation (9.19), and the result of noise-shaping is modeled using NTF. Hence, the following expression results for the in-band quantization noise power.

$$P_{in-band} = \int_{-\frac{f_b}{2}}^{\frac{f_b}{2}} \left[\frac{e^2}{f_s}\right] |NTF(f)|^2 \ df \tag{9.36}$$

Figure 9.14 shows the graph of the power spectral density for a first-order $\Delta\Sigma$ modulator. The "spike" occurs at the input signal frequency. Note the noise-shaping characteristics of the PSD. Also, the figure shows that the quantization noise can be approximated reasonably well as whitenoise, provided that the sampling frequency is much larger than the frequency of the highest spectral component.

Substituting:

$$z = e^{j2\pi f/f_s} \tag{9.37}$$

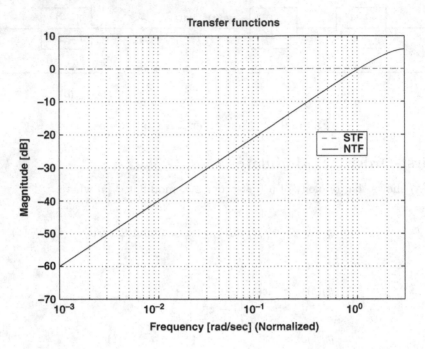

Figure 9.13: STF and NTF for the first-order $\Delta\Sigma$ modulator.

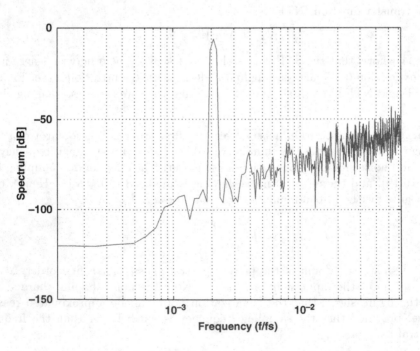

Figure 9.14: Power spectral density of the first-order $\Delta\Sigma$ modulator.

into

$$NTF = 1 - z^{-1} NTF = \left[1 - \cos 2\pi \frac{f}{f_s}\right] + j \sin(2\pi \frac{f}{f_s}) \tag{9.38}$$

results in

$$NTF = \left[1 - \cos 2\pi \frac{f}{f_s}\right] + j \sin(2\pi \frac{f}{f_s}) \tag{9.39}$$

$$|NTF|^2 = \left[1 - \cos 2\pi \frac{f}{f_s}\right]^2 + \sin^2 \left(2\pi \frac{f}{f_s}\right) = 4 \sin^2 \left(\pi \frac{f}{f_s}\right) \tag{9.40}$$

Hence, the total in-band quantization noise power can be predicted as:

$$P_{in-band} = \int_{-f_b/2}^{f_b/2} \left[\frac{e^2}{f_s} 4\right] \sin^2 \left(\pi \frac{f}{f_s}\right) df \tag{9.41}$$

$$= 4 \left[\frac{e^2}{f_s}\right] \int_{-f_b/2}^{f_b/2} \sin^2 \left(\pi \frac{f}{f_s}\right) df \tag{9.42}$$

$$= 4 \frac{e^2}{f_s} \left[\frac{1}{2}\left(f - \frac{f_s}{2\pi} \sin 2\pi \frac{f}{f_s}\right)\Big|_{-\frac{f_b}{2}}^{\frac{f_b}{2}}\right] \tag{9.43}$$

$$= 2 \frac{e^2}{f_s} \left[f_b - \frac{f_s}{\pi} \sin\left(\pi \frac{f_b}{f_s}\right)\right] \tag{9.44}$$

Using the approximation that:

$$\sin x \approx x - \frac{x^3}{3!} \tag{9.45}$$

and substituting it into Equation (9.44), the in-band quantization noise power for a single-bit quantizer becomes:

$$P_{in-band} = \frac{\pi^2}{9 \, OSR^3} \tag{9.46}$$

For a sinusoidal signal of amplitude A, the SNR of the first-order $\Delta\Sigma$ modulator is given by

$$SNR \; (dB) = 10 \log\left[\frac{\frac{A^2}{2}}{\frac{\pi^2}{9 \, OSR^3}}\right] = 10 \log\left[\frac{9 \, A^2 \, OSR^3}{2 \, \pi^2}\right] \tag{9.47}$$

Figure 9.15 is a graph of DR versus input amplitude for the first-order modulator and OSR = 32. The sampling frequency is normalized ($f_s = 1$), and the input signal frequency is $f_{in} = 1/512$. The DR is $\approx 34 \; dB$, and the highest SNR value is $\approx 46 \; dB$. The SNR increases with the amplitude of the input signal until reaching a value that saturates the modulator quantizer.

Figure 9.16 shows the dependency of the dynamic range of the first-order $\Delta\Sigma$ modulator on the OSR. The same conditions were used in Figure 9.15. DR is plotted for the OSR values 32, 64, 128, and 256. DR is 34 dB for OSR of 32 , 38 dB for OSR of 64, 42 dB for OSR of 128, and more than 50 dB for OSR of 256.

The next two subsections describe the implementation of first-order, $\Delta\Sigma$ modulators using PSoC's programmable, switched-capacitor block, and characterizes the impact of circuit nonidealities on the modulator performance.

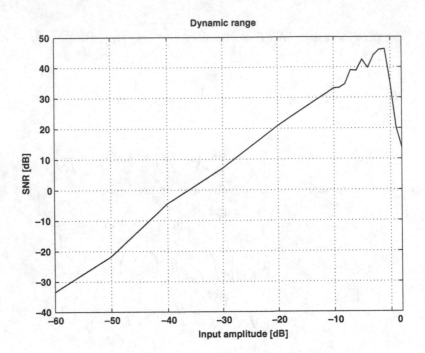

Figure 9.15: Dynamic range for first-order $\Delta\Sigma$ modulator.

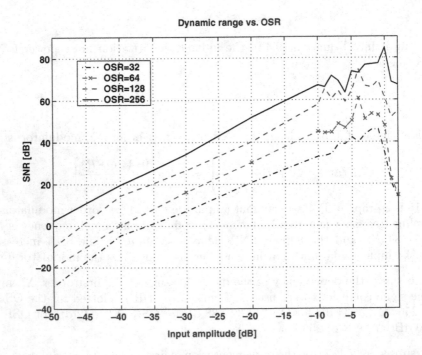

Figure 9.16: Dynamic ranges for first-order $\Delta\Sigma$ modulator and different OSR values.

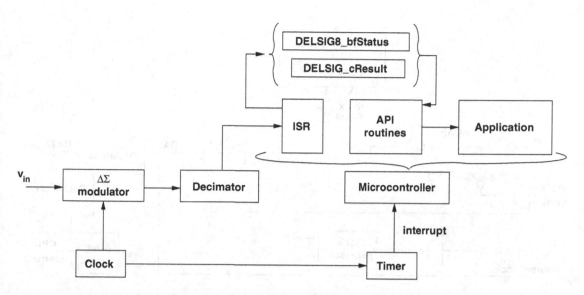

Figure 9.17: Block structure of the PSoC-based implementation of ΔΣ ADCs.

9.2.4   PSoC Implementation of First-Order ΔΣ Modulators

Figure 9.17 shows the block structure of the PSoC-based implementation of an eight-bit ΔΣ ADC [7]. The ADC's input signal is limited to the range:

$$V_{in} \quad \in \quad (-V_{ref}, +V_{ref}) \tag{9.48}$$

The voltage, V_{ref}, can be selected to be either $V_{DD}/2$, where V_{DD} is the chip's supply voltage, 1.6 $V_{bandgap}$ the bandgap voltage, or an external voltage is supplied to port 2 of the chip.

The oversampling ratio of the modulator is fixed at 64. The DR of the ideal modulator is shown in Figure 9.16. The value of the input signal is [7]:

$$V_{in} \quad = \quad \frac{n - 128}{128} \, V_{ref} \tag{9.49}$$

where n is the eight-bit output of the ADC.

The block structure shown in the figure is valid for both first-order and second-order ΔΣ ADCs. The structure is implemented in both hardware and software domains, and includes the following components.

- *ΔΣ modulator:* The modulator block implements the structure shown in Figure 9.12(a) and utilizes PSoC's programmable switched-capacitor blocks.

- *Decimator:* The functionality of the decimator block includes lowpass filtering of the high-frequency (outband) images of the sampled signal, and the downconversion of the signal by a factor equal to OSR. A sinc^2 filter [5] is used for lowpass filtering. The integration part of the filter is implemented in hardware and based on the PSoC decimator block. The differentiation part is handled in software. Signal downconversion is obtained by initializing the timer with the OSR value, and programming the timer to generate an interrupt at the end of the downcount process. The corresponding ISR routine implements the differentiation algorithm, and is executed once for each OSR input processed by the PSoC decimator block.

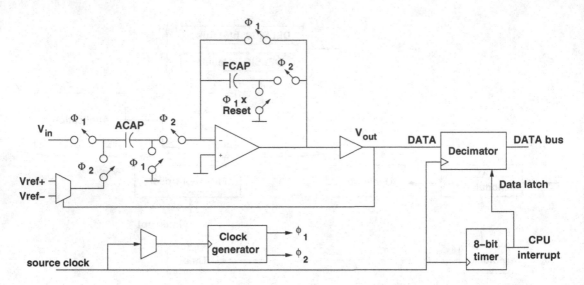

Figure 9.18: PSoC implementation of first-order $\Delta\Sigma$ ADC [7].

- *API routines*: The API routines handle starting/stopping the ADC block, starting/stopping the signal conversion procedure, managing the status information of the converter (including verifying that a new data byte is available), and accessing the data output of the converter. The application program relies on these API routines to access the ADC data by a polling mechanism.

- *Clocks*: Several clocks are needed for the circuit, including the clocks ϕ_1 and ϕ_2 for the switched capacitor circuits of the $\Delta\Sigma$ modulator, for the quantizer circuit (the clock operates at the frequency f_s), for the digital decimator block, and for the timer. The quantizer, digital decimator, and timer use the same clock frequency.

Figure 9.18 illustrates the implementation of a first-order $\Delta\Sigma$ ADC based on PSoC's programmable switched capacitor and digital blocks [7]. The implementation uses one programmable SC block configured as an integrator. The comparator of the SC block implements the quantizer of the modulator. In addition, the design uses two digital blocks: one Type 1 decimator block (cf. Subsection 4.4.3 in Chapter 4) and one programmable block configured as a timer. Figures 9.19 and 9.20 illustrate the measured power spectrum densities for a first-order $\Delta\Sigma$ modulator, with OSR = 32, and OSR = 64, respectively. To reduce the number of necessary clocks, all digital blocks are clocked with the same signal as the signal used for the SC block [7]. The frequency of this signal is four times higher then the required frequency. Therefore, the 8-bit timer generates an interrupt signal after $4 \times OSR$ clock cycles, which, in this design, is after 256 clock cycles. This explains the need for introducing an 8-bit timer.

The decimator block implements the following sinc^2 transfer function:

$$H(z) = \left[\frac{1}{OSR} \frac{1 - z^{-OSR}}{1 - z^{-1}} \right]^2 \tag{9.50}$$

Figure 9.21 presents the transfer function of the sinc^2 function for an OSR value of 64. Note that the zeros of the transfer function occur at the frequencies $f_b/2 + m\, a f_s/2$, which provides an efficient way of suppressing the images in the sampled signal. Another advantage is the

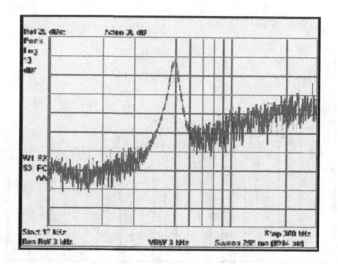

Figure 9.19: Power spectrum densities for a first-order ΔΣ modulator, with OSR = 32.

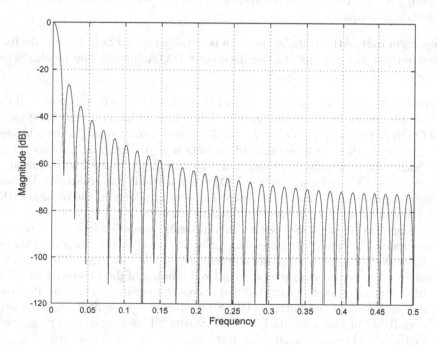

Figure 9.20: Power spectrum densities for a first-order ΔΣ modulator, with OSR = 64.

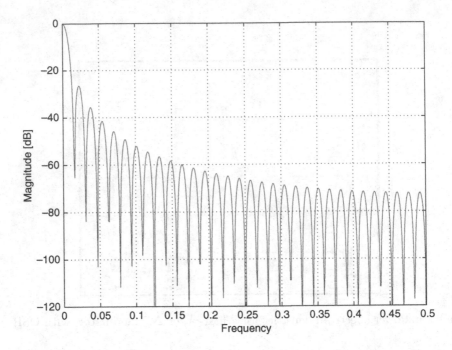

Figure 9.21: The sinc^2 function for an OSR value $= 64$.

simple implementation procedure of the transfer function, which includes few hardware blocks and simple software routines.

The integration part of the transfer function is obtained by a PSoC Type *1* decimator block. The quantizer output is connected to the decimator DATA input. The block operates at the sampling frequency of the modulator.

Differentiation is carried out in software, and is executed at the down-converted rate. Down-conversion is implemented by the 8-bit timer, which generates interrupts at time intervals equal to the value OSR/f_s (actually $4 \times OSR/f_s$ in this design because a single clock is used for all of the circuits). The ISR called performs the differentiation step. Figure 9.22 shows the pseudocode for the ISR which implements the FSM shown in Figure 9.22(a). State 1 is the initial state of the FSM. After receiving a timer interrupt, the FSM reads a 16-bit data from the digital decimator block implementing the double integration step. Data are available in the registers DEC_DL and DEC_DH of the decimator (cf. Subsection 4.4.3 in Chapter 4). Then, states 2-4 introduce a delay of four "steps" as required for double differentiation. State 5 accesses the data registers of the digital decimator block, and computes the difference between this value and the value found in State 1. State 5 may also include other processing of the converted values, such as finding their RMS values. If the ADC is to stop after a fixed number of data conversions State 5 counts the number of converted values. Figure 9.22(b) shows the pseudocode of the ISR routine. The converted value is placed in the variable DELSIG8_cResult by the ISR, and the status of the modulator is available in the variable DELSIG8_Status. These variables are accessible to the API routines called by the application programs.

In State 5, the ISR also determines whether the read value resulted in an overflow. This procedure consists of the following steps: (i) the bitstring of the modulator output is first converted to unipolar representation. This is needed because the input voltage range is centered around

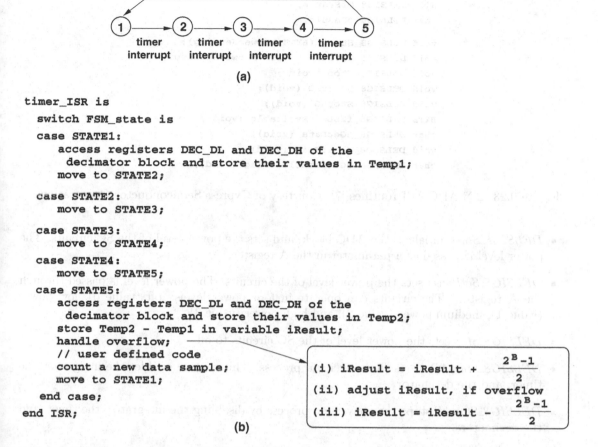

Figure 9.22: Timer interrupt service routine [7].

zero, that is $(-V_{ref}, +V_{ref})$. The conversion is achieved by shifting the input range up by the value V_{ref}, which is accomplished by adding the value

$$AD = \frac{2^B - 1}{2} \tag{9.51}$$

to the read value B being the number of bits of the ADC. (ii) If overflow has occurred, the read value is modified to the bit-representation of the largest (smallest, respectively) input value. (iii) The unipolar representation is converted back to the bipolar description, which corresponds to subtracting the value AD from the unipolar representation.

API routines: The API routines for the $\Delta\Sigma$ ADC include procedures for starting/stopping the ADC module, starting/stopping the data conversion process, accessing and clearing the converter's status information converter, and retrieving the converted data from the ADC. Figure 9.23 shows the prototypes for the existing API routines. In addition to these functions, variable *DELSIG8_Status* indicates the status of the ADC (the variable is zero unless new data are available), and variable *DELSIG8_cResult* holds the most recently converted byte.

The available API routines are [7]:

```
BOOL DELSIG8_bfStatus;
char DELSIG8_cResult;

void DELSIG8_Start (BYTE bPowerSetting);
void DELSIG8_SetPower (BYTE bPowerSetting);
void DELSIG8_Stop (void);
void DELSIG8_StartAD (void);
void DELSIG8_StopAD (void);
BYTE DELSIG8_fIsDataAvailable (void);
char DELSIG8_cGetData (void);
void DELSIG8_ClearFlag (void);
char DELSIG8_cGetDataClearFlag (void);
```

Figure 9.23: $\Delta\Sigma$ ADC API routines [7]. Courtesy of Cypress Semiconductor Corporation.

- *DELSIG8_Start* initializes the ADC block, and sets the power level of the SC circuits. The power level is passed as a parameter in the A register.

- *DELSIG8_SetPower* sets the power level of the circuits. The power level is passed through the A register. The circuits can operate in four power levels: off (value 0), low power (value 1), medium power (value 2), and high power (value 3).

- *DELSIG8_Stop* sets the power level of the SC circuits to off.

- *DELSIG8_StartAD* starts the conversion process. This routine enables the integrator, the timer, and the decimator.

- *DELSIG8_StopAD* stops the conversion process by disabling the integrator, the timer, and the decimators.

- *DELSIG8_fIsDataAvailable* confirms the availability of new converted data. This function returns a nonzero value, if data are ready.

- *DELSIG8_ClearFlag* clears the data availability flag.

- *DELSIG8_cGetData* returns the converted data.

- *DELSIG8_cGetDataClearFlag* returns the converted data and clears the data availability flag.

Figure 9.24 shows the Assembly language for the API routines.

ADC_AtoDCR2 and ADC_AtoDCR3 are the ASDxxCR2 and ASDxxCR3 registers of the SC block implementing the modulator. The ADC_T_DR1 and ADC_T_CR0 registers are DBBxxDR1 and DBBxxCR0 registers of the programmable digital block that implements the timer module. The ADC_ISR_REG register corresponds to the INT_MSKx register that enables the interrupt generation by the timer block. The DEC_CR1 register is the corresponding register of the decimator block. Finally, variable *ADC_fStatus* indicates the status of the ADC (i.e., whether new data are ready), and is set by the ISR. Variable *ADC_iResult* stores the converted data byte, and is set by the ISR.

```
macro ADC_RESET_INTEGRATOR_M
   IF ADC_NoAZ
     or    reg[ADC_AtoDCR2], 20H
   ENDIF
   or    reg[ADC_AtoDCR3], 10H
endm
```

```
ADC_fClearFlag:
_ADC_fClearFlag:
   mov    A, [ADC_fStatus]
   mov    [ADC_fStatus], 00H
   ret
```

```
ADC_fIsDataAvailable:
_ADC_fIsDataAvailable:

   mov    A, [ADC_fStatus]
   ret
```

```
ADC_cGetData:
_ADC_cGetData:

   mov    A, [ADC_iResult]
   ret
```

```
ADC_SetPower:
_ADC_SetPower:

   mov    X, SP
   and    A, 03H
   push   A
   mov    A, reg[ADC_AtoDCR3]
   and    A, ~03H
   or     A, [X]
   mov    reg[ADC_AtoDCR3], A
   pop    A
   ret
```

```
ADC_StartADC:
_ADC_StartADC:

   or     reg[DEC_CR1], C0H
   call   ADC_SetPower
   ADC_RESET_INTEGRATOR_M
   mov    reg[ADC_T_DR1], FFH
   or     reg[ADC_T_CR0], 01H
   ret
```

```
ADC_StopADC:
_ADC_StopADC:

   and reg[ADC_ISR_REG],~[ADC_ISR_MASK]
   ADC_RESET_INTEGRATOR_M
```

Figure 9.24: $\Delta\Sigma$ ADC API routines [7]. Courtesy of Cypress Semiconductor Corporation.

```
#include <m8c.h>
#include "PSoCAPI.h"
void main () {
  char cSample;
  M8C_EnableGInt;
  DELSIG8_Start (DELSIG8_HIGHPOWER);
  DELSIG8_StartAD();
  while (1) {
    if (DELSIG8_fIsDataAvailable()) {
      cSample = DELSIG8_cGetData();
      DELSIG8_ClearFlag();
    }
  }
}
```

Figure 9.25: Example of using the $\Delta\Sigma$ ADC API routines [7]. Courtesy of Cypress Semiconductor Corporation.

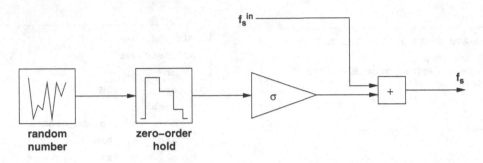

Figure 9.26: Modeling of jitter noise [2].

Example (Using the API routines) Figure 9.25 shows the C code for an application that uses the $\Delta\Sigma$ ADC API routines. First, the global interrupts are enabled. Then, the ADC block is started by calling the API routine, *DELSIG8_Start*, and the conversion process is initiated by the API routine, *DELSIG8_StartAD*. The program polls to determine the availability of new data by calling the *DELSIG8_fIsDataAvailable* routine. If new data are available, then they are returned by the *DELSIG8_cGetData* subroutine. The *ClearFlag* routine resets the ADC status flag before starting the next iteration of the while loop.

9.2.5 Impact of Circuit Non-Idealities on $\Delta\Sigma$ Modulator Performance

The performance of $\Delta\Sigma$ modulators is reduced by several types of circuit nonidealities [2, 3, 4, 5], for example (a) clock jitter, (b) switch thermal noise, (c) OpAmp noise, (d) OpAmp finite gain, (e) OpAmp slew rate, and (f) OpAmp saturation. This subsection estimates the performance degradation (e.g., in-band noise power) due to the different nonlinearity types. Performance degradation is characterized either in terms of analytical expressions if the mathematical computations are not too complex, or as simulation models.

A. *Impact of jitter noise.* Jitter noise is the noise introduced by variations in the sampling frequency f_s [3, 4]. The power spectral density for jitter noise is expressed by the following [3].

$$PSD_{jitter} = \frac{A^2}{2} \frac{(2\pi f_{in}\sigma)^2}{f_s} \tag{9.52}$$

where parameter A is the amplitude of a sinusoidal input signal, f_{in} the frequency of the signal, and σ the standard deviation of the sampling frequency.

The in-band noise power due to clock jitter is given by:

$$P_{in-band,jitter} = \int_{-f_b/2}^{f_b/2} PSD_{jitter}df = \frac{A^2}{2} \frac{(2\pi f_b\sigma)^2}{OSR} \tag{9.53}$$

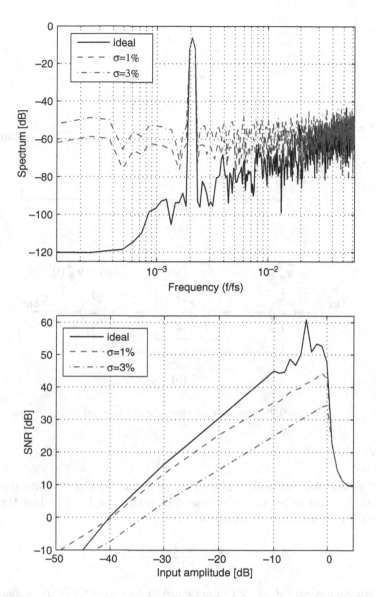

Figure 9.27: Effect of jitter noise on the PSD and DR of the first-order $\Delta\Sigma$ modulator.

This result shows that the in-band noise power decreases with the value of the oversampling ratio OSR, but increases with the signal amplitude A and frequency f_{in}.

Figure 9.26 shows the behavioral model used for modeling jitter noise [2]. The model adds the value obtained by the chain consisting of the random number generator, the zero-order hold block, and the block multiplying by the constant σ (the standard deviation) to the constant frequency f_s. Figure 9.27 illustrates the effect of jitter noise on the PSD and DR of a first-order $\Delta\Sigma$ modulator.

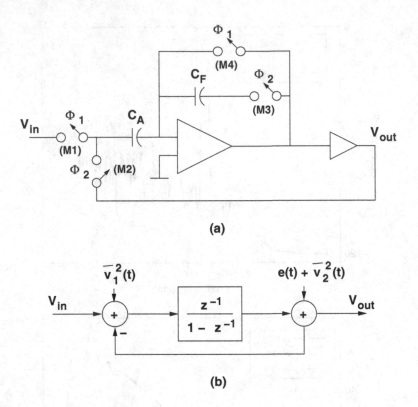

(a)

(b)

Figure 9.28: Impact of the switch thermal noise.

B. Impact of switch thermal noise. The switches in the SC circuits introduce thermal noise that increases the in-band noise power of the $\Delta\Sigma$ modulator [4, 5]. The thermal noise of a conducting MOSFET transistor M is expressed as

$$\bar{i}_M^2 = \frac{4kT}{R} \tag{9.54}$$

where k is Boltzmann's constant, T is temperature in Kelvin, and R the resistance of the conducting transistor.

Each of the switches in the SC modulator circuit is a noise source. Figure 9.28(b) shows the circuit including the noise sources for the circuit shown in Figure 9.28(a). The noise signal $\bar{v}_1^2$ represents the noise sources at the SC integrator's input, and the noise signal $\bar{v}_2^2$ the noise sources at the integrator output. Note that the impact of the noise signal $\bar{v}_2^2$ on the in-band noise power is much less than the impact of the signal $\bar{v}_1^2$. The reason is that the transfer function for signal $\bar{v}_1^2$ is the STF of the modulator, and the transfer function for $\bar{v}_2^2$ is the NTF of the modulator. Therefore, for simplicity reasons, this analysis considers only the noise source $\bar{v}_1^2$.

The in-band noise power, due to the switch thermal noise, is defined as:

$$P_{in-band,sw} = \int_{f_b/2}^{f_b/2} \bar{v}_1^2 \, |STF|^2 \, df \tag{9.55}$$

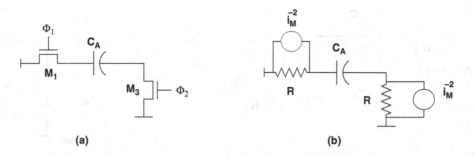

Figure 9.29: Modeling the impact of switching noise.

The noise source, $\bar{v}_1^2$, is due to the thermal noise of the switches controlled by the clock ϕ_1 (transistors M_1 and M_3), and the clock ϕ_2 (transistors M_2 and M_4). Hence, the noise, $\bar{v}_1^2$, includes the noise sources, $\bar{v}_{1,\phi_1}^2$, during clock ϕ_1 and $\bar{v}_{1,\phi_2}^2$ during clock ϕ_2:

$$\bar{v}_1^2 = \bar{v}_{1,\phi_1}^2 + \bar{v}_{1,\phi_2}^2 \tag{9.56}$$

The noise value, $\bar{v}_{1,\phi_1}^2$ represents the noise voltage across the capacitor C_A during the acquisition phase (the switches controlled by clock ϕ_1 are closed). This voltage generates the noise charge:

$$q\bar{c}_A{}^2 = C_A^2 \ \bar{v}_{1,\phi_1}^2 \tag{9.57}$$

The voltage, $\bar{v}_{1,\phi_1}^2$, is estimated using Figure 9.29(a), in which the conducting transistors M_1 and M_3 are replaced with their equivalent circuits, as shown in Figure 9.29(b). According to equation (9.54), the thermal noise of the two transistors is:

$$\bar{v}_{M1,2}^2 = \frac{4kT}{R_{M1}} \ R_{M1}^2 + \frac{4kT}{R_{M2}} \ R_{M2}^2 \tag{9.58}$$

If the two transistors are identical and hence, have equal on-resistances, R, then:

$$\bar{v}_{M1,2}^2 = 8kTR \tag{9.59}$$

The voltage $\bar{v}_{1,\phi_1}^2$ is given by:

$$\bar{v}_{1,\phi_1}^2 = \int_0^\infty \frac{8kTR}{1 + (2RC_A \ 2\pi f)^2} df = \frac{8kTR}{4\pi RC_A} \arctan(4\pi RCf) \ |_0^\infty = \frac{kT}{C} \tag{9.60}$$

The noise value $\bar{v}_{1,\phi_2}^2$ represents the noise voltage across the capacitor C_A during the transfer phase (the switches controlled by clock ϕ_2 are closed). Figure 9.30(a) represents the SC circuit during the charge transfer phase. After solving the nodal equations for the circuit, the following expression results for the voltage across the capacitor C_A:

$$\frac{v_{C_A}}{v_i} \approx \frac{A \ R_L \ X_{CA}}{2AR_LR + AR_LX_{CA} + X_{CA} + X_{CF}} \approx \frac{1}{1 + s(2R + \frac{1}{A})C_A} \tag{9.61}$$

where

$$X_{CA} = \frac{1}{s \ C_A} \tag{9.62}$$

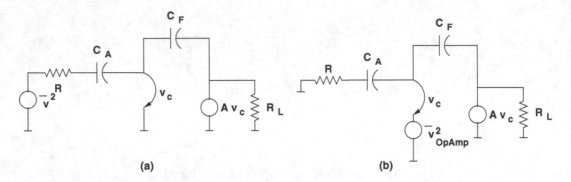

Figure 9.30: Analysis of the switch thermal noise and OpAmp noise.

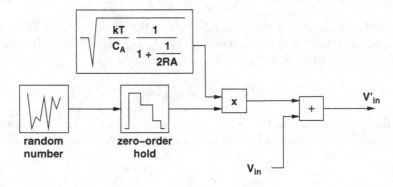

Figure 9.31: Modeling of the switch thermal noise [2].

and

$$X_{CF} = \frac{1}{s\, C_F} \tag{9.63}$$

are the impedances of the two capacitors, and the resistance R_L is very large. A is the gain of the OpAmp.

Using expression (9.61), the value of the voltage, $\bar{v}^2_{1,\phi_2}$, is:

$$\bar{v}^2_{1,\phi_2} = \int_0^\infty \frac{8kTR}{1 + [2RC_A\,(1+\frac{1}{A})\,2\pi f]^2}\,df = \frac{kT}{C_A}\left[\frac{1}{1+\frac{1}{2RA}}\right] \tag{9.64}$$

Substituting equations (9.60) and (9.64) into equations (9.55) and (9.56), the in-band noise power due to the switch thermal noise becomes:

$$P_{in-band,sw} = \int_{\frac{-f_b}{2}}^{\frac{f_b}{2}} \left[\frac{kT}{C_A} + \frac{kT}{C_A(1+\frac{1}{2R\,A})}\right]|STF|^2 df = \left[\frac{kT}{C_A} + \frac{kT}{C_A(1+\frac{1}{2R\,A})}\right]f_b \tag{9.65}$$

Alternatively, the effect of the switch thermal noise can be studied using the simulation model shown in Figure 9.31. Figure 9.32 illustrates the PSD and DR degradation due to the switch's thermal noise in a first-order $\Delta\Sigma$ modulator.

C. Impact of OpAmp noise. The modulator performance degradation due to the OpAmp noise can be estimated using the circuit in Figure 9.30(b) where OpAmp noise is denoted as v^2_{OpAmp}.

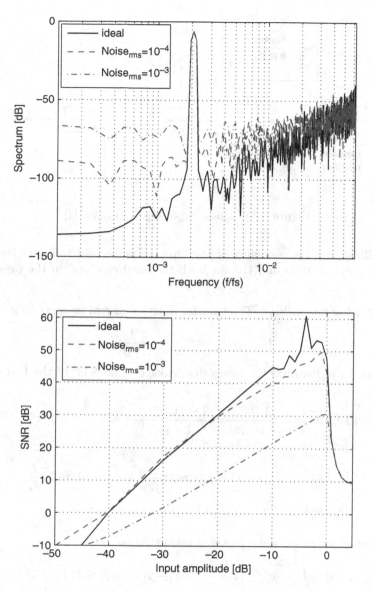

Figure 9.32: Impact of the switch thermal noise on the PSD and DR of a first-order $\Delta\Sigma$ modulator.

The voltage across the capacitor C_A is given by

$$\frac{v_{C_A}}{v_{OpAmp}} \approx -\frac{1}{1 + s(2R + \frac{1}{A})C_A} \tag{9.66}$$

The noise voltage across the capacitor C_A can be estimated by:

$$\bar{v}_{C_A,OpAmp}^2 = \int_0^{\infty} \frac{\bar{v}_{OpAmp}^2}{1 + (2RC_A\,(1 + \frac{1}{A})\,2\pi f)^2}df = \frac{\bar{v}_{OpAmp}^2}{8RC_A(1 + \frac{1}{A})} \tag{9.67}$$

Similarly, an estimate of the in-band noise power due to the OpAmp noise can be obtained from

$$P_{in-band,sw} = \int_{-f_b/2}^{f_b/2} \frac{\bar{v}_{OpAmp}^2}{8RC_A(1 + \frac{1}{A})}|STF|^2df = \frac{\bar{v}_{OpAmp}^2}{8RC_A(1 + \frac{1}{A})}f_b \tag{9.68}$$

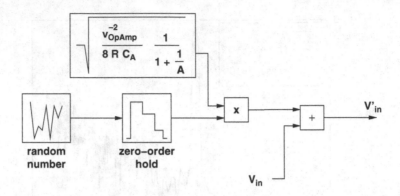

Figure 9.33: Modeling of OpAmp noise [2].

Figure 9.33 illustrates the modeling of the OpAmp noise [2]of a first-order $\Delta\Sigma$ modulator. Figure 9.34 is a graph of PSD and DR for both the ideal case and in the presence of OpAmp noise.

D. *Impact of integrator leakage.* The transfer function of real integrators is modeled by [3]:

$$H_{integ}(z) = \frac{z^{-1}}{1 - (1 - \mu)z^{-1}} \tag{9.69}$$

where the parameter μ describes the integrator leakage. The NTF of the first-order modulator is defined as

$$NTF(z) = \frac{1}{1 + H_{integ}(z)} = \frac{1 - (1 - \mu)^{z^{-1}}}{1 + \mu z^{-1}} \approx 1 - (1 - \mu)z^{-1} \tag{9.70}$$

and

$$|NTF|^2 = [1 - (1 - \mu)\cos 2\pi \frac{f}{f_s}]^2 + [(1 - \mu)\sin 2\pi \frac{f}{f_s}]^2 \tag{9.71}$$

$$|NTF|^2 = 1 + (1 - \mu)^2 - 2(1 - \mu)\cos 2\pi \frac{f}{f_s} \tag{9.72}$$

The total in-band quantization noise power is estimated by

$$P_{in-band} = \int_{-f_b2}^{f_b/2} e^2 f_s \left[1 + (1 - \mu)^2 - 2(1 - \mu)\cos 2\pi f f_s\right] df$$

$$P_{in-band} = \int_{-f_b/2}^{f_b/2} \frac{e^2}{f_s} (\mu^2 - 2\mu + 2) \, df - \int_{-f_b/2}^{f_b/2} \frac{e^2}{f_s} 2(1 - \mu)\cos 2\pi \frac{f}{f_s} \, df$$

$$P_{in-band} = \frac{\mu^2 - 2\mu + 2}{OSR} \left[\frac{\Delta^2}{12}\right] - 2(1 - \mu)\sin\left(2\pi \frac{f_b}{f_s}\right) \left[\frac{\Delta^2}{\pi \, 12}\right] \tag{9.73}$$

$$P_{in-band} \approx \frac{\mu^2 - 2\mu + 2}{OSR} \left[\frac{\Delta^2}{12}\right] - 2(1 - \mu)\left[\pi \frac{f_b}{f_s} - \frac{(\pi \frac{f_b}{f_s})^3}{3!}\right] \left[\frac{\Delta^2}{\pi \, 12}\right] \tag{9.74}$$

Thus, the total in-band quantization noise power can be estimated and is given by

$$P_{in-band} = \frac{\mu^2}{OSR} \left[\frac{\Delta^2}{12}\right] + (1 - \mu)\frac{\pi^2}{OSR^3} \left[\frac{\Delta^2}{36}\right]$$

$$= \frac{\pi^2}{OSR^3} \left[\frac{\Delta^2}{36}\right] + \frac{\Delta^2}{12} \left[\frac{\mu^2}{OSR} - \mu\frac{\pi^2}{3OSR^3}\right] \tag{9.75}$$

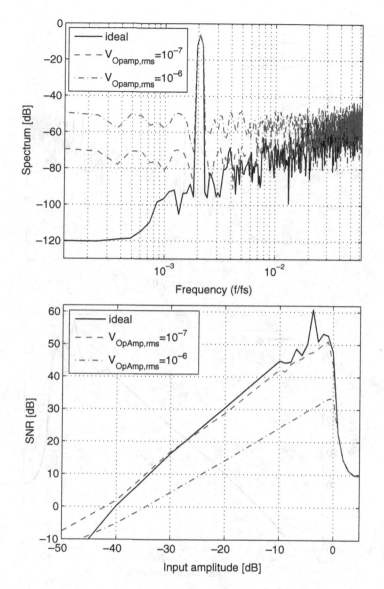

Figure 9.34: Effect of OpAmp noise on the PSD and DR of a first-order $\Delta\Sigma$ modulator.

The increase of the in-band quantization noise power due to the integrator losses is equal to the value

$$\Delta P_{in-band} = \frac{\Delta^2}{12}\left[\frac{\mu^2}{OSR} - \mu\frac{\pi^2}{3OSR^3}\right] \tag{9.76}$$

Hence, the effect of the integrator losses on the modulator accuracy is reduced by increasing the modulator's oversampling ratio, OSR. Figure 9.35 illustrates the degradation due to the integrator leakage of the PSD, and DR, of a first-order $\Delta\Sigma$ modulator.

E. Impact of OpAmp slew rate. Figure 9.36 shows the slew-rate model for the OpAmp of an integrator block. Finding analytical expressions that relate the modulator performance to the

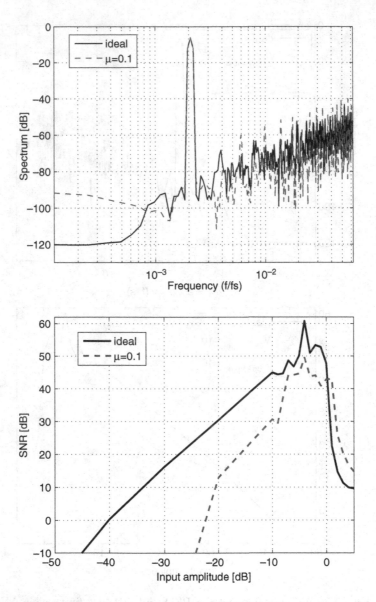

Figure 9.35: Impact of the integrator leakage on the PSD and DR of a first-order $\Delta\Sigma$ modulator.

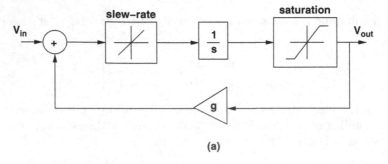

Figure 9.36: Modeling of OpAmp slew rate and saturation [2].

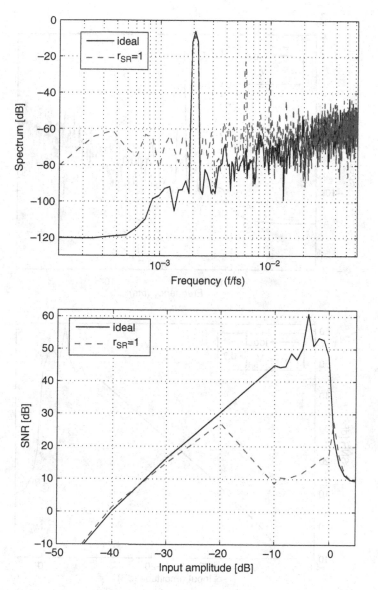

Figure 9.37: Effect of OpAmp slew rate on the PSD and DR of a first-order ΔΣ modulator.

OpAmp slew-rate is difficult. Figure 9.37 is a graph of the PSD and DR of a first-order ΔΣ ADC in the presence of OpAmp slew rate.

F. Impact of OpAmp saturation. The effect of OpAmp saturation on the modulator's performance is difficult to express in closed-form. Figure 9.36 shows the modeling of OpAmp saturation for an integration block [2]. Figure 9.38 is a graph of the PSD and DR of a first-order ΔΣ ADC in the presence of OpAmp saturation.

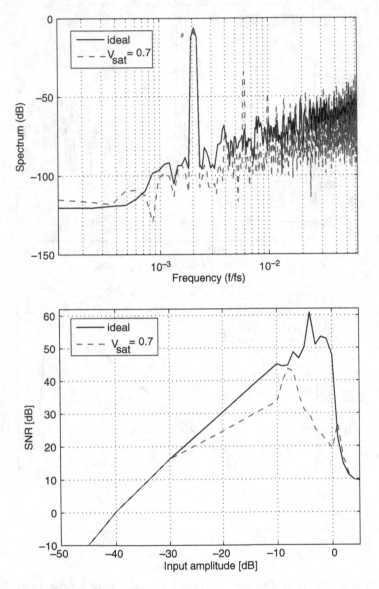

Figure 9.38: Effect of OpAmp saturation on the PSD and DR of a first-order $\Delta\Sigma$ modulator.

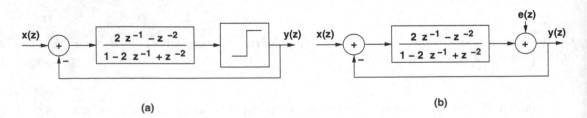

(a)

(b)

Figure 9.39: Second-order $\Delta\Sigma$ modulator.

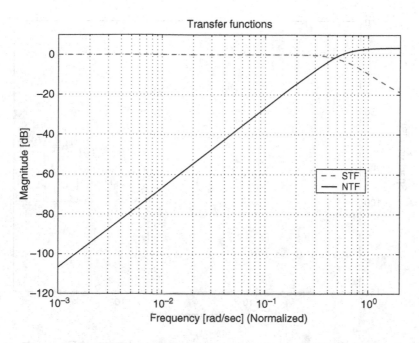

Figure 9.40: STF and NTF for the second-order $\Delta\Sigma$ modulator.

9.2.6 Second-Order $\Delta\Sigma$ Modulator

Figure 9.39 shows the structure of a second-order $\Delta\Sigma$ modulator. In this case, the transfer function H(z) is defined as [1, 5]:

$$H(z) = \frac{2\,z^{-1} - z^{-2}}{1 - 2\,z^{-1} + z^{-2}} \tag{9.77}$$

The modulator output is given by

$$y(z) = \frac{2\,z^{-1} - z^{-2}}{1 - 2\,z^{-1} + z^{-2}}\,[x(z) + y(z)] + e(z) \tag{9.78}$$

$$y(z) = [2\,z^{-1} + z^{-2}]\,x(z) + [1 - z^{-1}]^2\,e(z) \tag{9.79}$$

Hence, the modulator STF is

$$STF = 2\,z^{-1} + z^{-2} \tag{9.80}$$

and the NTF is

$$NTF = (1 - z^{-1})^2 \tag{9.81}$$

Figure 9.40 shows graphs of the STF and NTF. Note the noise-shaping characteristics of the modulator NTF. Figure 9.41 illustrates the power spectral density of the second-order modulator for OSR equal to 32, a normalized sampling frequency and an input signal frequency equal to 1/512 of the sampling frequency. This graph shows that the circuit operates as an ADC, because most of the signal power is concentrated at the input frequency. The quantization noise can be approximated reasonably well as white noise.

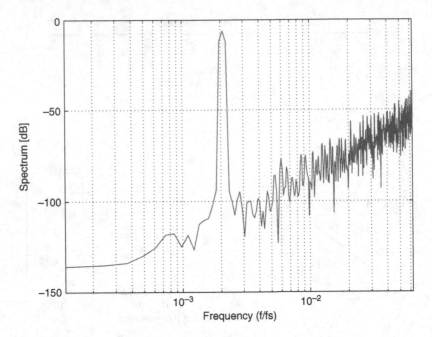

Figure 9.41: Power spectral density for the second-order $\Delta\Sigma$ modulator.

The in-band quantization noise power of the second-order modulator must be estimated in order to find prediction models for the modulator's performance. Substituting the expression for NTF, i.e., Equation (9.81) into Equation (9.36), yields the in-band quantization noise power as:

$$P_{in-band} = \frac{e^2}{f_s} \int_{-f_b/2}^{f_b/2} \left[1 - \cos\left(2\pi\frac{f}{f_s}\right) + \cos\left(4\pi\frac{f}{f_s}\right)^2 + 2\sin\left(2\pi\frac{f}{f_s}\right) - \sin\left(4\pi\frac{f}{f_s}\right)^2 \right] df \quad (9.82)$$

and

$$P_{in-band} = \frac{e^2}{f_s} \left[6 f_b - \frac{4 f_s}{\pi} \sin\left(2\pi\frac{f_b}{f_s}\right) + \frac{f_s}{2\pi} \sin\left(4\pi\frac{f_b}{f_s}\right) \right] \quad (9.83)$$

Using the approximation:

$$\sin x \approx x - \frac{1}{3!} x^3 + \frac{1}{5!} x^5 \quad (9.84)$$

yields,

$$\sin\left[2\pi\frac{f_b}{f_s}\right] = 2\pi\frac{f_b}{f_s} - \frac{1}{3!}\left[2\pi\frac{f_b}{f_s}\right]^3 + \frac{1}{5!}\left[2\pi\frac{f_b}{f_s}\right]^5 \quad (9.85)$$

$$\sin\left[4\pi\frac{f_b}{f_s}\right] = 4\pi\frac{f_b}{f_s} - \frac{1}{3!}\left[4\pi\frac{f_b}{f_s}\right]^3 + \frac{1}{5!}\left[4\pi\frac{f_b}{f_s}\right]^5 \quad (9.86)$$

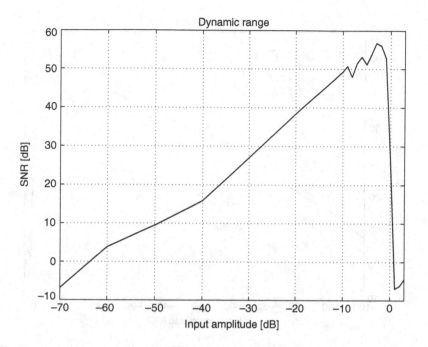

Figure 9.42: Dynamic range for the second-order ΔΣ modulator.

and the in-band quantization noise power is

$$P_{in-band} \approx \frac{2e^2}{OSR}\left[3 - 4\frac{2\pi\frac{f_b}{f_s} - \frac{1}{3!}(2\pi\frac{f_b}{f_s})^3 + \frac{1}{5!}(2\pi\frac{f_b}{f_s})^5}{2\pi\frac{f_b}{f_s}}\right.$$
$$\left. +\frac{4\pi\frac{f_b}{f_s} - \frac{1}{3!}(4\pi\frac{f_b}{f_s})^3 + \frac{1}{5!}(4\pi\frac{f_b}{f_s})^5}{4\pi\frac{f_b}{f_s}}\right] \tag{9.87}$$

$$P_{in-band} \approx \frac{\pi^4\, e^2}{5\, OSR^5} \tag{9.88}$$

Hence, if the modulator uses a one-bit quantizer, i.e., $\Delta = 2$, then

$$P_{in-band} \approx \frac{\pi^4}{15\, OSR^5} \tag{9.89}$$

For a sinusoidal input signal, the SNR of the second-order ΔΣ modulator is given by

$$\text{SNR (dB)} = 10\log\left[\frac{15\, A^2\, OSR^5}{2\pi^4}\right] \tag{9.90}$$

This result shows that doubling the OSR increases SNR by 15 dB corresponding to 2.5 bits, in precision.

Figure 9.42 shows the DR graph for OSR = 32 dB. The frequency of the input signal in this case was 1/512 of the sampling frequency. The maximum SNR is $\approx 56\ dB$, which is $\approx 10\ dB$ more than for the first-order ΔΣ modulator. The dynamic range is $\approx 65\ dB$, as compared to 34 dB for the first-order ΔΣ modulator. Figure 9.43 shows the DR graphs for four OSR values (32, 64, 128, and 256). Note that DR and the maximum SNR increase with the OSR value, as expected.

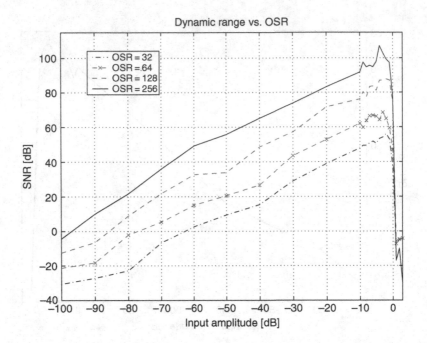

Figure 9.43: Dynamic range for a second-order $\Delta\Sigma$ modulator and different OSR values.

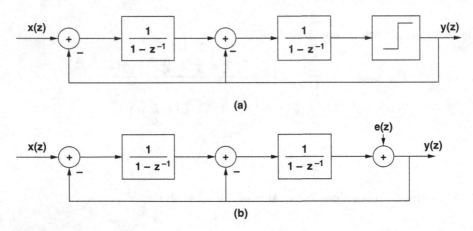

Figure 9.44: Topology and analysis model for a second-order $\Delta\Sigma$ modulator.

Also, for an input of -20 dB and OSR of 32, the resulting SNR is close to 39 dB. For the OSR of 256, the SNR increases to ≈ 84 dB, which is 15 dB higher, as predicted by Equation (9.90).

The PSoC implementation of the second-order $\Delta\Sigma$-ADC is similar to that of the first-order $\Delta\Sigma$ ADC shown in Figure 9.17. The only difference is the use of a second-order $\Delta\Sigma$ modulator instead of a first-order $\Delta\Sigma$ modulator. Figure 9.44 shows the topology for the second-order $\Delta\Sigma$ modulator [7]. Figure 9.44(a) shows the block structure of the topology, and Figure 9.44(b) illustrates the linear model used in a performance analysis. The circuit implementation of the modulator is shown in Figure 9.45. This implementation uses two programmable SC blocks to implement the modulator's two integrators. The digital part (the decimator and timer) and the related firmware routines are similar to those of the first-order modulator.

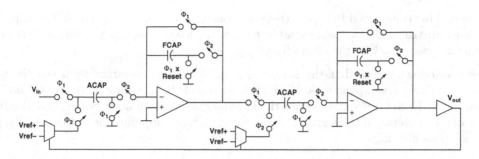

Figure 9.45: PSoC-based implementation of the second-order $\Delta\Sigma$ modulator [7].

9.3 Conclusions

This chapter has presented a discussion of the basic concepts of $\Delta\Sigma$ analog-to-digital convert-ers (ADCs), and detailed the implementation of first-order and second-order $\Delta\Sigma$ ADCs using the PSoC architecture.

ADCs are important subsystems in many embedded mixed-signal systems. They convert continuous-valued, continuous-time signals into discrete-valued discrete-time, data. The ADC op-eration is based on two steps, sampling and quantization. Sampling represents a continuous-time, continuous-valued signal as a discrete-time, continuous-valued signal. Quantization approximates continuous-valued signals as discrete-valued data.

The Nyquist sampling theorem was also introduced in this chapter and defines the sufficient condition for accurate reconstruction of the sampled signal from the sampling data. The Nyquist sampling theorem states that the signal can be correctly reconstructed if the sampling frequency is greater than twice the frequency of the highest spectral component (called Nyquist frequency). The sampled data included the original signal and images at negative/positive multiples of the sampling frequency. If the sampling frequency is greater than the Nyquist frequency, then the signal and the images do not overlap, otherwise aliasing occurs as a result of overlapping of the signal and its images. The original signal can be reconstructed by lowpass filtering. The transfer function of the filter is

$$H(f) = \text{rect}\frac{f}{f_s} \tag{9.91}$$

As shown, a B-bit quantizer approximates a continuous-valued signal as a bitstring of B bits. The introduced quantization error e_r is modeled as white noise with a power spectral density of $1/\Delta$, where Δ is the discretization range of the quantizer. The accuracy of the signal con-version process depends on the quantization noise power, which, for the white noise model for quantization noise, is expressed as:

$$\sigma_e^2 = \frac{\Delta^2}{12} \tag{9.92}$$

$$= \frac{1}{3(2^B - 1)^2} \tag{9.93}$$

Hence, the conversion accuracy improves if the number of quantization bits, B, is higher.

ADCs include a modulator and a decimator and to achieve the desired conversion accuracy, the modulator samples the input signal with the goal of minimizing the in-band quantization noise power. This is achieved by two orthogonal concepts, oversampling and noise-shaping. The decimator reconstructs the sampled signal by conducting lowpass filtering to eliminate the images, and downconversion back to the signal frequency band.

- Oversampling implies that the sampling frequency is much larger than the Nyquist frequency. The oversampling ratio (OSR) is the ratio of the sampling frequency to the Nyquist frequency. High OSR values reduce the in-band quantization noise power, and also increase the distance between the signal and its images, which simplifies the removal of the images by lowpass filtering.

- Noise-shaping selects the $\Delta\Sigma$ modulator transfer function such that the signal transfer function (STF) is an allpass filter, and the noise transfer function (NTF) is a highpass filter that eliminates the in-band quantization noise.

The performance of $\Delta\Sigma$ modulators in terms of accuracy of the signal conversion process is characterized by

- Signal-to-noise ratio (SNR) which is the ratio of the signal power to the in-band quantization noise power. Increasing SNR by 3 dB improves accuracy by 0.5 bits. Also, doubling OSR increases SNR by 3 dB.

- Dynamic range (DR) is the ratio of the output power of a sinusoidal input, with full-range amplitude, to the output power of the input signal, for which the SNR is zero.

The theory and implementation of first-order and second-order $\Delta\Sigma$ ADCs and the STF/NTF of associated modulators was discussed in detail. Analytical expressions and simulation results were derived for the power spectrum density, in-band quantization noise power, and SNR of the first-order and second-order $\Delta\Sigma$ modulators.

The PSoC-based implementation of an eight-bit first-order $\Delta\Sigma$ ADC was also discussed. The implementation included three parts, the modulator, decimator, and API routines:

- The modulator used PSoC's programmable SC blocks with an OSR of 64.

- The decimator, for lowpass filtering and downconversion, was based on a digital $sinc^2$ filter and a hardware/software design with the integration part of the decimator implemented using PSoC's customized Type 1 decimator blocks and the differentiation part of the decimator conducted in software.

  To keep the overhead low, downconversion was achieved by using a programmable digital block for generating the interrupts that activated the differentiation routine. The programmable block was configured as a timer, initialized with the OSR value, and produced an interrupt upon counting down to the value zero.

- The API routines discussed included subroutines for (i) starting/stopping an ADC, and the signal conversion process, (ii) managing (reading/clearing) the status of the converter, and (iii) reading the converted data values. This chapter also provided an example of reading data from an ADC by a polling mechanism.

Finally, analytical expressions and simulation models for estimating the ADC performance degradation due to nonidealities (e.g., clock jitter, switch thermal noise, integrator leakage, and OpAmp noise, finite gain, slew rate, and saturation) were presented and discussed in detail. Simulation graphs illustrated and quantified the degradation in performance, (e.g., power spectrum density and DR) as a result of the presence of certain nonidealities.

Bibliography

[1] G. Bourdopoulos, A. Pnevmatikakis, V. Anastassopoulos, T. Deliyannis, Delta-Sigma Modulators. *Modeling, Design and Applications*, London: Imperial College Press, 2003.

[2] P. Malcovati, S. Brigati, F. Francesconi, F. Maloberti, P. Cusinato, A. Baschirotto, Behavioral Modeling of switched-capacitor sigma-delta modulators, *IEEE Transactions on Circuits and Systems-I: Fundamental Theory and Applications*, Vol. 50, 3, pp. 352–364, March 2003.

[3] F. Medeiro, A. Perez-Verdu, A. Rodriguez-Vazquez, *Top-Down Design of High-Performance Sigma-Delta Modulators*, Kluwer Academic Publishers, Boston, 1999.

[4] S. Norsworthy, R. Schreier, G. Temes, Delta-Sigma Data Converters. *Theory, Design, and Simulation*, Piscataway, NJ: Nyork, IEEE Press, 1997.

[5] R. Schreier, G. Temes, *Understanding Delta-Sigma Data Converters*, New York: J. Wiley & Sons, 2005.

[6] PSoC Mixed Signal Array, Technical Reference Manual, Document No. PSoC TRM 1.21, Cypress Semiconductor Corporation, 2005.

[7] DelSig8 v3.2, 8 Bit Delta Sigma ADC, *Application Note*, Cypress Semiconductor Corporation, Oct. 3 2005.

[8] http://www.wikipedia.org.

Chapter 10

Future Directions in Mixed-Signal Design Automation

The goal of embedded system design is to develop customized hardware and software for an embedded application, so that the functional, cost, and performance needs of the application are met. This chapter introduces some of the current challenges related to design automation for analog and mixed-signal systems, as advanced topics. The topics include high-level specification of analog and mixed-signal systems, a simulation method for fast performance estimation and high-level synthesis of analog subsystems.

10.1 Top-Down Design and Design Activities

Figure 10.1 summarizes some of the popular modules in embedded applications, e.g., temperature and humidity sensing, statistical processing of the sensed data, data communication protocols, image processing, data encryption and decryption, control algorithms, and interfacing to a LCD. Most of these modules have been discussed in the previous chapters. In the following discussion, the implementations of the IDEA encryption routine and a face detection image processing procedure are discussed.

A typical top-down design flow for embedded systems starts from an abstract system specification for the functional, interfacing, cost, and performance requirements of the system. The design is incrementally refined by continuously adding new implementation details to the design. Each refinement step optimizes the design by conducting tradeoff analysis that considers different design solutions and their impact on the overall system cost and performance. The design flow should incorporate a performance evaluation mechanism to evaluate the quality of design decisions, and a modeling procedure to express the defining attributes of implementation modules.

The top-down design flow involves the following activities:

- *System specification:* The process of describing the defining properties of an embedded system.

- *Functional partitioning:* Reorganizing a system specification into modules, each having a specialized functionality, for example a certain interfacing function, or data processing.

A. Doboli, E.H. Currie, *Introduction to Mixed-Signal, Embedded Design*,
DOI 10.1007/978-1-4419-7446-4_10, © Springer Science+Business Media, LLC 2011

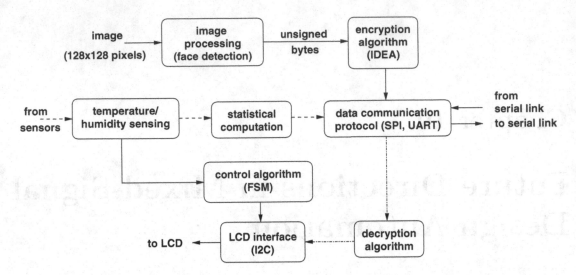

Figure 10.1: Signal and data processing in embedded applications.

- *System-level tradeoff analysis:* Mapping system-level performance and design requirements into requirements for the building blocks. Hence, this step produces a constraint transformation. Numerous design tradeoffs are analyzed (e.g., cost-speed, speed-power consumptions, accuracy-speed, etc.).

The outcomes of tradeoff analysis may include determining whether a module should be implemented in hardware or software (an activity referred to as hardware–software partitioning), the number of I/O ports and communication protocols to be used, mapping input and output signals to ports, allocating the hardware resources used in customization, mapping data to memory pages, and so on. Note that the system modules can be individually designed after a system-level tradeoff analysis.

- *Design of the individual modules:* Design of interfacing modules, data processing modules (for example video and audio processors), memory modules, and bus communication subsystem.

Modules can include hardware circuits and software routines. Complex modules undergo a separate top-down design process.

- *Analog circuit design:* Implementation of analog and mixed-signal circuits.

- *Digital circuit design:* Implementation of the customized digital circuits, for example interfacing circuits and customized data processing hardware.

- *Software development:* Creation of application software and system software, for example methods for data processing, control, and graphical user interfaces (GUIs).

- *Circuit modeling:* Characterizing the behavior of hardware circuits, for example propagation delay, power consumption, noise margins for digital circuits, and poles and zeros, bandwidth, harmonic distortion, and circuit noise for analog circuits.

- *Software characterization:* Development of the models that predict the performance of software routines, including speed, required memory, memory access patterns, and power consumption.

```
void IDEA () {
  WORD x[32][4];
  WORD y[32][4];
  WORD z[9][6];
  for (i=0; i<32; i++) {
    tempx1 = x[i][0];
    tempx2 = x[i][1];
    tempx3 = x[i][2];
    tempx4 = x[i][3];
    // eight rounds of transformations
    for (j=0; j<8; j++) {
      result1 = tempx1 * z[j][0];  // (1) multiply x1 and 1st subkey
      result2 = tempx2 + z[j][1];  // (2) add x2 and 2nd subkey
      result3 = tempx3 + z[j][2];  // (3) add x3 and 3rd subkey
      result4 = tempx4 * z[j][3];  // (4) multiply x4 and 4th subkey
      result5 = result1 ^ result3; // (5) XOR result of (1) and (3)
      result6 = result2 ^ result4; // (6) XOR result of (2) and (4)
      result7 = result5 * z[j][4]; // (7) multiply (5)and 5th subkey
      result8 = result6 + result7; // (8) add result of (6) and (7)
      result9 = result8 * z[j][5]; // (9) multiply (8)and 6th subkey
      result10 = result7 + result9; // (10) add result of (7) and(9)
      tempx1 = result1 ^ result9;  // (11) XOR result of (1) and (9)
      tempx3 = result3 ^ result9;  // (12) XOR result of (3) and (9)
      tempx2 = result2 ^ result10; // (13) XOR result of (2) and (10)
      tempx4 = result4 ^ result10; // (14) XOR result of (4) and (10)
    }
    // final transformation
    y[i][0] = tempx1 * z[8][0];  // (1) multiply x1 and 1st subkey
    y[i][1] = tempx3 + z[8][1];  // (2) add x2 and 2nd subkey
    y[i][2] = tempx2 + z[8][2];  // (3) add x3 and 3rd subkey
    y[i][3] = tempx4 * z[8][3];  // (4) multiply x4 and 4th subkey
  }
}
```

Figure 10.2: The IDEA algorithm.

- *Performance evaluation:* Analysis of the system performance that can be based on analytical expressions (e.g., system models, processor models, circuit macromodels), or on a simulation of simplified descriptions of the system.

10.2 Two Examples of Architecture Customization

This section provides a succinct description of two case studies for the design of embedded system implementations: (i) an IDEA encryption routine, and (ii) a face detection algorithm for image processing. In this discussion, particular stress is placed on top-down design (e.g., specification, tradeoff analysis, profiling or modeling, and performance evaluation).

10.2.1 IDEA Algorithm for Data Encryption

This subsection presents the PSoC-based implementation proposed by V. Sabino and J. Powers for the International Data Encryption Algorithm (IDEA) [30]. The design goal was to provide a time-efficient implementation of the algorithm using PSoC. The solution explores performance

```
; ASM_z = varx * vary
mov   A, [varx]            ; x_MSB
mov   reg[MUL_X], A        ; to MUL_X
mov   reg[MAC_CL0], A      ; clear accumulator
mov   A, [vary+1]          ; y_LSB
mov   reg[MAC_Y], A        ; to MAC_Y  ACC = (x_MSB * y_LSB)
mov   A, [varx+1]          ; x_LSB
mov   reg[MUL_X], A        ; to MUL_X  MUL = (x_LSB * y_LSB)
mov   A, reg[MUL_DH]       ; MUL_DH
tst   [varx+1], 0x80       ; MSBit of x_LSB is 1?
jz    .+4                  ; if not skip over add
add   A, [vary+1]          ; add to y_LSB
tst   [vary+1], 0x80       ; MSBit of y_LSB is 1?
jz    .+4                  ; if not skip over add
add   A, [varx+1]          ; add to y_LSB
mov   [ASM_z], A           ; to ASM_z-MSB
mov   A, reg[MUL_DL]       ; MUL_DL
mov   [ASM_z+1], A         ; to ASM_z_LSB
mov   A, [vary]            ; y_MSB
mov   reg[MAC_Y], A        ; to MAC_Y  ACC = (x_MSB * y_LSB) + (x_LSB * y_MSB)
mov   A, reg[ACC_DR0]      ; ACC_LSB
add   [ASM_z], A           ; add to ASM_z_MSB
```

Figure 10.3: Optimized 16-bit unsigned multiplication algorithm.

optimization by customization of PSoC's programmable, digital resources to address the performance needs of the algorithm. The solution follows the design and tradeoff analysis procedures illustrated in Chapter 4.

The IDEA algorithm generates four 16-bit unsigned integers that are the encryptions of the four 16-bit unsigned integers provided at the input. The outputs are stored in the y-matrix, and the inputs are in the x-matrix. Encryptions are computed in eight iterations and a final transformation, as shown in Figure 10.2. The algorithm uses fifty-two, 16-bit, key values that are stored in the z-matrix.

The profiling of the IDEA algorithm (described in the C language) indicated that the total number of clock cycles is $\approx$ 2.4 millions cycles. This corresponds to an execution time of $\approx$ 100 $msec$ for a clock frequency of 24 MHz. Hence, only 80 bytes can be encrypted per second, which is a serious constraint for a real-life application. Moreover, an analysis showed that $\approx$ 63% of the total execution time is spent performing multiplications, and $\approx$ 32% of the time is utilized for array indexing and memory paging. The total time required to execute the other instructions (e.g., addition, exclusive or, etc.) is less than 5%.

To speedup the IDEA execution time, an optimized algorithm, shown in Figure 10.3, was developed for fast multiplication of 16-bit unsigned integers. Note that the optimized algorithm's execution time is $\approx$ 112 clocks, as compared to $\approx$ 1400 clocks for integer multiplication in the C language. The algorithm uses PSoC's MAC block which is described in Chapter 4.

Array indexing and memory paging consumed approximately 1/3 of the total execution time of the algorithm. Table 10.1 details the total number of clocks for the instructions that are involved in accessing matrices x, y, and z for two possible addressing modes: indirect, by instructions MVI, and indexed. The setup time includes the overhead due to the instructions that establishes the

Table 10.1: Comparison of the indirect and indexed addressing modes.

	MVI instructions			Indexed addressing		
	# Clocks	# Loops	Total # Clocks	# Clocks	Loops	Total # Clocks
Setup	10	1	10	22	32	704
Memory access	20	256	5120	12	256	3072
Other	13	64	832	9	64	576
Other	6	32	192	7	32	224

values of the registers, for example the MVR_PP and MVW_PP registers used for reading and writing, based on indirect addressing. The table shows that MVI instructions have a lower setup overhead than indexed addressing. This is because the matrices x- and y-matrices can be stored in different pages (the x-matrix is read and the y-matrix is written), so that the MVR_PP and MVW_PP registers do not have to be modified. Indexed addressing requires modifying the IDX_PP register each time the memory page needs to be changed for addressing another data matrix, which increases the setup time significantly.

Nevertheless, the total number of clock cycles for indexed addressing is 4576 clock cycles, as opposed to 6154 clock cycles for indirect addressing. This is due to the fact that memory access is much slower for MVI instructions. In the final design, the IDEA implementation used indexed addressing in which the x and y-matrices were placed in different SRAM pages, and the z-matrix was stored in the local page. The resulting execution time for indexing and paging was $\approx$ 2457 clock cycles, which is almost 50% better than the unoptimized case.

The execution time for the optimized implementation was approximately 189, 241 clock cycles, less than the initial 2.4 millions clock cycles. The execution time was approximately 7.9 $msec$ for a clock frequency of 24 MHz. Hence, the optimized implementation was about12 times faster than the initial design. In the optimized design, $\approx$ 64% of the total time was required for multiplication, and only 1.3% for indexing and memory paging. The rest of the instructions, which represented less than 5% of the initial execution time, were $\approx$ 35% of the execution time. This result suggests that the performance-criticality of an instruction depends not only on the particular instruction but also on how the instruction was implemented. This is an important observation, in as much as further reduction of the overall execution time (i.e., below 189,241 clock cycles) would also require optimizing instructions other than multiplication, indexing, and memory paging.

10.2.2 Face Detection for Image Processing[1]

A simple algorithm for human face detection is discussed in this subsection as an illustrative example of an image processing algorithm. The goal of the algorithm is to identify human faces in an image in real-time. There are several kinds of face detection algorithms, such as

---

[1] Portions reprinted with permission from Y. Weng, A. Doboli, Smart sensor architecture customized for image processing. Proceedings of the 10th IEEE Real-Time and Embedded Technology and Applications Symposium, 396-403: 2004.

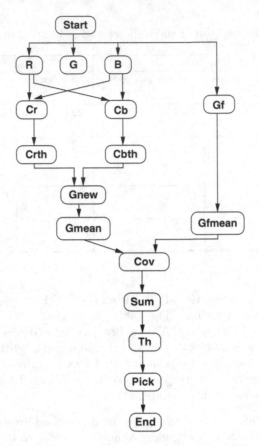

Figure 10.4: Task graph of the face detection algorithm [37] © 2010 IEEE.

knowledge-based, feature invariant, appearance-based, and template matching methods [38]. This subsection focuses on a simplified, template-matching algorithm for face detection [37].

The steps of the face detection algorithm are summarized in Figure 10.4:

- After the image is loaded, tasks R, G, B extract the red, green, and blue information from the image which can be extracted in parallel, as shown. The image size is 128×128 pixels.

- Steps Cr and Cb convert the RGB color space into the YCbCr color space. This step is executed because the skin can be detected more effectively due to the smaller variation in the skin chrominance [7]. The skin region is found using values Cr and Cb, as follows:

$$Cr \quad = \quad 0.5 \times R - 0.41869 \times G - 0.08131 \times B \qquad (10.1)$$
$$Cb \quad = \quad -0.16874 \times R - 0.33126 \times G + 0.5 \times B \qquad (10.2)$$

Figure 10.5 shows the pseudocode of the skin detection algorithm using values Cr and Cb.

- Steps $Crth$ and $Cbth$ find the skin region of an image by comparing the values Cr and Cb with predefined threshold values. The if instruction of the pseudocode in Figure 10.5 implements this step.

Note that the algorithm in Figure 10.5 was split into four tasks Cr, Cb, $Crth$, and $Cbth$ illustrating the parallelism of the tasks Cr and Cb, and tasks $Crth$ and $Cbth$, respectively.

Functional partitioning can reduce the algorithm's execution time by executing the tasks in parallel.

- Task *Gnew* leaves the *G* value of the face and skin regions of an image unchanged, and sets the colors of the rest of the image to black.

- Task *Gmean* calculates the mean *G* values of the image pixels.

- Task *Gf* finds the *G* color values of the face template image.

- Task *Gfmean* computes the mean of the *G* color values of the face template pixels.

- Task *Cov* computes the covariance of each pixel in the image and template.

- Task *Sum* adds up the covariances of a face region of the same size as the template.

Note that tasks *Cov* and *Sum* occur after the functional partitioning of the template matching algorithm in Figure 10.5.

The correlation coefficient is given by:

$$cor\_coef = \sum_{-temp\_length/2}^{temp\_length/2} \sum_{-temp\_length/2}^{temp\_length/2} (Template - Template\_mean)(B - B\_mean) \quad (10.3)$$

- Task *Th* filters out the regions that do not represent faces by comparing the correlation coefficient with predefined threshold values.

- Task *Pick* selects and marks the face region.

```
void skin_extraction() {
    for (int i = 0; i <= image_height; i++) {
        for (int j = 0; j <= image_width; j++) {
            Cr[i][j] = 0.5 * R[i][j] - 0.41869 * G[i][j] - 0.08131 * B[i][j];
            Cb[i][j] = -0.16874 * R[i][j] - 0.33126 * G[i][j] + 0.5 * B[i][j];
            if ((-35 < Cr[i][j]<40) && (Cb[i][j] < -68))
                skin[i][j] = true;
            else
                skin[i][j] = false;
        }
    }
}
```

```
void template_matching() {
    for (int i = 0; i < height; i++) {
        for (int j = 0; j < width; j++) {
            for (int k = -temp_length/2; k < temp_length/2; k++) {
                for (int l = -temp_length/2; l < temp_length/2; l++) {
                    cor_coef+=(Template[k+temp_length/2][l+temp_length/2-Template_mean)*
                              (B[k+temp_length/2][l+temp_length/2- B_mean);
                }
            }
        }
    }
}
```

Figure 10.5: Algorithms for skin detection and template matching [37] © 2010 IEEE.

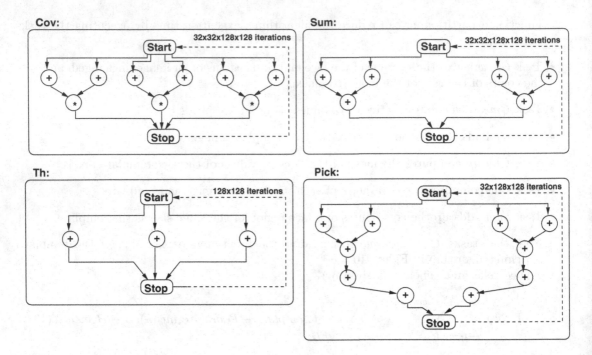

Figure 10.6: ADFGs for the time-intensive tasks of the face detection algorithm.

Implementation of the face detection algorithm in C++, and profiling the resulting code indicates that the execution time of the tasks *Cov*, *Sum*, *Th*, and *Pick* represents ≈ 99% of the total execution time. Therefore, speeding-up the execution mainly involves optimizing the implementation of the four tasks. The purpose of optimization is to identify the number and type of hardware resources that must be utilized by the general-purpose processor to achieve the maximum speed-up of the execution time as compared to the software-only implementation. The allocated hardware resources form an application-specific coprocessor, which can be based on the programmable digital circuits.

The data communication between processor and coprocessor is made possible by the use of shared memory. This communication time represents timing overhead that must be added to the task execution time, if the task is performed by a coprocessor. Assuming that the bus is B-bits, the timing overhead for a task "moved" to the coprocessor is expressed as

$$t_{com}^{in} = \frac{N_{input}}{B}(t_{com}) \tag{10.4}$$

$$t_{com}^{out} = \frac{N_{output}}{B}(t_{com}) \tag{10.5}$$

where t_{com}^{in} represents the time required by the task to read input data from the shared memory, time t_{com}^{out} is the time needed by the task to write output data to the shared memory, and N_{input} and N_{output} are the number of bits that are to be read and written, respectively. t_{com} is the time, assumed to be equal, required to read and write B bits to the shared memory.

The next step is to find the number and type of hardware circuits that form the application-specific coprocessor. The analysis uses acyclic dataflow graphs (ADFG) to describe the operators of a task and the data dependencies between operators. Recall that Chapter 1 introduced ADFGs, and Chapter 4 presented a case study that used ADFGs for developing the application-specific

Table 10.2: Speed-up results for different coprocessors

Case #	# adders	# multipliers	# logic (NAND)	Speed-up
1	1024	1024	512	2
2	512	512	512	1.9
3	512	128	512	1.56
4	128	512	512	1.33
5	128	128	512	1.15

coprocessor. Figure 10.6 shows the ADFGs for the four time-intensive tasks of the face detection task graph and, as dashed lines, the number of times each task needs to be performed.

Table 10.2 summarizes the execution time speed-ups that results for coprocessors with different resource sets. Five different cases are shown, each case being defined by the number of adders, multipliers, and NAND gates available in the coprocessor. The design procedure used is similar to the one discussed in Chapter 4. As expected, having coprocessors with more hardware resources leads to higher speed-ups, because more operations can be executed in parallel. Nevertheless, the difference in the resulting speed-ups can be very small for situations in which the amount of available hardware resources is very different. For example, for the first two cases, the speed-ups differ by less than 5%, even though the first case used twice the number of resources than the second case. Cases three and four show that additional adders are more beneficial than additional multipliers, due to the fact that a large number of addition operations were required by the original algorithm. Because multipliers are more complex than adders, thus allocating more complex hardware, and hence increasing cost, does not necessarily improve performance. In the fifth case, the speed-up is very small and is unlikely to justify the extra cost of a coprocessor.

Thus designing an application-specific coprocessor for an embedded application can lead to important performance improvements. However, allocating the coprocessor resources is not trivial, because more hardware does not necessarily result in substantial performance improvements.

10.3   Challenges in Mixed-Signal Design Automation

One current bottleneck in embedded, mixed-signal, system-on-chip (SoC) development is the design of the analog and mixed-signal IP cores, and the integration and verification of the final designs [14]. Existing research and CAD tools offer remarkable solutions to the synthesis of analog circuits, e.g., OpAmps, operational transconductors (OTA), comparators, etc. See [14] for a recent overview on this topic. Commercial tools, including Virtuoso NeoCircuit and Virtuoso NeoCell from Cadence and Circuit Explorer from Synopsys, are available for transistor sizing and layout design of analog circuits. Obviously, the next step is to address the problem of synthesis of more complex analog and mixed-signal systems, for analog-to-digital and digital-to-analog converters, phase-locked loop circuits, and transceivers. This is a significant undertaking and requires addressing three important problems:

- High-level specification for synthesis of analog and mixed-signal systems

- Fast simulation methods for analog and mixed-signal systems

- Automated analog and mixed-signal system synthesis

10.3.1   High-Level Specification of Analog and Mixed-Signal Systems[2]

According to Hosticka et al. [19], analog systems can be expressed at four levels of abstraction:

- *Concept level*: The system is defined as a network of modules (e.g., filters, analog-to-digital converters, digital-to-analog converters, programmable gain blocks, oscillators, etc.).

- *Algorithm level*: The system is presented as interconnected basic building blocks (e.g., adders, integrators, switches, multipliers, quantizers, etc.).

- *Macro level*: The system specification describes the interconnections of circuits (e.g., OpAmps, OTAs, comparators, etc.).

- *Component level*: The schematic of the system is defined as a network of devices (e.g., transistors, diodes, capacitors, inductors and resistors).

Depending on the amount of implementation detail present in a system specification, component and algorithm level descriptions are defined to be *high-level specifications*, because they contain little, or no detail about the implementation. In contrast, macro- and component-level descriptions are referred to as *low (physical)-level specifications*. This discussion focuses on high-level specification of analog and mixed-signal systems for automated design.

High-level specifications are developed using specification languages (e.g., VHDL-AMS [1, 8, 26], Verilog-AMS [26], SystemC [16], etc.) or graphical schemes, e.g., the notations in Simulink and PSoC Express. High-level specifications can serve different purposes, such as simulation, verification, and automated design (synthesis). Design methodologies, based on high-level descriptions, offer not only short design closures at lower costs, but are also capable of a more complete exploration of the possible design alternatives, than low-level specifications. However, low-level design environments (e.g., transistor sizing and layout tools) are arguably more mature than high-level design tools such as CAD tools for filters and ADCs. There are a relatively large number of design environments based on low-level specifications [9, 17, 22, 27, 28, 35] in contrast to the very few approaches for high-level synthesis [5, 13, 29, 33].

The focus in this discussion is on algorithmic high-level descriptions, because they are more approachable for automated design than concept-level specifications. Note that concept-level specifications do not indicate the nature of the modules in a specification, for example the type of filters or ADCs. This complicates the design flow because the nature of the modules also has to be selected. The most intuitive solution would be an expert system that identifies the best implementation option for each module depending on the performance requirements of the system from a predefined module library. For an effective selection, it is very important to have a comprehensive and accurate characterization (modeling) of the library modules. A limitation of this approach is the restricted selection of module implementations from the predefined library only. This issue can be addressed by providing an efficient mechanism for adding new module implementations to the library, and an automated procedure for modeling the implementations.

From an automated design perspective, algorithmic specifications must express the functionality, performance constraints and requirements of the implementation. This functionality defines

---

[2]This subsection is based on the paper - A. Doboli, R. Vemuri, "Behavioral modeling for high-level synthesis of analog and mixed-signal systems from VHDL-AMS, IEEE Transactions on Computer-Aided Design of Integrated Circuits and Systems, 22, (11), 1504-1520, November 2003, © 2010 IEEE.

mixed-domain relationships (e.g., relationships in the time and frequency domains) between the input, output, and state signals of a system. Performance descriptions denote relationships between the attributes of the signals and building blocks in a system.

Three different styles have been used to denote the functionality of analog systems:

- *Declarative specifications* define constraints among the signals in a system. Differential algebraic equations (DAEs) are typical declarative specifications in the time domain. The specification may include the initial state of the system. In the frequency domain, transfer functions are examples of declarative specifications. The system behavior results implicitly by solving the DAE set at consecutive instances, or computing the frequency response of the transfer function.

  Note that declarative specifications do not reveal the signal processing and flow structure of the implementation. However, this information is important for finding the system topology (architecture). Some type of symbolic processing must be employed to find the system structure, for example the well-known transformation rules for implementing a filter transfer function in observable, controllable, and ladder forms [5].

- *Algorithmic specifications* relate the input, output, and state signals based on the semantics of a simulation engine. The functionality is defined as a sequence of simulation steps. The system behavior can be determined by executing the simulation steps of the simulation engine. Similar to declarative specifications, they do not offer insight into the system topology.

- *Functional (denotational) specifications* define a system as a composition of basic functions. A pure functional specification has the property of referential transparency, which imposes the condition that the system functionality can be fully deduced from the semantics of its building blocks. Each basic function has inputs and outputs but no state. Signal flow graphs (SFGs) are an example of a functional specification notation. The system functionality results from explicitly showing the signal processing and flow. This information can be used as a starting point for finding different architectures for an analog system [11, 12, 33, 34].

  Figure 10.7 (a) illustrates the SFG for a fourth-order filter [11]. Figure 10.7(b) shows the high-level description of the SFG. The description reveals the architecture of the filter because it represents, at an abstract level, the filter's signal flow and processing. The macro construct defines one filter stage, including its inputs, outputs, and processing operations, for example addition, integration, and multiplication by a constant. The macro variables, e.g., variables m, n, p, and o, are introduced to express the signal flow between operators. The fourth-order filter includes two instantiations of the macro *stage_1*.

Descriptions of performance attributes may refer to the following two facets:

- Signal attributes refer to an electrical signal (e.g., voltage, current, frequency, and phase). Specifications may describe the value range or the punctual values of an attribute, such as maximum value, minimum value, and steady-state value. Sums and integrals can express the cumulative values of performance attributes. Performance attributes can also be related to the design parameters and nonidealities of a specific implementation.

- The constraints and relationships of the performance specifications can be expressed in declarative, algorithmic, or functional style. Specifications may also include partial derivatives to define sensitivities of certain system performance with respect to the signal and block attributes. Note that unlike the functionality specifications, performance descriptions

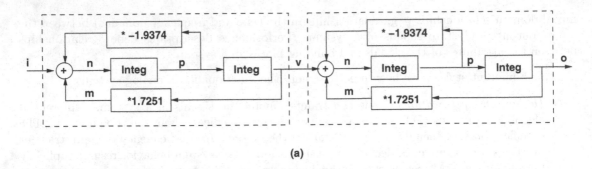

(a)

```
macro stage_1                        macro filter is continuous_time
    inputs i1;                           inputs i is voltage;
    outputs out;                         outputs o is voltage;
    arch controlable is                  arch two_stage_filter is
        variables m, n, p;                   variables v;
            o is array[2];
                                             v = stage_1.controlable (i);
            o[1] = i1 + m;                   o = stage_1.controlable (v);
            o[2] = -1.9374 * p;
            n = o[1] + o[2];             end arch;
            p = integ (n);           end macro;
            m = 1.7251 * integ (p);
            out = integ (p);
        end arch;
    end macro;
```

(b)

Figure 10.7: SFG and aBlox specification for a fourth order filter [11] © 2010 IEEE.

do not have to provide insight into the system topology. Moreover, algorithmic descriptions may be faster to simulate in as much as they do not require equation-solving, as do declarative and functional descriptions.

Figure 10.8 shows the SFG for a second-order $\Delta\Sigma$ modulator, and its high-level specification using a functional description. The attributes section of the specification defines a set of constraints that must be satisfied by the implementation of the modulator.

10.3.2   Fast Performance Estimation by Customized Simulation Code

It[3] has been repeatedly reported that one of the important challenges in mixed-signal design automation is the large number of optimization variables that must be simultaneously addressed [12, 17, 27]. Analog and mixed-signal circuit simulators are often too slow for use inside the SoC synthesis loop, experience stability problems, and are unable to exploit the specifics of circuits and systems. Behavioral circuit and system models that are used for speedingup mixed-signal

---

[3]This subsection is based on the paper by H. Zhang, S. Doboli, H. Tang, A. Doboli, Compiled code simulation of analog and mixed-signal systems using piecewise linear modeling of nonlinear parameters, Integration the VLSI Journal, Elsevier, 40 (2007), pp. 193–208 and the paper by H. Zhang, A. Doboli, Fast Time-Fomain Simulation Through Combined Symbolic Analysis and Piecewise Linear Modeling, in Proceedings of the 2004 IEEE International Behavioral Modeling and Simulation Conference, 2004, pp. 141–146.

system simulation are of two kinds: structural (physical) and mathematical models. (See [4] for a presentation of the most recent advances in behavioral modeling and simulation.)

The primary consideration when considering techniques for fast simulation of analog systems with nonlinear parameters is the development of optimized code for a simulator that is customized to each individual system. The code generation methodology relies on calculating symbolic expressions for the output voltages and currents, and the state variables of a system. Each kind of interconnection structure IST between two blocks b_i and b_j is captured as a separate C++ class $C_{IST}(b_i, b_j)$ with the related methods encapsulating the symbolic composition rule of the two blocks. All instances of the structure IST present in the system topology are formulated as objects of class C_{IST}, for which the identity of blocks b_i and b_j is set as the two blocks actually appearing in the structural instance. This representation is applied bottom-up, so that blocks b correspond to building blocks and composed blocks in the system.

Code generation utilizes detailed structural macromodels for the building blocks, for example OTA, OpAmp, comparators, including many nonidealities, such as finite gain, poles and zeros, CMRR, phase margin, and fall and rise time. Chapter 6 discussed one method for building structural OpAmp macromodels, but the technique can be utilized for other circuits also, for example operational transconductors. Code optimization identifies and eliminates loop invariants, that is code that has the same effect for all loop iterations, and propagates constant subexpressions present in the simulation loop. The technique is very fast and has few stability problems.

Figure 10.9 shows the proposed compiled-code simulation methodology and the structure of a single-loop $\Delta\Sigma$ modulator. The methodology produces C++ code for customized simulators generated from the system netlist used as input specification. The simulator code is organized as calling and called C++ methods, that reflect the hierarchical structure of a system. C++ methods include variables corresponding to the state variables of the system, and instructions for computing the values of output and state variables, that is voltages and currents at each time and frequency instance. Instructions are found using symbolic composition rules for each structural pattern that links multiport blocks. Blocks in structural patterns are either basic blocks, if they correspond to basic analog circuits, or composed blocks, if they are generated by block composition. Additional code is generated for postprocessing the simulation data to compute the required performance attributes of the system. For a $\Delta\Sigma$ ADC, the output voltage of the modulator is used to calculate typical performance figures, for example signal-to-noise ratio (SNR) and dynamic range (DR).

The remainder of this discussion centers on the basic circuit modeling for compiled-code simulation, block composition using symbolic rules, and provides some of the simulation methodology details.

A. Modeling of basic building blocks. A system architecture (netlist) is built out of interconnected multiport blocks. Blocks are either basic building blocks, such as operational transconductors, OpAmps, and comparators, or composed blocks, such as integrators and ADC stages. The behavior, as a function of time, of an N-port block is expressed as symbolic equations between voltages $V(t)$ and currents $I(t)$ at its N ports, derivates of its K state variables $V_s(t)$ denoting voltages across energy storing devices, (e.g., capacitors and inductors in the block) and M internal voltage and current sources $U_{internal}$, that is those used to express offset voltages of a circuit.

For example, Figure 10.10(a) shows the 4-port block for an OTA and the 4-port block for an integrator. The OTA block has two ports for across voltage $V_{idm} = V_{ip} - V_{in}$ and two ports for

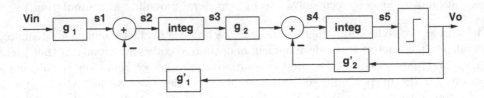

```
macro SD_converter is
  inputs vin is voltage;
  outputs vo is voltage with range GND-VDD;
  attributes
      define delta = max(vo.voltage) - min(vo.voltage);
      define Bandwidth(o) = Frequency.((o.voltage -
                                        o.voltage(DC) < 3dB) at 1);
      min(vin.voltage) in [-0.45*delta, 0.45*delta];
      max(vin.voltage) in [-0.45*delta, 0.45*delta];
      derivate (s3,time) > 176.0e3*delta;
      derivate (s5,time) > 176.0e3*delta;
      g4/(g1*g2) > 1.25;
      Bandwidth(s1) > 160kHz;
      Bandwidth(s2) > 160 kHz;
      Bandwidth(s3) > 160 kHz;
      Bandwidth(s4) > 160 kHz;
      Bandwidth(s5) > 160 kHz;
      g1 < g2;
  arch SFG is
  variables s1, s2, s4, s5;
                  s3 with range -0.2V-0.2V;
      s1 = g1 * vin;
      s2 = s1 - g3 * vo;
      s3 = integ (s2);
      s4 = g2 * s3 - g4 * vo;
      s5 = integ (s4);
      if s5 > VDD/2 then
          vo = VDD;
      else
          vo = GND;
      end if;
  end arch;
end macro
```

Figure 10.8: SFG description and high-level specification for $\Delta\Sigma$ modulators [7] © 2010 IEEE.

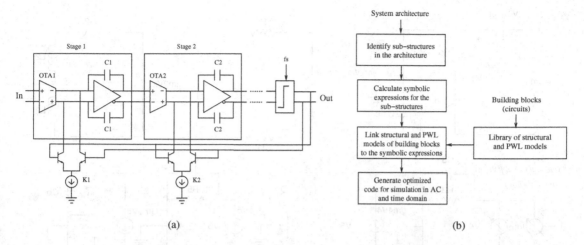

(a)   (b)

Figure 10.9: $\Delta\Sigma$ ADC structure and compiled-code system simulation methodology [39]. Reprinted from H. Zhang, S. Doboli, H. Tang, A. Doboli, Compiled Code Simulation of Analog and Mixed-Signal Systems using Piecewise Linear Modeling of Nonlinear Parameters, Integration the VLSI Journal, Elsevier, 40 (2007), pp. 193–208, with permission from Elsevier.

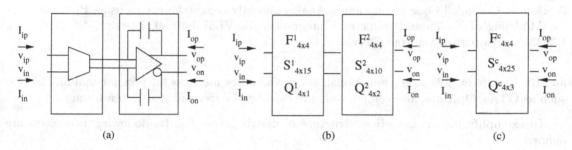

(a)   (b)   (c)

Figure 10.10: Two basic blocks and their composition into a $\Sigma - \Delta$ stage [39]. Reprinted from H. Zhang, S. Doboli, H. Tang, A. Doboli, Compiled Code Simulation of Analog and Mixed-Signal Systems using Piecewise Linear Modeling of Nonlinear Parameters, Integration the VLSI Journal, Elsevier, 40 (2007), pp. 193–208, with permission from Elsevier.

across voltage $V_{out} = V_{op} - V_{on}$. Voltages and currents at the four ports of the OTA block are related by the following symbolic expression.

$$[I_{ip}(t)\ I_{in}(t)\ I_{op}(t)\ I_{on}(t)]^T = [F_{ij}]_{4\times4}[V_{ip}(t)\ V_{in}(t)\ V_{op}(t)\ V_{on}(t)]^T +$$
$$[S_{ij}]_{4\times15}[V_{C_d}(t-1)\ V_{C_{cm,1}}(t-1)\ V_{C_{cm,2}}(t-1)\ V_{L_{1,1}}(t-1)\ V_{L_{1,2}}(t-1)\ V_{L_{2,1}}(t-1)$$
$$V_{L_{2,2}}(t-1)\ V_{C_{3,1}}(t-1)\ V_{C_{3,2}}(t-1)\ V_{L_{4,1}}(t-1)\ V_{L_{4,2}}(t-1)\ V_{C_{5,1}}(t-1)$$
$$V_{C_{5,2}}(t-1)\ V_{C_{o,1}}(t-1)\ V_{C_{o,2}}(t-1)]^T +$$
$$[Q]_{4\times1}V_{offset}$$

Matrix terms, F_{ij}, S_{ij}, and Q_{ij}, are computed by symbolically solving the nodal equations of the OTA structural macromodel in Figure 10.11 after replacing the derivates of state variables according to BEI (Backward Euler Integration) formula. The structural macromodel in the figure was obtained by extending the OTA model by Gomez[15] to the fully differential mode (DM) by duplicating the single end stage, the common mode stage, the intermediate and output stages,

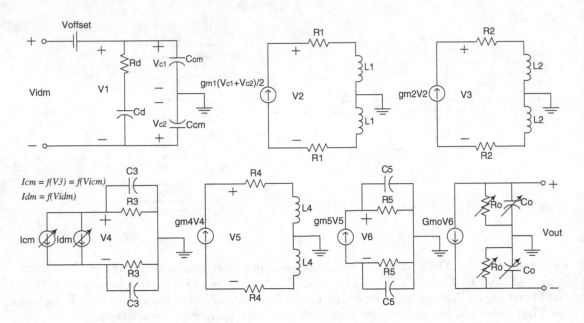

Figure 10.11: OTA structural macromodel [39]. Reprinted from H. Zhang, S. Doboli, H. Tang, A. Doboli, Compiled Code Simulation of Analog and Mixed-Signal Systems using Piecewise Linear Modeling of Nonlinear Parameters, Integration the VLSI Journal, Elsevier, 40 (2007), pp. 193–208, with permission from Elsevier.

and the dominant pole stage. Functional, state, and offset matrices for popular building blocks, such as OTAs, OpAmps, and comparators are calculated once, and stored in a library.

To exemplify the mathematical structure of matrix terms F_{ij}, the following two terms are defined:

$$F_{11} = \frac{C_{cm}}{h} + \frac{C_d}{h + C_d\,R_d} \tag{10.6}$$

and,

$$F_{31} = 8h\frac{G_m G_{m4} G_{m5} R_3 R_5 (L_4 + hR_4)}{(h + C_3 R_3)(h + C_5 R_5)} \tag{10.7}$$

Similarly, the terms S_{ij} and Q_{ij} have the following form:

$$S_{11} = S_{21} = -\frac{C_d}{h + C_d\,R_d} \tag{10.8}$$

$$Q_{11} = \frac{C_d}{h + C_d R_d} + \frac{C_{cm}}{h} \tag{10.9}$$

$$Q_{21} = -\frac{C_d}{h + C_d R_d} \tag{10.10}$$

B. Block composition. To produce the compiled-code simulator for the entire system, the simulation methodology merges small clusters of interconnected blocks into composed blocks with the same temporal behavior as the original clusters. The merging process is bottom-up, starting from the N-port representation of basic blocks until the port description of the entire system is found. For example, the two 4-port blocks in Figure 10.10(a) are merged to create the block for a $\Delta\Sigma$ stage, as shown in Figure 10.10(b). Then, as shown in Figure 10.10(c), the $\Delta\Sigma$ stage is described as a 4-port composed block with the same temporal behavior as the two connected blocks. Figure 10.12 shows the bottom-up merging of blocks for a third-order single-loop $\Delta\Sigma$ ADC. The topmost block corresponds to the entire ADC, and is simulated to find the time domain behavior of the system.

The behavior of composed blocks is found by symbolically settingup and solving the nodal equations for its composing blocks. Variables corresponding to the voltages and currents at the interconnection wires between the blocks are eliminated from the equation set. For example, the $\Sigma\Delta$ stage in Figure 10.10(a) consists of the OTA block linked to the OpAmp-C block by two wires. As explained in Figure 10.10(b), the OTA block has functional matrix F_{4x4}^1, state matrix S_{4x15}^1, and offset matrix Q_{4x1}^1. There are 15 state variables in the OTA structural macromodel. The symbolic functional matrix F_{4x4}^2, state matrix S_{4x10}^2, and offset matrix Q_{4x2}^2 describe the OpAmp-C block. The OpAmp-C macrocell has 10 state variables, and two offset voltages, V_{offset} and V_{odc}. Figure 10.10(c) presents the composed block for the stage, and described by matrices $F_{4\times4}^c$, $S_{4\times25}^c$, and $Q_{4\times3}^c$. The form of the matrix parameters was illustrated by showing the expression for element F_{11}^c. Also considered was the case where V_i and I_f (the DAC currents) are known, and I_i and V_o are unknown.

$$
\begin{aligned}
F_{11}^c \;=\; & \frac{(F_{22}^2 - F_{33}^1)(F_{11}^2 F_{11}^1 - F_{14}^1 F_{41}^2 + F_{11}^1 F_{44}^2)}{(F_{12}^2 + F_{43}^1)(F_{41}^2 + F_{34}^1) - (F_{11}^2 + F_{33}^1)(F_{11}^2 + F_{44}^1)} \\
& + \frac{(F_{21}^2 + F_{34}^1)(F_{12}^2 F_{11}^1 - F_{13}^1 F_{41}^2 + F_{11}^1 F_{43}^2)}{(F_{12}^2 + F_{43}^1)(F_{41}^2 + F_{34}^1) - (F_{11}^2 + F_{33}^1)(F_{11}^2 + F_{44}^1)} \\
& + \frac{F_{31}^1(F_{13}^1(F_{11}^2 + F_{44}^1) - F_{14}^1(F_{12}^2 + F_{33}^1))}{(F_{12}^2 + F_{43}^1)(F_{41}^2 + F_{34}^1) - (F_{11}^2 + F_{33}^1)(F_{11}^2 + F_{44}^1)}
\end{aligned}
\tag{10.11}
$$

The parameters, F_{ij}^c, S_{ij}^c, and Q_{ij}^c, are described by similar expressions.

The *structural pattern* for a given cluster of interconnected blocks is defined by (i) the number of blocks in the cluster, (ii) number of ports, state variables, internal voltage and current sources of each block, and (iii) the connection structure, that is the number of wires between blocks. Structural patterns do not describe the symbolic descriptions of the F, S, and Q matrices of their interconnected blocks. For example, the composed blocks for the three single stages of the third-order $\Delta\Sigma$ modulator in Figure 10.12 have the same structural pattern, even though each stage is based on different circuits.

The same symbolic expressions are used to calculate matrices F^c, S^c, and Q^c for all composed blocks corresponding to the same structural pattern. For example, all matrices for the single stages in Figure 10.12 are found using the same symbolic expressions, obviously involving the parameters of the corresponding circuit macromodels. If the symbolic expressions of the composed block matrix elements are interpreted as functions of the composing macromodel parameters, then these functions are the same for identical structural patterns. This property results from the fact that the nodal equations are the same for identical structural patterns. Thus, after eliminating the same unknown voltages and currents at the internal interconnection wires, the same symbolic expressions are obtained.

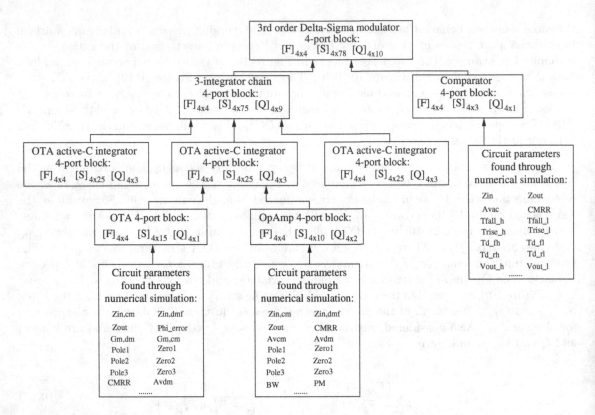

Figure 10.12: Structural patterns in third order single-loop $\Delta\Sigma$ ADC [39]. Reprinted from H. Zhang, S. Doboli, H. Tang, A. Doboli, Compiled Code Simulation of Analog and Mixed-Signal Systems using Piecewise Linear Modeling of Nonlinear Parameters, Integration the VLSI Journal, Elsevier, 40 (2007), pp. 193–208, with permission from Elsevier.

A *Symbolic Composition Rule* (SCR) is the set of symbolic expressions for all the F^c, S^c, and Q^c matrix elements that characterize a given structural pattern. SCRs relate the symbolic functional, state, and offset matrices of composed blocks to the symbolic matrices of their composing blocks. As explained, SCRs are found by setting up the nodal equations for the blocks in the structural pattern, and eliminating the unknown voltages and currents at the internal wires of the connected ports.

C. Compiled-code simulation methodology. Figure 10.9(b) shows the proposed compiled-code simulation methodology. The first step identifies the structural patterns (substructures) that connect the building blocks in the system architecture (netlist). Several structural patterns can usually be identified for complex systems, and Figure 10.12 shows the structural patterns in the third -order single-loop $\Delta\Sigma$ ADC. The 3-integrator chain macromodel is obtained by composition of the three structurally identical macromodels for OTA active-C integrators.

The next step computes the symbolic expressions for the SCR of each structural pattern. Next, the parameters in the symbolic expressions are linked to the building block macromodel parameters. Finally, optimized code is generated for customized simulation. Code is generated for each unit in the system architecture, for example basic blocks, structural patterns, and composed blocks. The code for a unit is a sequence of assignment statements that numerically calculate the elements in the F, S, and Q matrices of the unit. Code generation carefully identifies any redundant subexpressions. For example, for the OTA macromodel, subexpression $h + C_d R_d$ was identified as common to all matrix elements. Hence, the subexpression was named as a new variable, and its value reused, in all instances.

The simulation algorithm implements a simulation loop for the entire time range to be simulated. The time increment is h, the parameter also used in the BEI formula. At each instance in time, the algorithm calculates only a subset of all voltages and currents in the system netlist, including output signals, state variables, and across voltages and by currents of nonlinear devices. For nonlinear devices, the simulation algorithm also identifies the correct linearized region. The identification step calculates the voltages and currents for nonlinear devices assuming that the linearized regions at the current instance remain the same as for the previous time. If the assumption is false, then the algorithm recalculates the voltages and currents of the nonlinear devices after switching these to the closest linearized regions. The identification step ends when the closest feasible regions are found.

Figure 10.13 presents the pseudocode of the C++ main program for a simulator, customized for a given system netlist. The program includes local variables defined for all internal nodes of the system. Each variable includes a voltage and a current component. Then, objects are instantiated for each block of the system depending on the kind of the block. These classes embed the structural model of the circuit and the symbolic composition rules (SCR) to compute its functional, state, and offset matrices. For example, similar objects are created for all identical OTA circuits using the class defined for these OTAs in Figure 10.14. Similar classes exist for OpAmps, comparators, etc. Objects are instantiated for each block composition depending on the specific partitioning of the system architecture. For the $\Delta\Sigma$ ADC in Figure 10.9(a), and its decomposition shown in Figure 10.12, all active OTA-C integrator models are obtained by using the same composition rule of the OTA objects and the objects describing capacitors. Objects are connected by the variables describing inputs, outputs, and nodes. Next, the system is initialized by setting all its state variables to their initial values. Finally, the code for time domain simulation calls the system simulation method at successive instances separated by h, the time integration step.

Figure 10.14 presents the pseudocode for the C++ class corresponding to OTAs. Any other circuit class has a similar structure. Protected variables correspond to the two input nodes representing the differential inputs, the two output nodes for the differential output of the OTA, and the arrays used to store the values for the circuit functional, state, and offset matrices. The class methods set the numerical values of the model components, e.g., the values of capacitors, resistors, and constant current and voltage sources, initialize the state matrix, calculate the state and functional matrix elements, update the value of the state sub-matrix for the next instance of time, and simulate the OTA circuit over time. Classes for SCRs have similar structures.

The methods for computing functional, state, and offset matrices and for updating the state are optimized by sharing common sub-expressions. In the present case, the functional matrix F of the OTA class, e.g., has elements F_{31} and F_{32} defined by the following symbolic expressions:

$$F_{31} = \frac{8hG_m G_{m4} G_{m5} G_{mo} R_3 R_5 (L_4 + hR_4)}{(h + C_3 R_3)(h + C_5 R_5)} \tag{10.12}$$

and

$$F_{32} = \frac{-8hG_m G_{m4} G_{m5} G_{mo} R_3 R_5 (L_4 + hR_4)}{(h + C_3 R_3)(h + C_5 R_5)} \tag{10.13}$$

```
struct node { double v, i; };
void main (void) {
    struct node internal signals;

    for all blocks in the system architecture do
        create an instance of the class corresponding to the block type;

    for all symbolic composition rules (SCR) in the architecture do
        create an instance of the class corresponding to the SCR type;

    connect all created objects to inputs, outputs, and internal signals;
    set the initial state of the system;

    for all blocks in the architecture do
        calculate the functional, state, and offset matrices of the block;

    for the time interval of interest at successive instances with time step h do
        simulate the system (current time instance);
};
```

Figure 10.13:   Structure of the C++ main program for system simulation [39]. Reprinted from H. Zhang, S. Doboli, H. Tang, A. Doboli, Compiled Code Simulation of Analog and Mixed-Signal Systems using Piecewise Linear Modeling of Nonlinear Parameters, Integration the VLSI Journal, Elsevier, 40 (2007), pp. 193–208, with permission from Elsevier.

The simulation code calculates the numerical value of element F_{31} and then assigns its negative value to element F_{32} without actually computing the expression of F_{32}. Common expression sharing is very efficient, especially for code that is repeatedly executed inside a loop statement, for example the simulation loop. The method *update state matrix*, called at each new instance of the simulated time range, computes sub-expressions, such as $h/(h + C_d R_d)$, $hG_m R_3/(h + C_3 R_3)$, and so on. These sub-expressions are constant, and thus represent invariants for the simulation loop. Invariants are extracted outside the simulation loop, and executed only once before the loop execution starts. As experiments show, loop invariant elimination is an important source for speedingup simulation speed, especially for the methods of SCR classes.

Figure 10.15 shows the pseudocode of the system simulation algorithm. At each instance of time, the algorithm identifies the current linearized region for the nonlinear devices. The identification step first calculates the voltages and currents for the nonlinear devices by traversing, bottom-up, the structure hierarchy of the system and assuming that the linearized regions at the current instance remain the same as those at the previous time. If this assumption is false, then the algorithm reiterates the calculation of the voltages and currents after switching to the closest PWL region, and so on. The iteration process stops when the closest feasible PWL region is found. The nonlinear component, linearized region checking and adjusting process, are inside each basic module and are called before state matrices are updated. Changing the current region of a module also results in updating of the functional and offset matrices for this module and all composed modules that contain this module.

Experiments demonstrate the importance of circuit nonidealities, (e.g., poles, zeros, input and output impedances etc.) on the accuracy of ADC simulation. Figure 10.16 shows the SNR and DR plots for the ADC. The maximum SNR is 64dB and DR is 67dB. Similar values resulted from a Spectre simulation. This confirms the correctness of the symbolic method. The figure also shows the importance of using detailed circuit models, for example circuit models that include

```
class OTA {
    public:
        friend class OTA_ActiveC_Integrator;
        OTA (parameter list);
        set_input(input signals);
        simulate();
        get_output();
    protected:
        Node in_p, in_n; // input ports
        Node out_p, out_n; // output ports
        double F[4][4]; // functional matrix
        double S[4][15]; // state matrix
        double Q[4]; // offset matrix
        set_initial_state();
        update_state();
    private:
        calculate_functional_matrix();
        calculate_state_matrix();
        calculate_offset_matrix();
};
```

Figure 10.14:   C++ code for OTA class [39]. Reprinted from H. Zhang, S. Doboli, H. Tang, A. Doboli, Compiled Code Simulation of Analog and Mixed-Signal Systems using Piecewise Linear Modeling of Nonlinear Parameters, Integration the VLSI Journal, Elsevier, 40 (2007), pp. 193–208, with permission from Elsevier.

```
void simulate the system (current time instance) {
    while current linearized segments are not correct do
        for all blocks following the bottom-up structure of the architecture do
            calculate the port signals of the blocks;
        for all blocks in the architecture do
            find the corresponding linearized segments and check if they are correct;
            if the guessed linearized segments are not correct then update the segments;

    for all blocks following the bottom-up structure of the architecture do
        update the functional, state, and offset matrices;
};
```

Figure 10.15:   Pseudocode of the system simulation method [39]. Reprinted from H. Zhang, S. Doboli, H. Tang, A. Doboli, Compiled Code Simulation of Analog and Mixed-Signal Systems using Piecewise Linear Modeling of Nonlinear Parameters, Integration the VLSI Journal, Elsevier, 40 (2007), pp. 193–208, with permission from Elsevier.

poles and zeros, rather than ideal models. In the right-hand side of Figure 10.16, the three plots with dotted lines correspond to simulations, which used circuit macromodels with one pole and two poles. In the first two cases, the system still worked as an ADC, but the SNR was reduced by $\approx$ 5 dB and 13 dB, and the DR by $\approx$ 4 dB and 12 dB, respectively, due to the poles. In the third case, the poles prevented the system from functioning correctly. This example confirms that using detailed circuit models is compulsory.

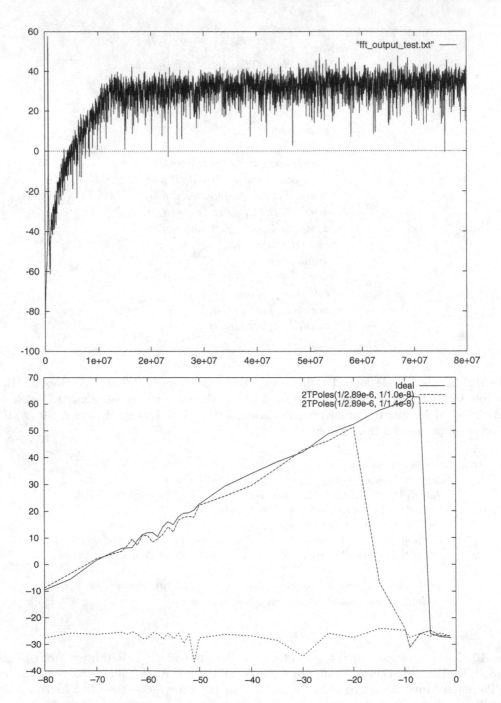

Figure 10.16: SNR and DR plots for $\Delta\Sigma$ ADC [39]. Reprinted from H. Zhang, S. Doboli, H. Tang, A. Doboli, Compiled Code Simulation of Analog and Mixed-Signal Systems using Piecewise Linear Modeling of Nonlinear Parameters, Integration the VLSI Journal, Elsevier, 40 (2007), pp. 193–208, with permission from Elsevier.

procedure *mapping (signal-flow, cur block, OpAmp nr)* **is**
  for $\forall$ *sub-graph* $\in$ *signal-flow* that has *cur block* as its output block **and** can be
    mapped to a library component; in decreasing order of the number of blocks in *sub-graph* **do**
      if sharing is possible **and** library component for *sub-graph* exists in *netlist* **then**
        make the required connections for *sub-graph* in *netlist*;
        if *signal-flow* was completely mapped **then**
          call GA for constraint transformation & component synthesis, and save solution if best so far;
        else
          signal = select an input signal of *sub-graph*;
          mapping (*signal-flow*, block $\in$ *signal-flow* with output *signal*, *OpAmp nr*);
        end if
      end if
      if (*OpAmp nr* + nr of OpAmps for *sub-graph*) * *MinArea* < *current best* **then**
        allocate hardware component for the mapping of *sub-graph*, and add it to *netlist*;
        if *signal-flow* was completely mapped **then**
          call GA for constraint transformation & component synthesis, and save solution if best so far;
        else
          signal = select an input signal of *sub-graph*;
          mapping (*signal-flow*, block $\in$ *signal-flow* with output *signal*, *OpAmp nr* + nr of OpAmps for *sub-graph*);
        end if
      end if
    end for
end procedure

Figure 10.17: Algorithm for architecture generation, A. Doboli, N. Dhanwada, A. Nunez-Aldana, R. Vemuri, A Two-Layer Library-based Approach to Synthesis of Analog Systems from VHDL-AMS Specifications, ACM Transactions on Design Automation, 9(22), pp. 238–271, April 2004, © 2010 ACM, Inc. reprinted here by permission.

10.3.3   High-Level Synthesis of Analog Subsystems

This[4] subsection focuses only on architecture generation, and does not discuss other related steps such as constraint transformation [12] and system parameter optimization [11, 12, 34]. A modern way of synthesizing $\Delta\Sigma$ modulator topologies is discussed in [33].

Architecture generation produces multiple mappings of an SFG to netlists of library components. It is difficult to predict what the best implementation architecture might be for a system, given the large diversity of possible performance requirements. The performance of an architecture is evaluated only after distributing the system-level constraints to the components, and instantiating the topologies and sizing the transistors of the OpAmps in the netlist. Although the problem of architecture generation is NP-hard, it can be addressed, for small netlists, by a branch-and-bound algorithm [18]. The SFG representations for many applications (e.g., signal

---

[4]This subsection is based on the paper A. Doboli, N. Dhanwada, A. Nunez-Aldana, R. Vemuri, A Two-Layer Library-based approach to synthesis of analog systems from VHDL-AMS specifications, ACM Transactions on Design Automation, 9, (2), 238–271, April 2004, © 2010 ACM, Inc. reprinted by permission.

conditioning systems) filters and ADC, are small, so that it is practical to search for all corresponding mappings.

Figure 10.17 shows the architecture generation algorithm. It maps the SFG structure, denoted by variable *signal-flow*, onto the netlist indicated by variable *netlist*. Variable *OpAmp nr* represents the number of OpAmps in a partial mapping solution. This variable is equal to the number of programmable analog blocks used, for architectures similar to that of PSoC.

To minimize the area, or equivalently, the number of programmable analog blocks, of the implementation, architecture generation attempts two hardware sharing possibilities between blocks in different signal paths, and between blocks of the same signal path. Blocks in distinct SFG paths can share the same library component if they have identical inputs and perform the same operation. A set of blocks of the same SFG path can share a component if the component implements the functionality of the entire set. The algorithm analyzes all possible mappings, because the two sharing options might conflict with each other.

The following three elements are specific to architecture generation.

- The *branching rule* (line 2 of the code), describes how distinct mapping solutions are produced for a partial solution. It lists all of the SFG block structures, pointed to by the variable *sub-graph*, that have *cur block* as their output block, and that can be mapped to library components. The branching rule contemplates two kinds of SFG transformations. *Functional transformations* replace a particular block structure with a distinct, but semantically equivalent structure. For example, for improving bandwidth, an OpAmp is replaced by a chain of two OpAmps with lower gains, or two noninverting amplifiers are substituted by two inverting amplifiers. These structural changes permit improved gain-bandwidth tradeoff explorations during constraint transformation. Transformations pertaining to circuit interfacing introduce additional circuits, for example follower circuits, or various input/output stages for reducing the loading/coupling effects between connected components.

- The *bounding rule* (line 13 of the code) eliminates a partial solution if it finds that its minimum possible area is greater than the area of the best solution found thus far (variable *current best*). The minimum area for a partial mapping is estimated using value *MinArea*, the area of an OpAmp with transistors sized to the minimum dimensions. This rule is important for eliminating architectures with an unreasonably large number of OpAmps. Such architectures could result from repeatedly replacing singular blocks with chains of OpAmps, or using multiple OpAmp-based components as the basis for a single block. For example, an adder block could be mapped to a summing amplifier with additional gain stages for each of the inputs. Even though these architectures provide the necessary signal flow and processing, they do not have any potential for significant performance improvements, and their parameters are hard to optimize, given the large number of parameters.

- The *sequencing rule*, a heuristic, decides the order of traversing the branching alternatives. A good sequencing rule can dramatically improve the speed of the algorithm. The bounding rule becomes efficient if a high-quality solution is found early. The proposed sequencing rule first considers branching alternatives, which map a higher number of blocks to one library component. Also, the algorithm analyzes the case in which blocks of SFG sub-graph share the existing components in the partial netlist, and then try mapping the sub-graph to dedicated components. The two strategies help finding an architecture with fewer OpAmps, early.

Figure 10.18(a) shows an SFG and a fragment of the decision tree for its mapping to library circuits. Each node in the decision tree relates to a partial solution and a specific value for variable *cur block*. Arcs are annotated by the corresponding mapping decisions. The number of OpAmps used is indicated for each complete mapping of the SFG. The algorithm uses a library of patterns, which relate SFG structures to library circuits. A block structure, referred to as *comp1*, and its corresponding library circuit are shown in Figure 10.18(b). This example uses similar patterns for the block that multiplies an input voltage by a constant and is mapped to a gain stage denoted as *comp2*, and the block that adds two input voltages and is mapped to a summing amplifier named *comp3*. The algorithm first attempts to map the output summing *block 1* of the SFG to component *comp3* (summing amplifier), followed by mapping blocks 2, 3, and 4 to a second summing amplifier, and *block 5* to a gain stage. The resulting architecture uses 3 OpAmps. A distinct solution shown in the decision tree is generated by mapping *block 2* alone to a summing amplifier, followed by mapping blocks 3, 4, and 5 to three different gain stages. The resulting architecture has 5 OpAmps. The branching rule also introduces an additional block

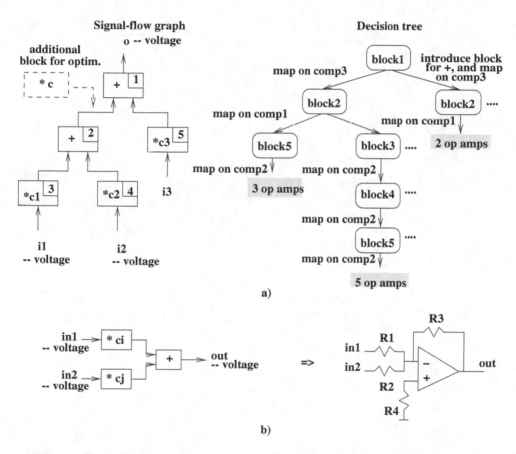

Figure 10.18:   Example for architecture generation with branch-and-bound, A. Doboli, N. Dhanwada, A. Nunez-Aldana, R. Vemuri, A Two-Layer Library-based Approach to Synthesis of Analog Systems from VHDL-AMS Specifications, ACM Transactions on Design Automation, 9(22), pp. 238–271, April 2004, © 2010 ACM, Inc. reprinted here by permission.

into the SFG, shown as a dashed box. This results in an architecture with two OpAmps. Blocks 1 and 5 are mapped to one summing amplifier, followed by mapping blocks 2, 3, and 4 to a second summing amplifier.

This brings to an end the discussions in this textbook that were designed to introduce the reader to some of the many facets of embedded, mixed-signal, system design. Although the presentation is by no means comprehensive, the authors hope that by now the reader is sufficiently motivated to continue her or his quest to master more and more complex embedded mixed-signal systems.

Bibliography

[1] *IEEE Standard VHDL Language Reference Manual* (Integrated with VHDL-AMS Changes), IEEE Std. 1076.1.

[2] PSoC Express, Version 2.0, Cypress Semiconductor, 2006.

[3] PSoC Mixed Signal Array, Technical Reference Manual, *Document No. PSoC TRM 1.21*, Cypress Semiconductor Corporation, 2005.

[4] *IEEE Transactions on CADICS*, Special Issue on Behavioral Modeling and Simulation of Mixed-Signal/Mixed-Technology Circuits and Systems, 22, (2), February 2003.

[5] B. Antao, A. Brodersen, ARCHGEN: Automated synthesis of analog systems, *IEEE Transactions on Very Large Scale Systems*, 3, pp. 231–244, June 1995.

[6] O. Bajdechi, G. Gielen, J. Huijsing, "Systematic Design Exploration of Delta-Sigma ADCs", *IEEE Transactions on Circuits & Systems - Part I*, 51, (1), pp. 86–95, January 2004.

[7] D. Chai, K. N. Ngan, Locating facial region of a head-and-shoulders color image, *Proceedings of the Third International Conference on Automatic Face and Gesture Recognition*, pp. 124–129, 1998.

[8] E. Christen, K. Bakalar, VHDL-AMS - A hardware description language for analog and mixed-signal spplications, *IEEE Transactions on Circuits and Systems - Part II*, 46, pp. 1263–1272, October 1999.

[9] J. Cohn, D. Garrod, R. Rutenbar, L. Carley, KOAN/ANAGRAM II: New tools for device-level analog placement and routing", *Journal of Solid-State Circuits*, 26, pp. 330–342, March 1991.

[10] G. De Micheli, *Synthesis and Optimization of Digital Circuits*, New York: McGraw-Hill, 1994.

[11] A. Doboli, R. Vemuri, Behavioral modeling for high-level synthesis of analog and mixed-signal systems from VHDL-AMS, *IEEE Transactions on Computer-Aided Design of Integrated Circuits and Systems*, 22, (11), pp. 1504–1520, November 2003.

[12] A. Doboli, N. Dhanwada, A. Nunez-Aldana, R. Vemuri, A two-layer library-based approach to synthesis of analog systems from VHDL-AMS specifications, *ACM Transactions on Design Automation*, 9, (2), pp. 238–271, April 2004.

[13] K. Francken, G. Gielen, A high-level simulation and synthesis environment for delta-sigma modulators, *IEEE Transactions on Computer-Aided Design of Integrated Circuits and Systems*, 22, (8), pp. 1049–1061, August 2003.

[14] G. Gielen, R. Rutenbar, Computer-aided design of analog and mixed-signal integrated circuits, *Proceedings of the IEEE*, 88, pp. 1824–1854, December 2000.

[15] G. J. Gomez et al., "A Nonlinear Macromodel for CMOS OTAs", *Proceedings of International Symposium on Circuits and Systems (ISCAS)*, pp. 920–923, 1995.

[16] T. Grotker, S. Liao, G. Martin, S. Swan, *System Design with SystemC*, Boston: Kluwer, 2002.

[17] M. Hershenson, S. Boyd, T. H. Lee, Optimal design of a CMOS op-amp via geometric programming, *IEEE Transactions on Computer-Aided Design of Integrated Circuits and Systems*, 20, (1), pp. 1–21, January 2001.

[18] E. Horowitz, S. Sahni, *Fundamentals of Computer Algorithms*, New York: Computer Science Press, 1985.

[19] B. J. Hosticka, W. Brockherde, R. Klinke, R. Kokozinski, Design methodologies for analog monolithic circuits, *IEEE Transactions on Circuits & Systems - Part I*, 41, pp. 387–394, May 1994.

[20] X. Huang et al., Modeling nonlinear dynamics in analog circuits via root localization, *IEEE Transactions on Computer-Aided Design of Integrated Circuits and Systems*, 22, (7) pp. 895–907, 2003.

[21] V. Kremin, DTMF Detector, *Application Note*, AN 2122, Revision A, Cypress Semiconductor Corporation, August 3 2004.

[22] W. Kruiskamp, D. Leenaerts, DARWIN: CMOS opamp synthesis by means of genetic algorithm, *Proceedings of the Design Automation Conference*, 1995, pp. 433–438.

[23] F. Leyn et al., Analog small-signal modeling. behavioral signal-path modeling for analog integrated circuits, *IEEE Transactions on Circuits & Systems - Part II*, 48, (7), pp. 701–711, 2001.

[24] F. Medeiro, A. Perez-Verdu, A. Rodriguez-Vazquez, *Top-down design of high-performance sigma-delta modulators*, Boston: Kluwer, 1999.

[25] A. Moscovici, High Speed A/D Converters - *Understanding Data Converters through SPICE*, Boston: Kluwer, 1999.

[26] F. Pecheux, C. Lallement, A. Vachoux, VHDL-AMS and Verilog-AMS as alternative hardware description languages for efficient modeling of multidiscipline systems, *IEEE Transactions on Computer-Aided Design of Integrated Circuits and Systems*, 24, (2), pp. 204–225, February 2005.

[27] R. Phelps, M. Krasnicki, R. Rutenbar, L. Carley, J. Hellums, Anaconda: simulation-based synthesis of analog circuits via stochastic pattern search, *IEEE Transactions on Computer-Aided Design of Integrated Circuits and Systems*, 19, pp. 703-717, June 2000.

[28] B. N. Ray, P. P. Chaudhuri, P. K. Nandi, Efficient synthesis of OTA network for linear analog functions, *IEEE Transactions on Computer-Aided Design of Integrated Circuits and Systems*, 21, pp. 517–533, May 2002.

[29] E. Sanchez-Sinencio, J. Ramirez-Angulo, AROMA: An area optimized CAD program for cascade SC filter design, *IEEE Transactions on Computer-Aided Design of Integrated Circuits and Systems*, CAS-14, pp. 296–303, 1985.

[30] B. Schneier, *Applied Cryptography*, Hoboken, NJ: John Wiley, 1996.

[31] D. Seguine, Capacitive Switch Scan, *Application Note AN2233a*, Revision B, Cypress Semiconductor Corporation, April 14 2005.

[32] R. Sommer, I. Rugen-Herzig, E. Hennig, U. Gatti, P. Malcovati, F. Maloberti, K. Einwich, C. Clauss, P. Schwarz, G. Noessing, From system specification to layout: Seamless top-down design methods for analog and mixed-signal applications, *Proceedings of the Design, Automation and Test in Europe Conference (DATE)*, 2002.

[33] H. Tang, A. Doboli, High-level synthesis of delta-sigma modulators optimized for complexity, sensitivity and power consumption, *IEEE Transactions on Computer-Aided Design of Integrated Circuits and Systems*, 25, (3), pp. 597–603, March 2006.

[34] H. Tang, H. Zhang, A. Doboli, Refinement based synthesis of continuous-time analog filters through successive domain pruning, plateau search and adaptive sampling, *IEEE Transactions on Computer-Aided Design of Integrated Circuits and Systems*, 25, (8), pp. 1421–1440, August 2006.

[35] G. Van Der Plas et al., AMGIE - A synthesis environment for CMOS analog integrated circuits", *IEEE Transactions on Computer-Aided Design of Integrated Circuits and Systems*, 20, pp. 1037–1058, September 2001.

[36] Y. Wei, A. Doboli, Systematic development of analog circuit structural macromodels through behavioral model decoupling, *Proceedings of Design Automation Conference (DAC)*, 2005.

[37] Y. Weng, A. Doboli, Smart sensor architecture customized for image processing, *Proceedings of the 10th IEEE Real-Time and Embedded Technology and Applications Symposium*, pp. 396-403, 2004.

[38] M.-H. Yang, et al., Detecting faces in images: A Survey, *IEEE Transactions on Pattern Analysis and Machine Intelligence*, 24, (1), January 2002.

[39] H. Zhang, S. Doboli, H. Tang, A. Doboli, Compiled code simulation of analog and mixed-signal systems using piecewise linear modeling of nonlinear parameter, *Integration the VLSI Journal*, New York: Elsevier, 40, (3), pp. 193–209, 2007.

[40] H. Zhang, A. Doboli, Fast Time-Fomain Simulation Through Combined Symbolic Analysis and Piecewise Linear Modeling, in Proceedings of the 2004 IEEE International Behavioral Modeling and Simulation Conference, 2004, pp. 141–146.

Index

A. Doboli, E.H. Currie, *Introduction to Mixed-Signal, Embedded Design*,
DOI 10.1007/978-1-4419-7446-4, © Springer Science+Business Media, LLC 2011